Photographing Canada *from* Flying Canoes

S. BERNARD SHAW

Published by

GSPH GENERAL STORE PUBLISHING HOUSE

Box 28, 1694B Burnstown Rd., Burnstown, Ontario, Canada K0J 1G0
Telephone (613) 432-7697 or 1-800-465-6072

ISBN 1-894263-42-1
Printed and bound in Canada
Layout and cover design by Derek McEwen
Printing by Custom Printers of Renfrew Ltd.

Also by the author:

The Opeongo: Dreams, Despair and Deliverance.
Canoe Lake, Algonquin Park: Tom Thomson and Other Mysteries (with Gary Long).
Lake Opeongo: Untold Stories of Algonquin Park's Largest Lake.

National Library of Canada Cataloguing in Publication Data

Shaw, S. Bernard
Photographing Canada from flying canoes

Includes bibliographical reference and index.
ISBN 1-894263-42-1

1. Aerial photography in geography–Canada–History.
I. Title.

TA592.6.C3S53 2001 526.9'82'0971 C2001-900398-6

Front cover photograph:
Crew of Canadian Vickers Viking flying boat G-CYET on return to Victoria Beach, Winnipeg, from the Reindeer Lake Expedition, 14 Aug. 1924. (L-R) Cpl. A.J. Milne, S/L B.D. Hobbs, R.D. Davidson, DLS, P/O D.J.R. Cairns.
NAC C25910

Back cover photograph:
Laminated elm structure of Canadian Vickers Vedette G-CYFS awaiting mahogany planking applied with glue and copper nails in the canoe building tradition. Canadian Vickers, 19 September 1925.
DNDPU RE11710-13

Contents

"Prior to the war of 1914–18, air photography was practically non-existent, and air survey flights were flights of the imagination only; a type of wishful thinking sometimes indulged in by surveyors where the going along the ground became particularly tough, wherein they saw aircraft as the beautifully simple solution of all their technical and transportation problems."

Robert D. Davidson, B.A. Sc., DLS, Army Survey Establishment, Ottawa.
The Canadian Surveyor, July 1957, p. 499.

Bob Davidson was one of the first surveyors loaned from Topographical Survey to the RCAF as navigator on photographic flights, notably the 2,800 mile, month-long 1924 expedition in a sole Vickers Viking over unmapped territory to Reindeer Lake on the northern Manitoba-Saskatchewan border.

"All maps are crude and unfinished as compared with the living terrain when seen from an airplane. As you float through the air with a gently swaying motion, you see spread out before you the work of that perfectly finished draughtsman, Nature; a map revealing the minutest detail and every conceivable shade of colour that is in use to express the handiwork of the Creator of our world."

A.O. Wheeler, DLS, after a reconnaissance flight over unmapped territory in the Rocky Mountains north of Mount Robson with Capt. J.H. Tudhope, Canadian Air Board, in a D.H.4 biplane, July 1922.
Canadian Alpine Journal, 1923, p. 146.

"In those days [1923], when you looked down from an aircraft, you couldn't recognize the countryside from your map, even around Winnipeg! It was that bad!"

Alex J. Milne, *Canadian Aviation Historical Society Journal*, winter 1983, p. 106.
Alex Milne was a flight mechanic.

"A tribute must be paid to the operating staffs of the air services of Canada. Many Canadian pilots have records which, if they had been made in the full glare of publicity, would have made them famous. In Canada such work is done, not for its publicity value, but because it is urgently required in the development of the country. It is part of the day's work and it seems to pass unrecognized. Its value to the state and to those interests which have fostered it is none the less great."

Report on Civil Aviation, 1929, p. 6.

"Will Canada remember that the Royal Canadian Air Force first removed the veil of secrecy from our treasure-laden Northland to mark the opening of an epoch-making period of development? Or that immeasurable wealth was conserved by aerial forest protection and that invaluable photographic and survey work was accomplished with greater dispatch than was hitherto dreamed possible?"

Editorial, *Canadian Aviation*, March 1934, p.3.

Foreword

MANY BOOKS have been written about the adventures of civilian airmen flying in Northern Canada in the pioneering days of the 1920s and '30s. This is not the case with the Canadian Air Force, a large part of which, by government policy, was employed specifically on non-military flying sponsored by other departments desiring to expedite and expand their work in opening up the Canadian North to civilization. These activities are described in various published histories of the Air Force to a reasonable degree, but are limited by publication space. One activity having high priority was aerial photography to provide accurate maps, initially to facilitate travel through the North and eventually to produce accurate surveys comparable to those of Southern Canada.

Bernard Shaw has chosen, because of personal interest, to concentrate on the story of the aerial photography role of the Canadian Air Force and he recounts the life stories of some men representative of those employed in this work. His information has been gathered from official Air Force and Defence Department records, archives, histories and miscellaneous sources to give a rare insight into the conditions in which the Photo Detachments operated. By patient research he has also located relatives of these pioneer airmen and obtained by mail or by personal interview many stories, photographs and personal histories.

I flew in the North during the period and knew and worked with several of the individuals profiled. I am sure this book will be helpful to researchers, historians and to anyone interested in the pioneering days of Canadian aviation and its important role in the growth of Canada.

Ralph E. McBurney

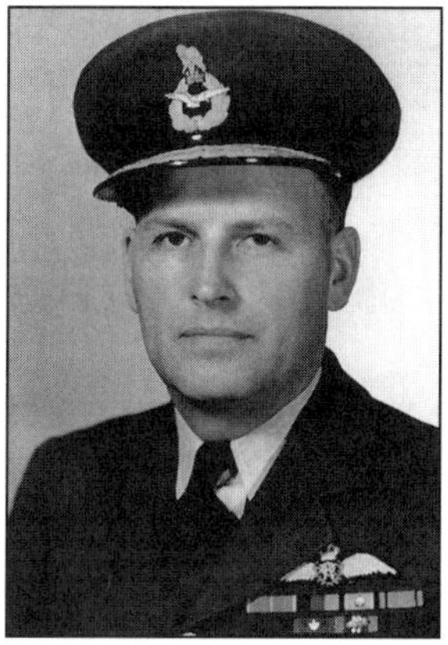

Air Vice-Marshal Ralph E. McBurney, CBE, CD

Ralph McBurney joined the RCAF as a provisional pilot officer in 1924. He obtained his wings in 1926 and spent the next decade on photographic, fire patrol and general duties, mostly in Northern Canada, before taking over as director of signals at Air Force Headquarters. Posted to 6 Group of Bomber Command in 1942, he commanded stations in Britain and became senior air staff officer of 6 Group in 1944. Returning to Canada in 1945, he formed and commanded Maintenance Command and was president of the RCAF War Crimes Trials in 1946. He spent two years as RCAF liaison officer at the Canadian High Commission in London before returning to head Air Materiel Command until his retirement in 1952.

Introduction

THE FIRST KNOWN aerial photograph was taken in 1858 from a tethered balloon near Paris, France, by Gaspard-Felix Tournachon, better known as Nadar. Union forces during the US Civil War (1861–65) are believed to have photographed the Confederate-held city of Richmond from a tethered balloon. Worldwide interest in the potential for aerial photography soon developed and a Russian, R. Yu Tile, is given credit for the first practical multi-camera system for topographical mapping, photographing a stretch of the Pripyat River with a horizon-to-horizon, seven-camera system from a balloon in 1898. The French fascination with photography generated many experiments with balloons and kites. In Germany, Julius Neubronner patented a pigeon-borne camera with a swivel lens and weighing only seven grams in 1903, and Alfred Maul demonstrated a gyroscopically stabilized parachute system launched by rocket in 1912. G.R. Lawrence photographed the cinders of San Francisco after the 1906 fire with a huge kite-borne camera system heavier than the all-up weight of the Wright brothers' 1903 *Flyer*.

The art—it was not yet a science—of aerial photography was advanced by the introduction of gelatin/silver-bromide plates, removing the necessity for on-site development of the exposed negative. Replacement of wet

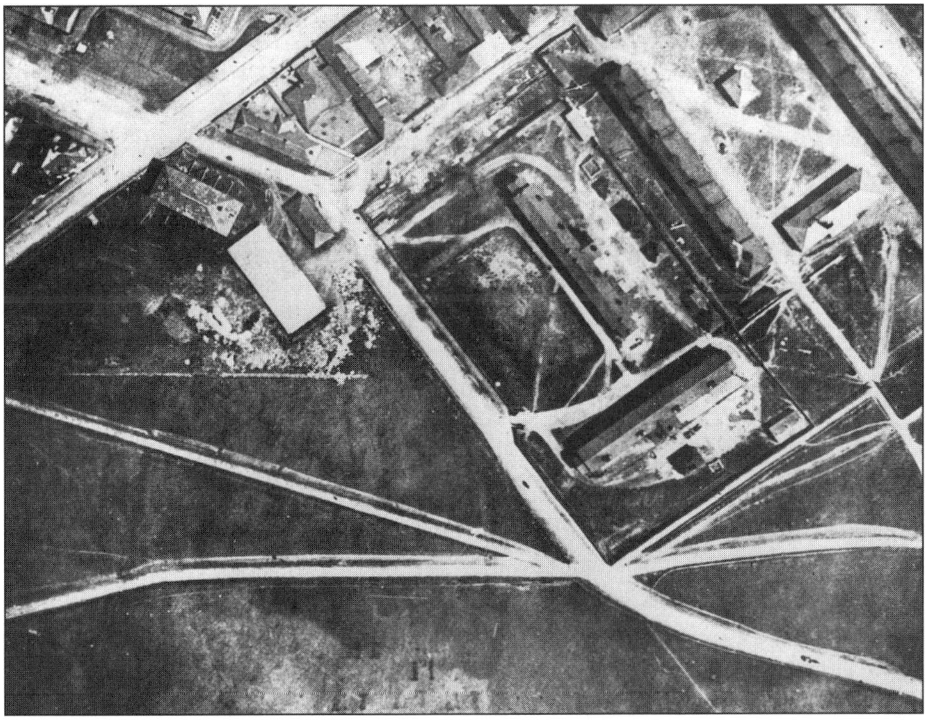

The earliest recorded aerial photograph in Canada was taken from a balloon-borne camera at about 1,450' over the Halifax Citadel in August 1883. The camera was operated from the ground by Capt. Henry Elsdale, Royal Engineers.
NAC PA 172326

Unable to interest the British authorities in aerial photography, Elsdale conducted experiments at his own expense while he was on a two-year tour of duty in Canada. He attached a camera actuated by a clockwork mechanism to a balloon of about 1,000 cubic feet capacity tethered to a windlass. He subsequently returned to duty in England and was promoted major in charge of the Balloon Establishment, 24th Field Company of the Royal Engineers.

photographic plates by faster, dry film was also of great practical value. There was a clear need for an aerial platform free from the limitations of balloons and kites. The first recorded photograph from a powered aircraft, however, was not made until 1909 when Wilbur Wright's passenger took motion pictures of the military airfield at Centocelli near Rome. In the same year, L.P. Bonvillian, a Pathé cameraman, took aerial photographs near Le Mans, France. British, French and German military authorities recognized the intelligence gathering potential of the aerial camera; Zeppelins, for example, were carrying cameras at the outbreak of World War I. During this conflict, considerable refinement of aerial photographic techniques and equipment flowed from early "snapshots" with hand-held personal cameras. By Armistice Day in 1918, no fewer than 4,000 cameras of special design were in use on the British Front with another 5,000 used for training in England.

When World War I staggered to its conclusion in 1918, only some navigable waterways and fragments on the fringes of Canada had been mapped with any degree of accuracy. Much of the rest remained unseen by Europeans, and the extent of a vast storehouse of natural resources could only be guessed at.

War-surplus military aircraft donated by Britain and the United States of America were the foundation of civil aviation in Canada. Rejecting military missions, returning Canadian airmen set out to prove the value of aviation in a variety of civil roles. Forest patrol, firefighting, transport of Indian treaty payments, immigration patrols, customs patrols, crop dusting, spore collection, air mail, pilot training, passenger route development, aerial sketching and aerial photography were all undertaken at various times during the next two decades.

The economic benefit of aerial photography was the consistent thread that justified the funding of a national air force during the Depression years. Thousands of photographs of "unexplored territory" were taken from the air and transformed into maps essential for forests to be inventoried; mineral resources discovered; immigrants settled; land drained and irrigated; hydroelectric power sites identified; roads, canals and highways built; villages, towns and cities planned; and borders established. Today, we take it for granted that an accurate topographical map of any location in Canada is readily available, and virtually everyone born in Canada can find an aerial photograph of his or her birthplace.

None of this would have come about so efficiently without the skilled and dedicated men who made it happen. They are the heart of this story—the bemedalled heroes who returned from one world war to start an air force and train a new generation that successfully led the RCAF through a second major conflict. And the fitters, riggers, mechanics, photographers and clerks— the less-visible "can-do" types who made everything work between the wars—were the backbone of the British Commonwealth Air Training Plan that operated 97 bases in Canada and supplied 131,553 aircrew for the Allied war effort.

This book is divided into three parts. The first covers organizational aspects of aerial photography from 1919 to 1939, with particular reference to the conditions experienced by the participants. The interrelated mix of government and commercial operations is covered in the formative years. By 1927, however, some provinces were taking greater responsibility for their own operations and an expanding private sector was venturing into diverse areas while aerial photography for mapping purposes continued to be, in the main, the responsibility of the Air Force. Accordingly, after 1926, this account concentrates on the RCAF's aerial photography role, only touching on other activities where pertinent to this central theme.

The second part profiles the lives of 14 men, representative of those who undertook such a huge task with minimal material resources, but were liberally endowed with skill, dedication, courage, and that most important characteristic, a sense of humour. These heroes made a vital—and largely unrecognized—contribution to the development of Canada. A direct line can be traced from these pioneering aerial photographers to Canada's present leading position in aerial survey and remote sensing from world-girdling satellites. Similarly, kick-start contracts for photographic aircraft placed by the Air Force in the early 1920s led to Canada's current pre-eminent position in specialized aerospace design and manufacture now amounting to some $10 billion annually, most of it exported. An argument could also be made that aerial forest fire patrol and firefighting techniques initiated by these men in the 1920s were the basis of worldwide operations today.

The final part of the book describes the aircraft used on photo operations—the early flying boats reminiscent of traditional cedar-strip canoes with wings and engine added. Different sources give marginally different figures for dimensions and performance than those listed, which are from RCAF specifications.

The government fiscal year runs from 1 April to 31 March. The civil aviation annual report for 1925, for example, covers the period 1 April 1925 to 31 March 1926. This is not particularly important during the 1920s, as little flying was undertaken during the winter months, but should be borne in mind when the introduction of improved equipment in the 1930s made it possible to extend the traditional summer–fall flying season through the winter into March of the following year.

Events described occurred before the contentious metrification of Canada. Imperial measurements, then standard, are retained. Similarly, place names continue in the style of the time. For example, the Quebec cities now known as Sept-Îles and Trois Rivières were then known by the Air Force as, respectively, Seven Islands and Three Rivers. I have escaped the modern gender problem, as all the personnel of the photographic detachments during the period were men.

Much of the information in this book is drawn from the annual reports of the successive organizations, Canadian Air Board, Canadian Air Force and Royal Canadian Air Force. Additional sources included DND files now in RG24 at the National Archives of Canada; biographical files at the DND Directorate of History; a review of air force activities by G/C Fred H. Hitchins, PhD, CD, compiled when he was air historian of the RCAF; archives of the National Aviation Museum, National Archives of Canada, National Library of Canada, National Air Photo Library, DND Photo Unit, Hudson's Bay Company, Glenbow Museum, Western Canada Aviation Museum and Alpine Club of Canada; and the extensive personal records of aviation historian John A. Griffin, who provided most of the information for part three. I could not have located many photographs without the friendly guidance of NAC archivist Peter Robertson. Other information was gleaned from a wide variety of sources listed in the bibliography, and from private records and recollections noted in the text. A particular effort was made to locate the individuals profiled, or their descendants, in order to both garner information and ensure its accuracy. I have done my best to resolve the contradictions that inevitably result from tapping such a diversity of information sources, some of them oral, and any errors that may have crept into this account are because I did not listen closely enough to the experts, notably Larry Dunlap, John Griffin, Ralph McBurney and the families of the heroes profiled. The idea for the book originated with John Griffin and it would never have been completed without his patient guidance. I am also indebted to editors Susan Code and Jane Karchmar who shattered my self-satisfaction with their constructive criticism and convinced me to write a better book.

Part I

THE ORGANIZATION

1919-1920

CONCEIVED IN FRANCE during World War I, Canada's air force had a difficult birth. Apprehensive of expensive operation and maintenance costs, preoccupied with demobilization, and with little conviction that the Dominion needed to be defended from the air, the government reluctantly accepted about one hundred war-surplus aircraft and other equipment offered by Britain in 1919. More than 22,000 Canadians had served with the British air forces and those who returned (1,563 died) were "air-minded," and had a vision of the role aircraft could play in developing Canada. They influenced the government to accept the Imperial Gift, conservatively valued at $5 million, and included some cameras, probably of Thornton-Pickard manufacture as used by the RAF, among an extensive collection of dirigibles, vehicles and support equipment.

Most of the British aircraft were new, still in their original crates, the result of wartime production momentum. The 62 reliable two-seater Avro 504K biplanes were put to work immediately refreshing and training pilots. Five of the Avros, designated 504L, were equipped with floats to enable them to operate away from the few airfields then existing. The 504Ls played a minor role in subsequent photographic operations, but were too small for effective work. Some of the 11 Felixstowe F.3 flying boats included in the Imperial Gift were put to work on photographic duties. Several of the 12 two-seat D.H.4s were modified to D.H.4Bs and used for photography. Probably the most important acquisition from a photographic perspective, however, was from the US Naval Flying Corps. Twelve Curtiss HS-2L flying boats used for submarine patrol off Halifax harbour since August 1918 under the command of Lieut. Richard Evelyn Byrd, later an admiral and claimed to be the first man to fly over both the North and South Poles, were donated to Canada. Other facilities at the two US Navy bases, North Sydney and Dartmouth, Nova Scotia, were purchased by Canada to give the air force its first toehold on the east coast.

The bill to establish "a Board on Aeronautics" was given royal assent on 6 June 1919, "to supervise all matters connected with aeronautics," including responsibility for the gifts from Britain and the USA. To carry the organization through to 31 March 1920, $250,000 was provided, although only $109,464 was actually expended. Chairman of the Canadian Air Board (CAB) was the Hon. A.L. Sifton, PC, KC, minister of Customs, and the vice-chairman was Col. A.M. Biggar, KC, judge advocate general. Members were the Hon. S.C. Mewburn, CMG, minister of Militia and Defence; the Hon. C.C. Ballantyne, minister of Naval Service; Dr. R.M. Coulter, CMG, deputy postmaster general; Mr. John Armistead Wilson, assistant deputy minister, Department of Naval Service; and Mr. E.S. Busby, chief inspector of the Department of Customs and Inland Revenue. There was some criticism of these appointments as none was a pilot, "who had seen service on the battlefields of Europe."[1] This omission was remedied on 15 July when the CAB outlined an organization and requested the Civil Service Commission to make appointments and to provide office facilities. Key positions, however, were not filled until November and permanent office accommodation in Ottawa was not made available until 24 December.

The next echelon of management appointed by the CAB in 1919 to head the three branches of the new aviation organization were veterans, and virtually all air force senior positions were subsequently filled by qualified pilots. The key players were as follows: Lt.-Col. Robert Leckie, DSO, DSC, DFC, as superintendent of Flying Operations at $4,500 per annum; Lt.-Col. J. Stanley Scott, MC, AFC, as Superintendent, Certification Branch at $3,600; and Maj. A.M. Shook, DSO, DSC, AFC, Croix de Guerre, as secretary, and responsible for administration, at $3,000. Also appointed was Capt. F.R. Smith as medical officer at $2,400. These men were well aware of the possibilities offered by aircraft. Disregarding, or perhaps because of, wartime experiences, they concentrated on identifying non-military roles, such as forest inventories, fire detection, anti-smuggling patrols, fisheries inspection, environmental protection, police support, mapping, and transportation assignments, including carrying treaty-paying parties to Indian reserves. It was more than a decade before any Canadian military aircraft was armed. One of the CAB's first tasks was to co-opt a Department of the Interior committee that had already been formed to consider the possibilities of aerial survey.

The Imperial Gift was uncrated and assembled by a hastily organized crew at the army's Camp Borden, near Barrie, Ontario. From this rudimentary organization evolved the fourth largest Allied air force in World War II with almost a quarter of a million men and women in uniform, 93,844 serving overseas.

The CAB commenced presenting annual reports with a slim volume, *Report of the Air Board for the Year Ending March 31, 1920*, covering events from June 1919. The emphasis was on establishing a basic organization; aerial photography included only by association with the identified need for aerial navigation maps. Recommendations were made to Parliament for the

1 F.H. Hitchins, *Air Board, Canadian Air Force, Royal Canadian Air Force*, Paper No. 2, Mercury Series (Ottawa: Canadian War Museum, 1972), 3.

formation of a Canadian air force, "which the very large number of ex-officers and airmen of the Royal Air Force in Canada are expected to enthusiastically support."[2] A decision was postponed, however, "until the general policy with regard to Canada's future sea, land and air forces was determined."[3] It was generally believed by the politicians that line officers and airmen could maintain their skills by undertaking one month's training every second year: "The air force as a whole could be maintained in a sufficient state of efficiency if one twenty-fourth of its total non-professional strength was always on duty."[4] They were paid only for the time served.

The identification of Canada's forest resources was, and still is, an important part of aerial survey. The first aerial timber survey, and the first civil use of airborne radio, in what is now Canada was carried out by the H.V. Green Aerial Survey Company Ltd. of Canada in Labrador during the summer of 1919. Although it is difficult to confirm claims that it was the first such venture in the world, it was certainly an ambitious and innovative enterprise—and expensive, at an estimated cost of $200,000. *The Halifax Herald*, 10 July 1919, stated that a million and a quarter acres were to be surveyed: "Here is an undertaking that stands unique and without previous example."

Capt. Daniel Owen, leader of the expedition, was still officially in the RAF. He went to France with the Canadian Expeditionary Force, transferred to the Royal Flying Corps and was shot down behind the German lines in a battle that cost him an eye and took him to prison camp. Exchanged with an injured German pilot because of his wounds, he returned on sick leave to Annapolis Royal, where his father, J.M. Owen, was a judge. Owen hung up his shingle as a barrister and became president of North American Securities Corporation. It was his legal expertise coupled with his flying experience that brought him a directorship with The Southern Labrador Pulp and Paper Co. Inc. This company was associated with The Belle Island Straits Lumber and Pulp Company Limited, Boston, which had spent five years trying to survey its enormous logging rights in southern Labrador. Owen convinced the company that he could provide an accurate survey of an area 75 by 42 miles covering the watersheds of three large rivers in one month plus a week on each end for travelling. The prime motivation, but not the only one, was access to timber estimated to be worth $26 million. In addition to pilots, surveyors, photographers, a forester and "other specialists necessary to comfortable living in the wilderness," he proposed to take along a mineralogist and a naturalist. They were also directed to search for other exploitable natural resources and to locate possible hydroelectric power generation sites.

The corporate structure behind what became known as the Owen Labrador Expedition was somewhat complicated as described in newspaper clippings collected by Owen's mother, Isabella Anne, and held by the National Archives of Canada (NAC).[5] Mr. H.V. Green of Boston invested in the venture and the expedition was officially known as The H.V. Green Aerial Survey

2 Canadian Air Board, *Report of the Air Board for the Year Ending March 31, 1920* (Ottawa: Dominion of Canada, 1921), 7.
3 Hitchins, op. cit. p. 8.
4 Hitchins, op. cit., p. 10.
5 National Archives of Canada. Scrapbook of Isabella Anne Owen. (MG30, A22).

Company Ltd. of Canada, a company incorporated in both Canada and the USA. Transportation was the responsibility of yet another participant, Annapolis Transportation Company owned by Boer War veteran and Boston dentist with a taste for adventure, Dr. Murdock M. Graham. He went with the expedition in charge of the thirty personnel involved. In the heterogeneous group were two US Navy pilots who had seen service in France—Ensigns William S. Cormack and William P. Smith; Lieut. Henry Irving Twiss, a medical doctor who had served in France with the US Army and was called "an explorer of note who has spent many years in Australia and South Africa";[6] photographer Jack Wade from Boston; and two mechanics, William Beehler, a warrant officer in the US Navy, and Joseph Halpin from the RAF.

The expedition was given a great send-off on 7 June 1919 by the mayor of Annapolis with a band playing *The Maple Leaf Forever* and *The Star-Spangled Banner* before a thousand spectators in St. Ann's Park. The 5,000-ton steamer *Granville* was chartered to transport the expedition to Battle Harbour, a small community in southern Labrador with a radio station, mission hospital and a population of about 100. Three Curtiss JN-4 biplanes were carried, two on wheels and one on floats. The aircraft were transported, with considerable difficulty, six miles up the Alexis River to a relatively flat area utilized as an "aerodrome," where canvas Besonneau hangars were erected.

Owen established stringent safety procedures hinging upon airborne radios, and there were no untoward events during the survey of the area inland of St. Louis Inlet and Alexis Bay. The first of his *General Instructions To Pilots and Observers* read, "Safety first—no stunting."[7] He also required two days' rations and hand compasses to be carried on every flight, and he specified actions to be taken in the event an aircraft was forced down—wireless message, smudge fire, flares and walk towards the nearest waterway. The task was successfully completed in ten days of flying by aircrews in shifts during the long daylight hours. The Canadian-built JN-4s were probably the first aircraft to fly over Labrador. Photography was carried out from 2,000' to 9,000' and 15,000 negatives were rushed by Owen to Boston where he arrived on 26 August.

Less enthusiastic when the expedition returned, Owen compared his experience unfavourably with flying over the Western Front in France. He moved to Halifax and resumed his law practice until his death in 1939. The H.V. Green Company apparently found no further work and the *Granville* was advertised for sale in the St. John's *Daily News*, 20 August 1919.

Ellwood Wilson, chief forester of the Laurentide Company Limited, an important paper manufacturer in the St. Maurice River watershed north of Three Rivers, was an early proponent of aerial survey. One of the few foresters of the period to recognize the need for forest management, he introduced the use of nursery-grown trees in the reforestation of logged land in 1911. He also advocated the growing of spruce specifically for the aircraft industry—a need that became critical in 1918 and again during World War II.

6 *Halifax Herald*, 10 July 1919, p. 3.
7 NAC, op. cit.

Wilson explored the possibilities of forest fire detection from balloons as early as 1908. He was not, however, able to get his ideas off the ground with any success until he heard of the US Navy's post-war gift to Canada of HS-2L flying boats. Through the St. Maurice Forest Protective Association, six companies with mutual interest in the area, he applied for the loan of two HS-2Ls from the federal government and financial assistance from the Quebec government to operate them. The Canadian government was not in a position to operate the aircraft, but welcomed the opportunity to gather information and complied with the sole condition that comprehensive reports be submitted to Maj. Claire MacLaurin at the CAB in Ottawa.

Stuart Graham, who had flown a Curtiss HS-1L in Britain, was taken on as pilot and Walter (Bill) Kahre as flight engineer. Their first task was to fly the aircraft from Halifax to the St. Maurice Forest Protective Association's headquarters at Grand Mère, north of Three Rivers, then considered a dangerous overland flight for a flying boat. The aircraft operated from Lac-à-la-Tortue, three miles from Grand Mère, which thereby became the first official "air harbour" in Canada. There, the wooden hulls of the HS-2Ls soaked up water to the extent that they became too heavy to take off; a ramp had to be built in great haste so they could be hauled out and drained before they sank at their moorings.

Ellwood Wilson's enthusiasm was not dampened when his first flight, with Stuart Graham at the controls, concluded with a forced landing just short of Lac-à-la-Tortue. The flying boat came to rest with a dead engine, but without serious damage, in a pasture filled with tree stumps and surprised cattle behind the Tremblays' farmhouse. With the help of mechanic Roméo Vachon, the HS-2L was dismantled and towed on a stone boat back to its natural element. Vachon later married one of the Tremblay daughters, Georgette, in the small church at Lac-à-la-Tortue.

The first aerial photographic survey in Canada proper was undertaken in October 1919. Pilot Stuart Graham and engineer-photographer Bill Kahre, in one of the HS-2L flying boats borrowed from the CAB by the St. Maurice Forest Protective Association, made a flight to demonstrate the value of aerial photography to the pulp and paper industry in the valley of the St. Maurice River, north of Three Rivers. In the *CASI Journal*, September 1970, historian Ken Molson identified this region as the birthplace of Canada's air transport industry.

A hole was cut in the bow of one HS-2L for a "pot" vertical camera mount and sealed by a tube projecting above the water line while on the water, or permanently by a plug when not on photo ops. A similar system was installed in most of the subsequent photographic flying boats. Several cameras were tried, but none was satisfactory until an Eastman Kodak K-1 with a roll film magazine was located late in the season. The pioneers exposed about 380 9"x7" negatives of varying topography as ice formed on

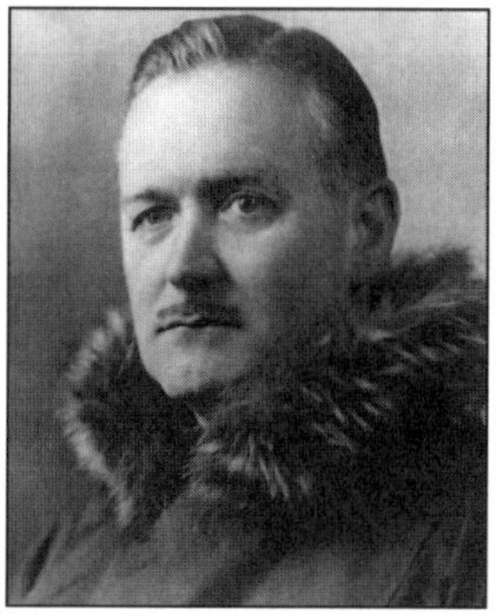

Stuart Graham, 1940.

Graham was wounded at Ypres in 1915 and spent a year in hospital. He transferred to the RNAS and was awarded the AFC for attacks against German submarines. Graham is recognized as Canada's first bush pilot, received the OBE and McKee Trophy, and was elected to the Aviation Hall of Fame. *ROBERT GRAHAM*

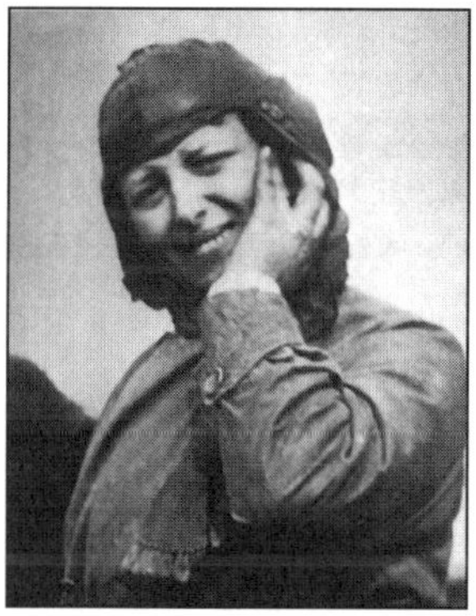

Stuart Graham's wife Marguerite Marie (Madge) went along on both HS-2L deliveries to Lac-à-la-Tortue as unofficial navigator. She was photographed by engineer Bill Kahre in the bow cockpit of the ex-US Navy HS-2L, *La Vigilance*, on 8 June 1919, just after being drenched by an errant wave at Rivière du Loup, Quebec, en route to Three Rivers. *ROBERT GRAHAM*

Joseph Pierre Roméo Vachon, born in 1898 to French-Canadian parents in Sainte-Marie-de-Beauce, Quebec, had a powerful drive to excel in what was then an English-dominated profession. He left the family farm in 1917 for the Royal Canadian Navy, leaving as a trained mechanic with the very rare capability of technical expertise in both the French and English languages. He joined the CAF at Camp Borden in 1920, qualifying for aero-mechanic certificate No.93, which, at his request, was printed in French.

While he was a flight engineer with Laurentide Air Services Ltd. from 1921 to 1925, Vachon qualified as a pilot at the Wright Flying School in Dayton, Ohio. He pioneered airmail and passenger routes; helped establish the British Commonwealth Air Training Plan during World War II; and participated in the formulation of Canadian and international air transport development policies.

Roméo Vachon received Canada's premier aerospace award, the McKee Trophy, in 1937 for his efficient organization of air transport services and outstanding safety record. At a time when very few Canadians of French origin were involved in aviation, he served as a role model for many compatriots who followed his inspiring example. He died in 1954.

the lakes and blizzards blew them off course. From 5,000' they were able to photograph an area three-quarters of a mile wide and 60 miles long with 100 exposures on one roll of film. Results exceeded expectations and their report notes, "Watersheds, and their drainage, swamps and burnt over lands are all accurately shown. The hardwood is readily distinguished from the coniferous and the difference in size of growth is readily noted. Pine is easily spotted from other coniferous types and as our experience increases in the interpretation of aerial forest photos . . . innumerable other valuable features will undoubtedly be disclosed."[8] A forecast that turned out to be very accurate.

Despite the positive report, the economics were not encouraging. One study estimated that patrol of a certain forest area would cost $23,000 per season, plus the capital cost of aircraft and base facilities. With the same amount of money, it was argued, 40 fire wardens could be hired. All the association members, except for the Laurentide Company, decided that the cost of aerial fire patrol was too great. No doubt due to the persuasive arguments of Ellwood Wilson, Laurentide acknowledged the value of aircraft for not only fire patrol, but also for firefighting, mapping, photography and the transport of personnel. The end result was that the two HS-2Ls, still on loan from the CAB, were turned over to Laurentide.

Returning war heroes William Barker and Billy Bishop, each with a VC and many other medals, cashed in on their popularity to launch a commercial venture, Bishop-Barker Company Limited, in November 1919. Their charter allowed them to do virtually everything in aviation including carrying passengers and freight, designing and manufacturing aircraft, operating flying schools, and repairing, leasing and selling aircraft. Their first purchase was an ex-RAF Curtiss JN-4, which generated the first positive cash flow by photographing, with war-surplus British cameras, Queen's University at Kingston, new construction along the Toronto waterfront, and the city of Stratford. This work resulted in the production of scenic postcards and led to a variety of commissions to photograph factories, golf courses, industrial sites, housing developments and individual homes. This was, however, intermittent work dependent on the weather, and the two somewhat anti-social colonels were not good businessmen. The company went out of business after an exciting couple of years.

8 K.M. Molson. "A History of Laurentide Air Service—Canada's First Scheduled Air Service," *CASI Journal*, 16, no. 7 (September 1970): 271.

1920-1921

THE 1920 CAB report covered the government fiscal year, 1 April 1920 to 31 March 1921, a pattern followed in all subsequent reports. Important changes at the senior levels of the CAB were noted. Having completed the preliminary work of organization for which it was formed, the founding board resigned and a new board was appointed on 19 April. The new chairman was the Hon. Hugh Guthrie, MP, KC, recently appointed minister of Militia and Defence, and Col. Biggar returned as vice-chairman. The three department heads were appointed to the board: Lt.-Col. Leckie as director of Flying Operations; Lt.-Col. Scott as controller of Civil Aviation; and Sir Willoughby Gwatkin,

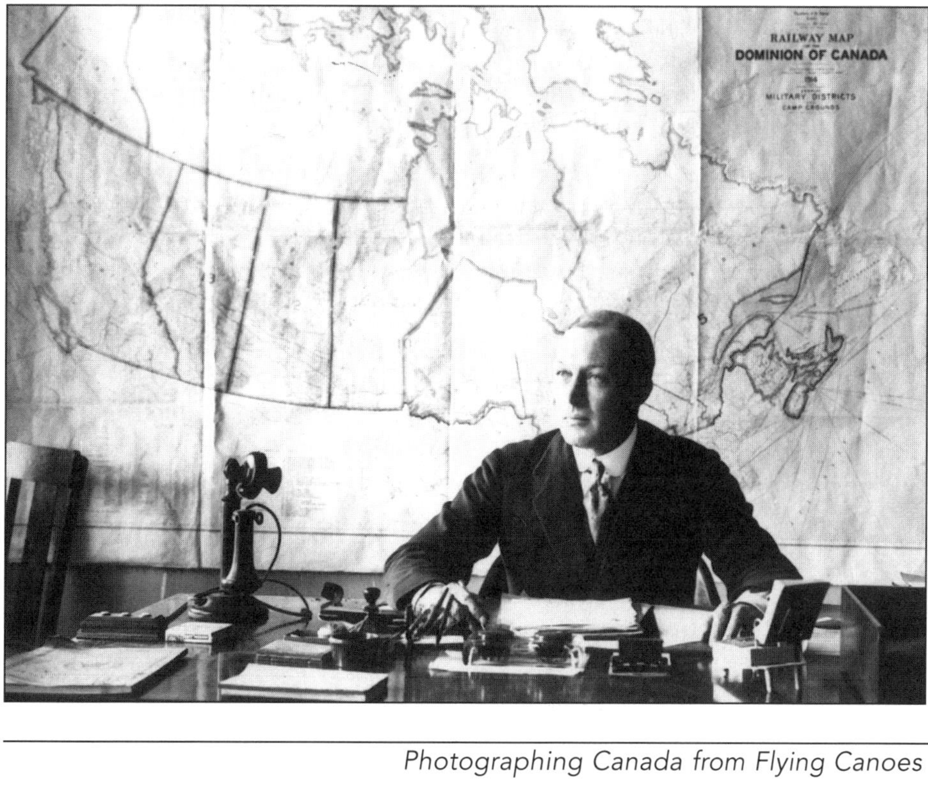

John Armistead Wilson took over the key post of secretary, CAB, when Maj. A.M. Shook resigned due to illness in 1920. He is shown in his Ottawa office in 1922.
NAC PA64578

Wilson, born in Scotland in 1879, went to India in 1902 before arriving in Canada in 1905 to work for the Canada Cement Company. He moved to the federal government in 1909 where he subsequently organized the RCNAS in 1918. He relinquished his position as assistant deputy minister in the Department of the Naval Service to transfer to the CAB. He was an energetic proponent of the Air Force and a trans-Canada passenger service, playing what was probably the most significant individual role in advancing early Canadian aviation.

KCMG, CB, former chief of staff, as inspector-general with the rank of air vice-marshal. The two advisory member positions were filled by Capt. W. Hose, CBE, director of Naval Service, and Dr. Édouard Gaston Daniel Deville, LL.D, DLS, the surveyor general, who became a forceful proponent of aerial photography.

The CAB received a 1920–21 appropriation of $1.9 million for both civil and military aviation. Canadian Air Force (CAF) headquarters were officially opened at 529 Sussex Street in Ottawa on 17 May by the commanding officer, A/C A.K. Tylee, OBE, with a provisional staff of one wing commander, one flight lieutenant, one warrant officer, one flight sergeant, and one sergeant. In June, the CAF received provisional authority to hire 1,340 officers and 3,905 airmen—a "non-permanent, non-professional air force."[9] The CAF uniform was cut on the army pattern: dark blue serge with silver buttons, officers' ranks denoted with silver stars and crowns on shoulder straps, other ranks by stripes on the sleeve. All officers wore white shirts with turndown or wing collars and carried canes. Squadron leaders and above had a row of silver leaves on the cap brim and could wear riding breeches and field boots. The motto *Sic Itur Ad Astra* (This Is the Way To the Stars) featured on caps, collars, wings and buttons. The service enjoyed a curious duality of traditional army ranks and the new ranks adopted by the RAF. A major-general was an air vice-marshal (then the highest rank); a brigadier-general an air commodore; a colonel a group captain; a major a squadron leader; a captain a flight lieutenant; a lieutenant either a flying officer or a pilot officer; and a sergeant major a flight sergeant. The more familiar army ranks were in common use, but the new CAF titles were required in official communications.

Buildings constructed for the RFC and RAF during the war at Camp Borden were officially transferred from the Department of Militia and Defence to the CAB on 5 July 1920. The Department of Naval Service bases at Kelly Beach, North Sydney, and Baker's Point, Dartmouth, were consolidated at Dartmouth under the authority of the CAB, and ten ex-USN HS-2Ls were taken out of storage and reassembly commenced. A new base was established at Jericho Beach, part of the old Naval Reserve at English Bay, Vancouver, British Columbia, over the objections of the Jericho Golf and Country Club, which had enjoyed a lease on the property for many years. Two HS-2Ls were shipped by rail from Dartmouth and reassembled at Jericho Beach, and one was flown to Ottawa. Work also started on new bases near the village of Morley, on the Stoney Indian Reserve west of Calgary, Alberta, and at Roberval, on Lac Saint-Jean in central Quebec. Two HS-2Ls were flown from Dartmouth to Roberval under the command of Capt. W.R. Kenny, primarily for forest surveillance on behalf of the Quebec Department of Lands and Forests. Operations were also initiated with a single HS-2L from Victoria Beach, located at the terminus of the Canadian National Railway on the eastern shore of Lake Winnipeg, and photographs were taken of the Winnipeg

9 Canadian Air Board, op. cit., p. 11.

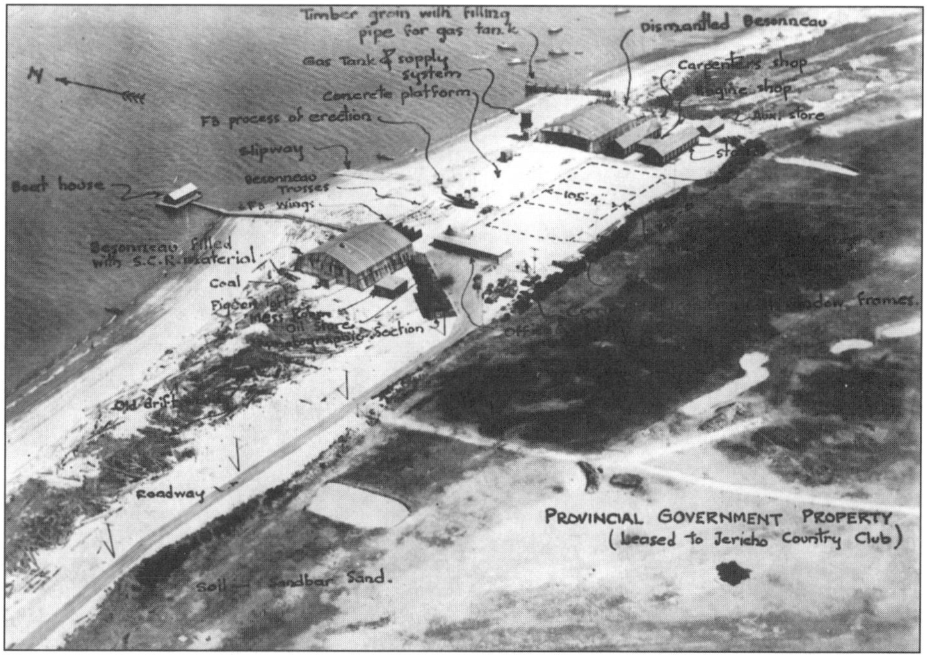

Vancouver Air Station at Jericho Beach under construction, 1921. (left)
DNDPU PL143834

Jericho Beach, Vancouver, looking south, 24 July 1931. The waterfront is now a public park and two private clubs. (below)
NAPL A3832-96

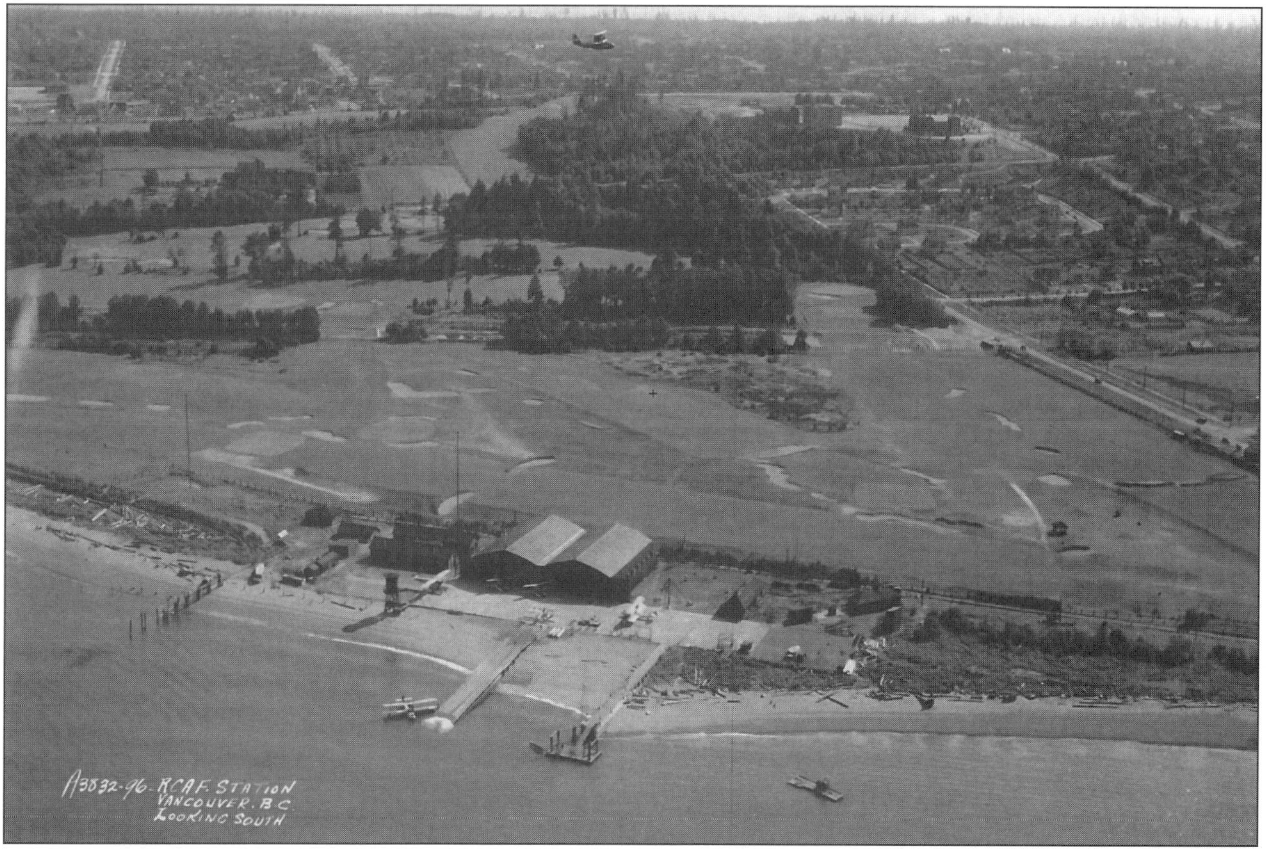

The first flying boat, HS-2L G-CYBA, at Vancouver, 28 October 1920, being loaded on a flatcar for transport through the Rockies. Maj. C.C. MacLaurin is on the left. *DNDPU PL143820*

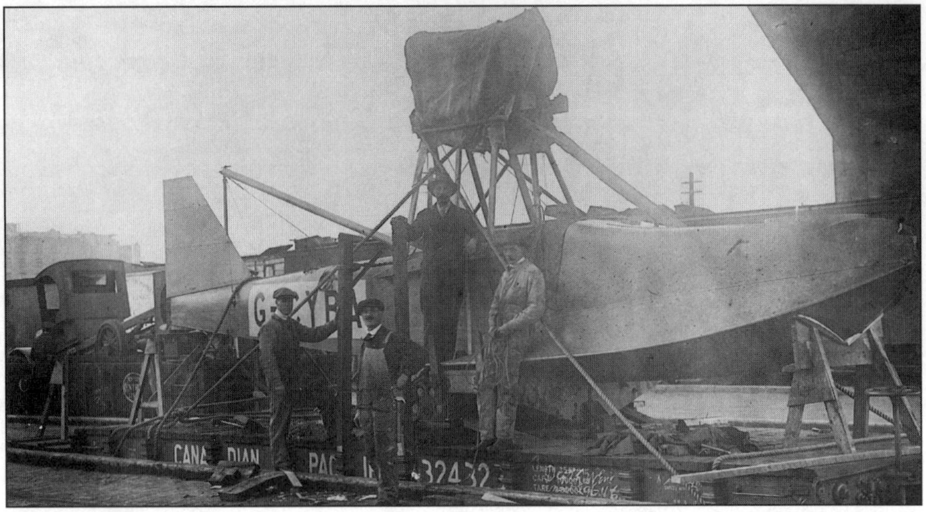

Until 1928, federal government-owned aircraft operated with five-letter civil-type registrations, G-CY**. Private and commercial aircraft were registered G-CA**. In 1928, the RCAF introduced a numerical system for registering its aircraft and, starting in 1929, Canadian civil aircraft were registered CF-***.

River for the Water Powers Branch, Department of the Environment. Most of these bases utilized canvas Besonneau hangars, part of the Imperial Gift, until permanent buildings could be erected.

Jericho Beach CO, Maj. Claire C. MacLaurin, was keen to demonstrate the civil applications of aircraft. The provincial government had criticized the expenses involved in constructing the base so he made a quick trip to Victoria to "show the flag" as soon as the first HS-2L was assembled. The aircraft was dispatched in September to the Slave River area in northeast Alberta to transport a Dominion entomologist for inspection and photography of swamps where mosquitoes were thought to be breeding in large numbers.

In the closing days of 1920, 'BA was flown to Victoria, Esquimalt and Sooke for coastal defence reconnaissance, and demonstration flights were

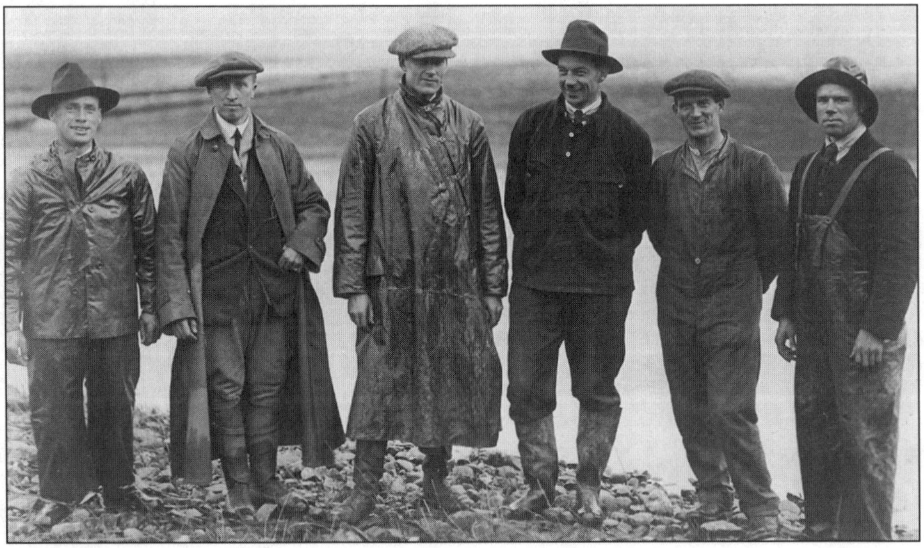

HS-2L G-CYBA was transported in October by rail to a siding near Sicamous on the Thompson River for a series of flights from Sushwap Lake as far west as Kamloops, returning by rail to Jericho Beach on November 15. Photographed at Kamloops in November, pilot MacLaurin, third from left, was accompanied on these pioneering flights by photographer Lieut. Clarence J. Duncan, second from left, and mechanic A.L. Hartridge. *DNDPU PMR82-174*

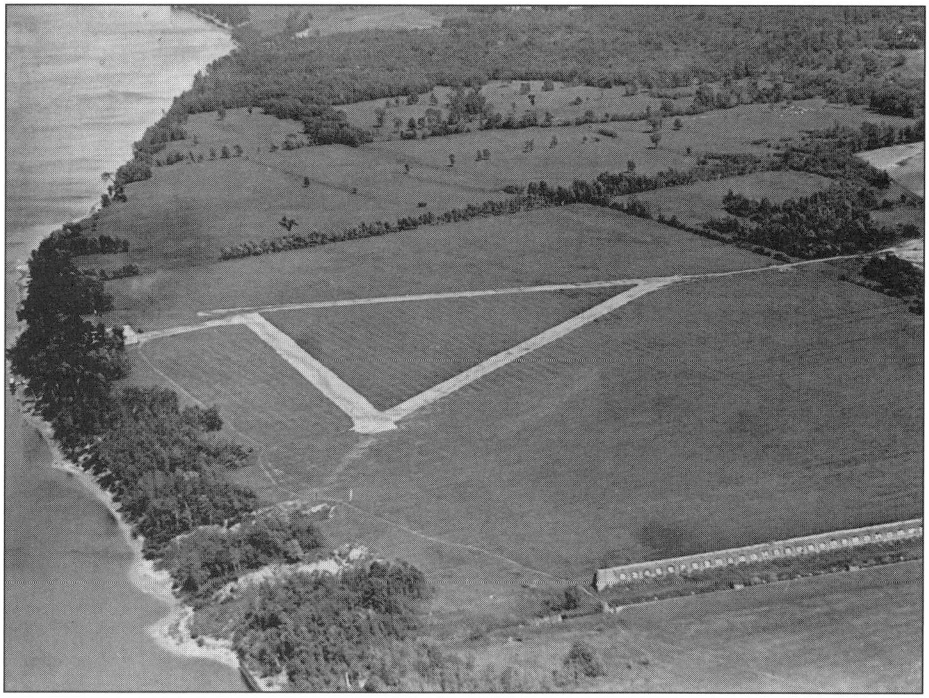

Arrangements were made with the Department of Militia and Defence in 1919 for the use of vacant land by the Ottawa River "in the danger area behind the butts at Rockcliffe range"11 for the use of an aerodrome. This site also offered landing facilities in the Ottawa River for float planes and flying boats, the only combined land-and-water air force base until Trenton, Ontario was activated a decade later. Ninety-one hours and thirty minutes were flown from Rockcliffe by a variety of aircraft in its first year of operation, primarily for experimental work with cameras and developing photographic techniques. "Ottawa Air Station" is shown here on 2 June 1921 with the rifle range butts in the foreground. The packed-dirt runways were grassed over until paved in the late 1930s. *NAC PA117853*

made from Jericho Beach. Representatives of the federal and provincial departments involved were impressed and "expressed themselves fully satisfied that the use of aircraft will increase the efficiency of their service and render possible much work which cannot be undertaken to advantage at present."[10]

The enthusiastic and energetic MacLaurin encouraged the use of HS-2Ls for non-military applications, often carrying RCMP officers in the front cockpit to enforce laws against smuggling, illegal fishing and illegal immigrants, in addition to routine photographic, fire patrol and transportation duties.

The inter-departmental committee formed in 1919 to investigate "photo-topographic methods of surveying" had made positive recommendations and action was soon under way. Ottawa was identified as the favoured location for test and evaluation, as the terrain was varied and provided areas of land and water with diverse elevations within a small radius of operation. Another inducement was the availability of government land for an airfield.

Howard Frederick John Lambart, DLS, the project manager for the Rockcliffe operation, had extensive experience using photography as an aid to map-making in mountainous areas—called photo-topography—a technique that had first been used in the Rockies in 1886 during surveys for the Canadian Pacific Railway by J.J. McArthur and W.S. Drewry, both dominion land surveyors. It is not surprising that these mountain surveyors founded the Alpine Club of Canada. They carried surveying gear and cameras to high

10 Ibid., p. 8.
11 Ibid., p. 9.

"triangulation points," precisely located as key references to plot maps from photographs taken from these locations. The same technique had been used to some advantage in establishing the US–Canada border along the 49th parallel of latitude and the Yukon-Alaska boundary. Arthur Oliver Wheeler, DLS, then commissioner for British Columbia on the British Columbia–Alberta Boundary Commission, worked closely with Lambart and the two were among the first surveyors to recognize the tremendous value of aircraft for both photography and transportation, "where the highly accented contours of the terrain rendered the application of ordinary methods too laborious and expensive, or else altogether impractical."[12] In a detailed description of the process in the *Canadian Alpine Journal*, 1919, Wheeler commented on ground-based photo-topography: "The method is best adapted to boldly accented areas such as are found in mountainous regions, and can be applied no matter how exalted the area may be, depending only on the ability of the operators to reach the elevated positions used as photographic stations with their instruments—camera and mountain theodolite—and while there to experience weather conditions suitable for photography."[13]

Concerns in 1920 centred on the practicality of compiling a mosaic of many photographs, each subject to variations dependent on aircraft altitude, position and attitude at the time it was taken. Correction for crosswinds would skew the photographs, and changes in ground elevation also introduced variations that made it difficult to match adjacent photographs. Each photograph had to be painstakingly "rectified" to make it line up with the adjoining images. In later years equipment was developed that eased this onerous task.

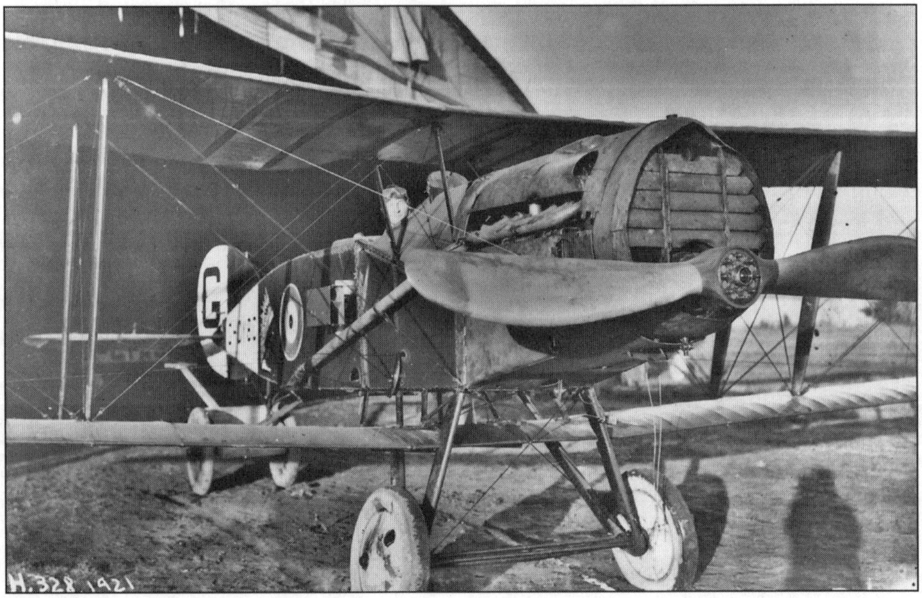

Bristol F.2B Fighter used on early photo research at Rockcliffe with Capt. C.M. McEwen in the cockpit, 1921.
NAC PA62276

"Black Mike" McEwen had 27 victories in World War I and went on to command the famed Canadian 6 Group of Bomber Command in World War II.

12 A.O. Wheeler. "Application of Photography to the Mapping of the Canadian Rocky Mountains," Canadian Alpine Journal, 11 (1920): 77.
13 Ibid., p. 80.

Two aircraft, both two-seaters, were used on the tests from Rockcliffe, a 130 hp Clerget-powered Avro 504K and a Bristol F.2B Fighter with a 270 hp Rolls-Royce Falcon III engine. An ex-RAF LB (Mk 1) semi-automatic camera with 4" x 5" plates and a focal length of 8 1/2", built by Williamson Kinographic Co., London, was used and photographs were taken with Ilford Panchromatic Polychrome and Wratten Panchromatic plates from 6,000' and 10,000' altitudes. Considerable difficulty was experienced in isolating the camera from aircraft vibration, partially alleviated by the addition of sponge rubber gaskets and spring mountings.

Because of its limited ceiling and excessive vibration, the Avro was unsuitable for vertical air photography. The Bristol, although of adequate performance, had controls so sensitive that the pilot experienced difficulty in maintaining straight and level flight for prolonged periods. Effective communication between pilot and cameraman was essential, but usually proved impossible due to engine noise and lack of any intercom. Crews had to rely on hand signals, supplemented by handwritten messages, or, on occasion, pulling on a string tied to the cameraman's wrist.

While difficulty was experienced in matching the edges of the photographs to construct a mosaic of Ottawa, the tests demonstrated that rectification of aerial photographs was possible. Photographs could, it was argued, be accurately transformed into conventional topographical maps, although a need was identified for improved stabilization of the aerial camera. Tests were continued over the next two years with a D.H.4B aircraft equipped with a Steinheil Orthostigmat Series E camera with a circular photographic plate 11" in diameter, and a camera devised by Prof. H.L. Cooke of Princeton University. Cooke arrived on the scene following a visit to Princeton by Dr. Deville who offered co-operation in support of Cooke's theories on topographical survey based on aerial photography. The second of Cooke's cameras had a Georz-Hypergon lens of 2.95" focal length and a circular field 12" in diameter giving an angular coverage of 128 degrees and incorporating a unique feature. Slow film speed at that time necessitated a minimum exposure of one second so the camera was rotated backwards to compensate for aircraft forward movement. Although reasonable results were obtained, the system was complex and cumbersome and it never went beyond the experimental stage.

Laurentide's two HS-2Ls were civilly registered as G-CAAC and G-CAAD, and the company purchased a smaller Curtiss Seagull flying boat, which it was thought would prove more economical for photographic operations. Graham and Kahre completed 70 flights and 150 hours' flying time during 1920, but both resigned at the end of the season.

Aviation in Canada received a publicity boost in October 1920, when CAF aircraft completed the first coast-to-coast flight. Six aircraft were used over ten-and-a-half days, covering 3,341 miles in 49 hours and 7 minutes at an average speed of 68 mph. Painfully slow by today's standards, but the achievement triggered the attention of the populace, who began to appreciate the potential benefits of aviation in the vast reaches of Canada. Not as visible, but perhaps more significant in the long term, was the appointment in the same month of Lt.-Col. Ernest W. Stedman, MEIC, AMICE, ARCSc, AFR, as Director of Technical Services for the CAF. A dedicated officer from the National Physical Laboratory at Teddington, England, Stedman guided the development of military aviation in Canada for the next quarter century and was promoted air vice-marshal in 1942.

Immediately on his appointment, Stedman initiated airworthiness tests of the HS-2Ls. Stress tests of the wings showed that six-foot extensions installed by the US Navy to increase payload and range of the flying boats were inadequate for the anticipated fully-loaded duties in Canada. New sections were designed and manufactured under the direction of R.G. (Bob) Ford in a rented workshop on Sparks Street, Ottawa, and installed, much to the relief of the aircrews.

Hand-held cameras were initially used in the HS-2Ls for oblique photography with resulting inconsistencies. The machine-gun ring was soon replaced with a camera mount and the cockpit enlarged to accommodate the cameraman and his gear. The first mount was a concertina design with a universal joint screwed to the cockpit floor. However, it was difficult to extend in the still-cramped cockpit, and was subject to vibration under the pressure of buffeting winds. This design was replaced by a simple trolley attached to the cockpit lip and stabilized by a brace attached to a floor-mounted universal joint. In his recollections, Stedman gave credit to Alan Ferrier and Jack Greenhalgh for much of this work. He also noted a great debt to the mechanics whose skill kept the machines operating safely at a time when funds were limited: "The high standard of skill they introduced was handed down to the younger men they trained."[14] Official records of these men are scanty, but Stedman recalled a number of names—Black, Boyd, Ford, Gorham, Grepson, Graham, Hems, McConnell, Moore, Rabnett and Staveley.

14 E.W. Stedman, From Boxkite to Jet (Toronto: University of Toronto Press, 1963), 5.

1921-1922

BY 1921, the CAB, now with the Hon. G.P. Graham, minister of Militia and Defence, as chairman, was beginning to flex its muscles. Notice was given in the annual report that surplus World War I aircraft were due for replacement. Note was made that the $1.6 million air force allocation was 14% less than the previous year, only one-third of the navy's allocation, and the militia was to receive 14 times as much. British Columbia, however, contributed $20,000 to the CAF's coffers in 1921 for aerial operations, including photographic surveys of Vancouver Harbour and the Fraser Delta, conducted on its behalf from Jericho Beach.

The relatively large Felixstowe F.3 flying boats with twin Rolls-Royce Eagle VIII engines were unpopular with the pilots because of their unwieldy handling characteristics, both in the air and on the water. Designed for wartime service by the Royal Naval Air Service (RNAS) and part of the Imperial Gift, they were difficult to maintain and expensive to operate. The smaller, single engine, ex-US Navy Curtiss HS-2Ls performed valiantly, and 12 more were purchased during 1921 at $1,000 each. Cheap to purchase and economical to operate, they were the best the CAF could afford, but underpowered for the duties expected of them.

Primitive by today's standards, the single-engine, three-seater wood, wire and canvas HS-2Ls had hulls fabricated from elm frames covered with cedar planks, similar to a traditional cedar-strip canoe. The CAF's aircraft were modified for photographic operations and proved versatile workhorses. Their performance, however, was best in the spring before the wooden hull soaked up many gallons of water. Held together by elaborate wire bracing, they gave rise to the story that a rigging check was accomplished by releasing a pigeon from the cockpit: if the bird escaped, a wire was missing. The HS-2L, with a ceiling of only 6,500', suffered, as conceded in the CAB annual report, "a disadvantage" operating near mountains 10,000' high.

In the early days at Jericho Beach, pigeons had a much more practical role as the sole means of emergency communication from aircraft in distress.

Superintendent MacLaurin placed high priority on establishing a pigeon service, and a pigeon loft was one of the first buildings constructed at Jericho Beach. He hired local pigeon expert Alexander Dickie as a carpenter, a trade for which he was qualified, but his prime role was to establish a strain of homing pigeons for use at all CAF stations. MacLaurin no doubt had support from Lt.-Col. Leckie, director of Flying Operations, who, in 1917, had been rescued from a harrowing three days and nights adrift in the North Sea as the result of a message delivered ashore by a pigeon. Retired A/C Earl Leslie MacLeod recounted his west coast experiences with pigeons in the CAHS Journal, Winter 1974. The young pilot made an emergency landing in December 1921 when the magneto in his HS-2L failed about 50 miles from base. His released pigeon with a message tied to its leg made a swift trip to the loft at Jericho Beach where "pigeon man" Dickie dispatched the standby HS-2L to rescue MacLeod. In 1924, S/L A.E. Godfrey was rescued in a similar manner, and in 1925 P/O A.T. Gowling was saved after dispatching a pigeon SOS. (Rumour has it that several hungry downed pilots weighed the benefits of a bird in the hand versus a bird in the bush—to the pigeon's disadvantage!)

The use of homing pigeons for carrying messages has a long history. The Greek poet Anacreon mentioned carrier pigeons in the fifth century B.C., the Romans used pigeons, the Crusaders found them in use by the Saracens, and they are credited with delivering many messages to and from besieged Paris during the 1870–71 Franco-German war.

MacLaurin tackled the difficult problem of releasing pigeons without harm from aircraft moving at over 60 mph in a series of experiments with different types of wrapping. The most successful was a partially torn paper bag that protected the bird from the initial blast of the slip stream and released it without injury. Cpl. Dickie was ordered to establish a pigeon farm at Camp Borden in 1926, but bought himself out of the service rather than move from the west coast. By 1928, the RCAF was operating eight lofts in Canada, the largest at Rockcliffe with over 2,000 birds, many from Dickie's original stock.

The HS-2Ls from Jericho Beach photographed the Fraser River estuary and Cowichan and Chadsey Lakes. An aerial survey near Chilliwack was

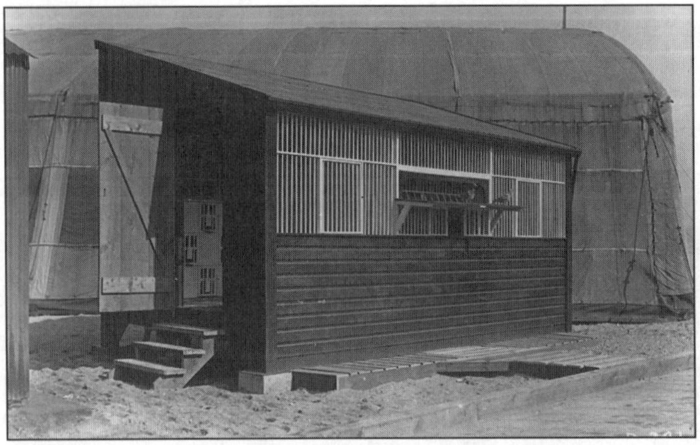

Pigeon lofts were a permanent feature at Air Force stations. In September 1931, the one at Jericho Beach was in better shape than the weary World War I Besonneau hangar in the background. *DNDPU PL 143818*

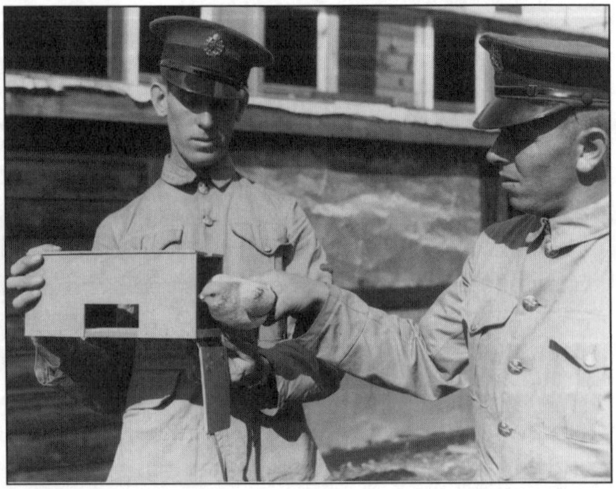

Personnel inserting pigeons into carrying case, Rockcliffe, 21 September 1931. *NAC PA62838*

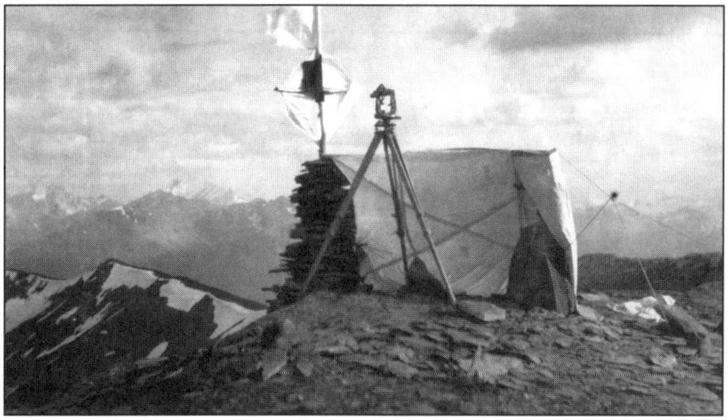

Survey triangulation point in the Rocky Mountains, c.1921.
NAC PA202765/EVELYN LAMBART COLLECTION

The Geodetic Survey Branch had a long-term program to establish basic triangulation points across the high mountain ranges in southern British Columbia. These points provided the basis on which all maps would be plotted. They had to be accurately located with no visual obstruction between each point and the next, ranging from five to 40 miles apart. This task would require time-consuming treks and laborious trial-and-error ascents of rugged and heavily treed mountains.

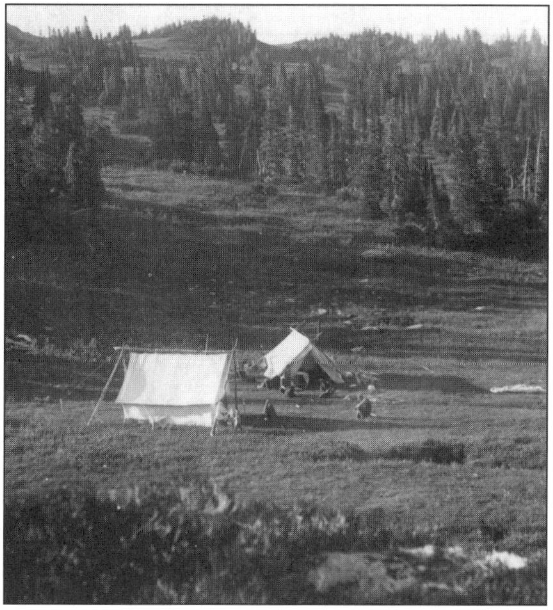

Survey party camp in the Rocky Mountains, 1922.
NAC PA 202782/EVELYN LAMBART COLLECTION

initiated that eventually led to the draining of Sumas Lake and the addition of some 10,000 acres of farming land to the Fraser Valley. A mosaic of Vancouver Harbour had to be made with extraordinary accuracy as enough film was allocated to allow only a five percent overlap, instead of the 60 percent later considered necessary for optimum results. Only one small gap, in the centre of Stanley Park, appeared when all the photographs were assembled, but Earl MacLeod recalled that it was neatly filled to everyone's satisfaction with a slice of trees from a photograph of the Seymour Creek area.

H.F. Lambart, DLS returned to Vancouver on 15 July 1921 from a frustrating two-week trip by boat in deplorable weather to triangulation stations at Bruce, near Sidney, and Constitution, on Orcas Island. He immediately sought the assistance of Maj. MacLaurin at Jericho Beach. Only too willing to demonstrate the value of his aircraft, MacLaurin took him, a colleague and camping equipment in an HS-2L to Alouette Lake. The two surveyors climbed to their selected peak the next day and an enthusiastic Lambart reported, "Under ordinary means that same journey would have meant leaving by train for Port Hardy and thence by pack train over a trail to the lake, a journey of at least two days." He noted the densely forested and steep rocky slopes where the slash and underbrush was, "damnible [*sic*], that is the only way to describe it, it's a fight through the woods, not a walk, at one mile per hour."[15]

Always conscious of costs, Lambart noted the economy of the HS-2L which, he estimated, used 35 gallons of gasoline and one gallon of oil per hour,

15 H.F. Lambart, Collection MG30, C243 Vol. 3 (Ottawa: National Archives).

High River, south of Calgary, was opened in 1921 to replace the base at Morley, Alberta, about 100 miles west, closed due to hazardous flying conditions caused by turbulent winds in the lee of the nearby Rocky Mountains.
NAPL CA47-2

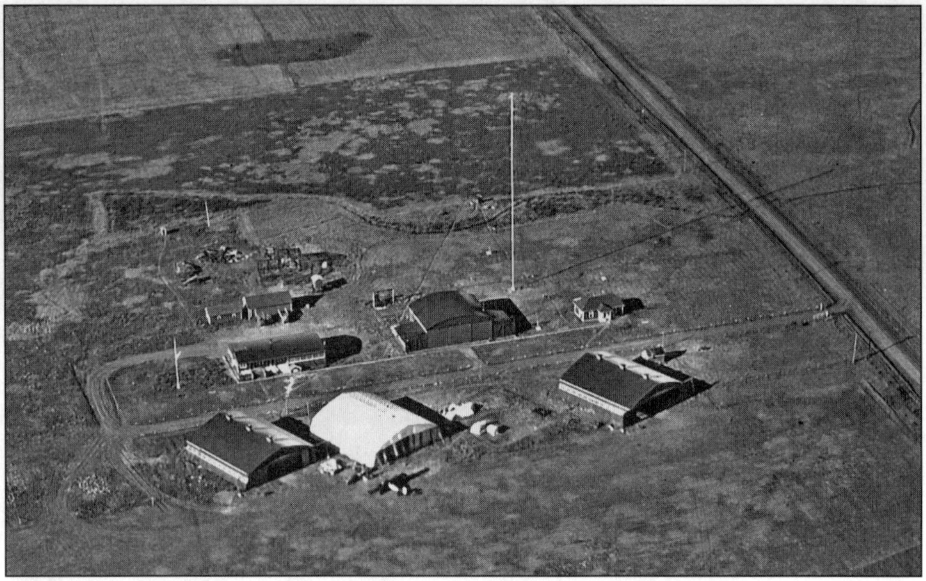

Besonneau hangar, tractor and mobile workshop, all part of the 1919 Imperial Gift, Victoria Beach, 27 September 1921.
NAC PA53268

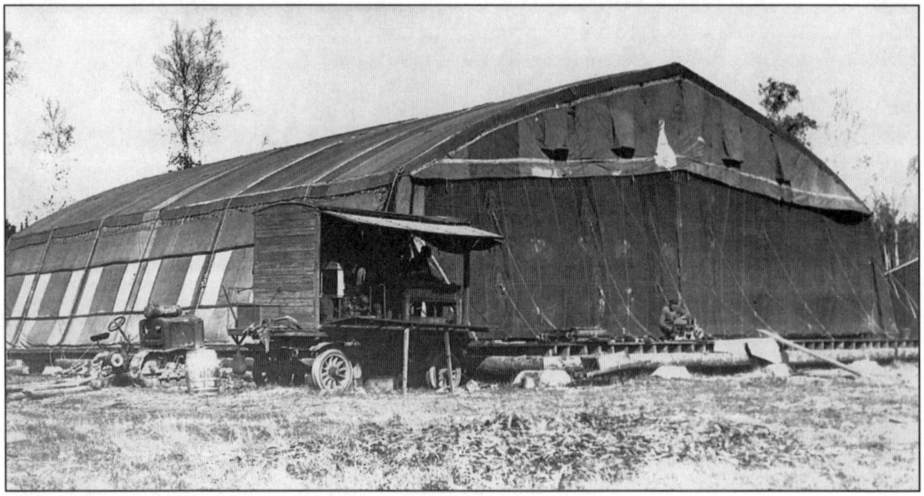

making the hourly running cost about $12. The only other aircraft on station, a twin-engine F.3, he expected to be "vastly more expensive." Fifteen HS-2L and one F.3 flights were made in the 1921 season for a total of 28.24 hours covering 2,026 miles at an average speed of 71 mph. Lambart believed this venture to be the first application of the aeroplane to primary triangulation reconnaissance. He summarized the cost at $4,700 for the survey plus $2,300 for the CAF, which he compared with $19,000 and three seasons "under ordinary methods." His conclusion: "Triangulation survey in difficult mountainous country can ill-afford to do without the services of an aeroplane."[16] On his return to Ottawa, he and Col. Stedman laid down their specification for a photographic and patrol flying boat that materialized as the Canadian Vickers Vedette in 1924.

16 Ibid.

With a total of 710 flying hours, High River was the busiest station in 1921, primarily on forest patrol. Time was found, however, to take Col. Maynard Rogers, Jasper Park superintendent, on three flights over virtually every valley that could be used for a trail in the park. Col. Rodgers commented, "It appears almost incredible that one could have covered such enormous distances in so short a time and with such absolute comfort and enjoyment."[17] A photographic survey was made on behalf of the International Joint Commission of the Waterton Lakes District and the courses of international waters at the Alberta–Montana border.

In addition to forestry patrol, aircraft from Victoria Beach photographed along the Winnipeg River for the Water Powers Branch, and provided transportation for survey parties and their supplies in the district east of Lake Winnipeg.

Extensive aerial forest surveys were conducted with two HS-2Ls from a temporary base at Sioux Lookout, northern Ontario, on behalf of the Ontario government that contributed $15,000. The nearby railroad inspired a novel alternative to living in tents. Known as the Northern Ontario Mobile Unit, four railway boxcars were converted to sleeping quarters, cookhouse and mess, operations office, and workshop under the command of Capt. A.W. Carter. Sub-bases were established at Minaki, Allanwater and Banning to facilitate the survey of an area 200 x 60 miles. From 20 May until 25 October, when the lakes started to freeze, 188 sorties were made for a total of 329 flying hours.

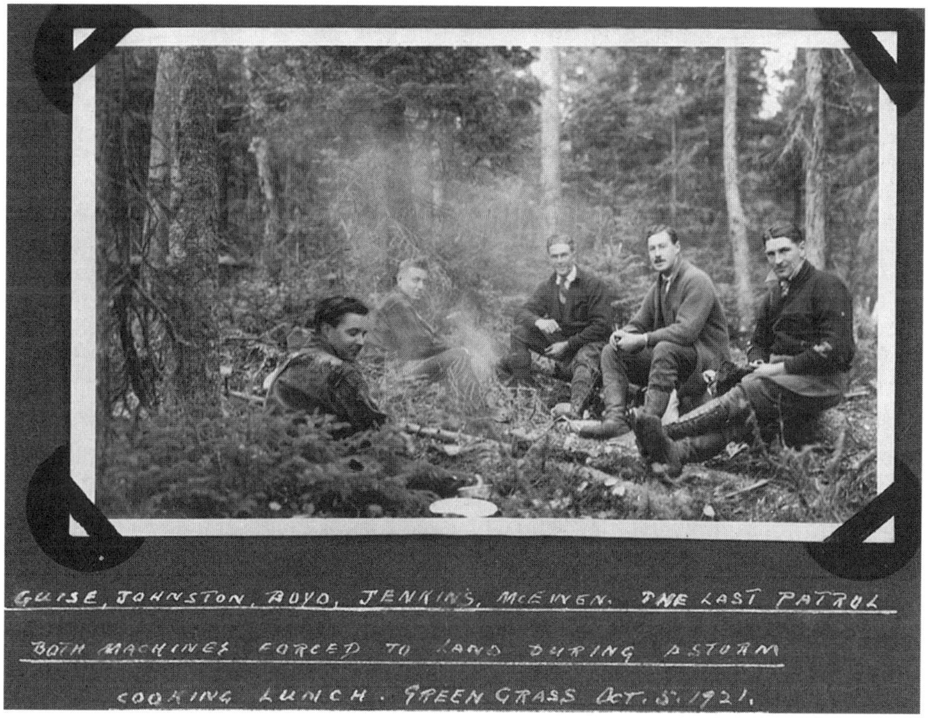

Stretching the flying season, two HS-2Ls were forced to wait out a storm, 5 October 1921. "Black Mike" McEwen on the right.
DNDPU RE15414

17 Hitchins, op. cit., p. 62.

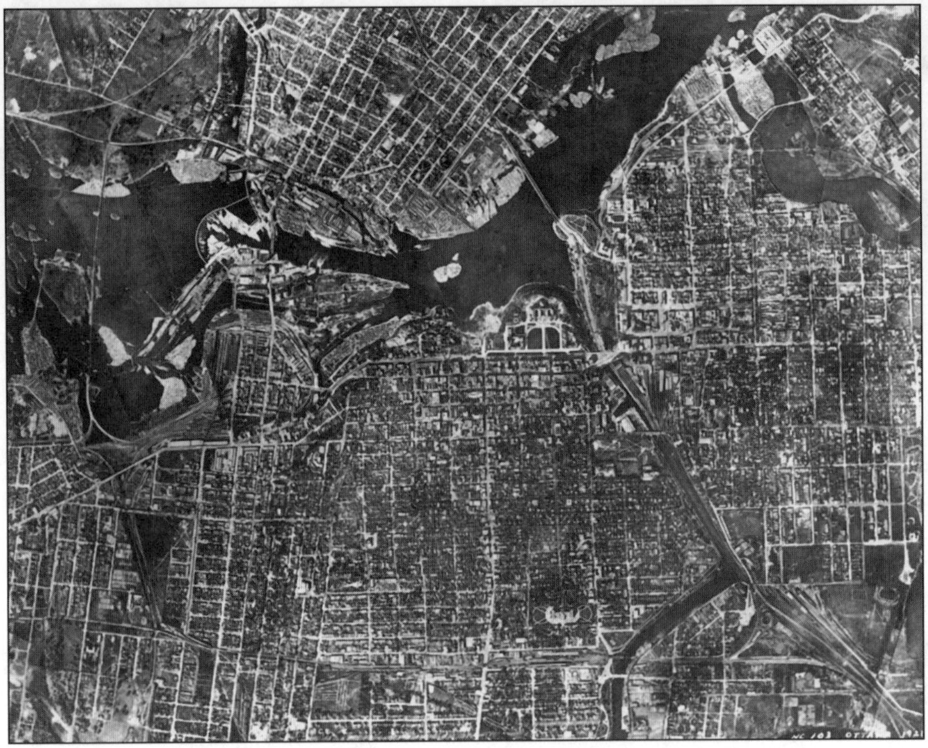

Early teething problems with matching the edges of photographs to form mosaics were solved by "rectified" prints of each negative that compensated for their individual differences. This mosaic of Ottawa with the Parliament Buildings in the centre was completed in 1921. *NAC PA64580*

The operation was a success and provided valuable experience for future detached photographic operations.

The Ottawa Station continued its experimental photographic work during 1921 with the cameras developed by Prof. Cooke. Photographs were taken of the Gatineau region, Quebec, and mosaics were completed of London and the Welland Canal, Ontario.

The London operation was an experiment to establish what could be accomplished with large-scale mosaic photographs and was under the direction of Maj. Douglas Henry Nelles, DLS, supervisor of Topography in the Geodetic Survey Branch. The primary objective was to create a topographical map on which details could be filled in later by a ground survey party. Some urgency was attached to the mission as it was considered important to photograph the city "before the leaves come out in the trees and bushes."[18] Capt. Hubert L. Holland, Rockcliffe superintendent, visited London to make arrangements with civic officials and he identified an appropriate landing field at Merwin Heights. Meanwhile, a D.H.4B was modified at Camp Borden by enlarging the photographer's cockpit to accommodate a gimbal mount for an Eastman Kodak K-1 camera with a 20" lens and 100-exposure film magazine. Holland and Air Photographic Inspector Lieut. Edward R. Owen exposed six rolls of film from 5,000' on 28 April to photograph 42 square miles with a 33 percent overlap.

18 Ibid., p. 57.

Rain delayed departure of the aircraft from London for two days and its propeller was broken when landing on the waterlogged airfield at Leaside, near Toronto. Owen continued to Ottawa by train with the exposed film; Holland remained to have the propeller repaired, completed the flight to Rockcliffe on 5 May. The CAF sent a message to the chairman of the London Chamber of Commerce expressing "sincere appreciation of your valuable assistance . . . without your hearty co-operation the work could not have been brought to so successful a conclusion."[19]

Thoughts were turning in 1921 to "the practicality of providing a deep waterway from the Great Lakes to the sea."[20] In May, L.J. Burpee, secretary of the US–Canada International Joint Committee appointed to look into the matter, requested that he be flown along the waterway from Kingston to Montreal: "A view from the air of the river and surrounding countryside would make it possible to interpret more intelligently the information contained in maps."[21] His request was granted and resulted in 13 photographic flights by Capt. Holland and Lieut. Owen between 5 July and 9 August in, as available, an HS-2L or D.H.4B. Numerous vertical and oblique photographs were taken and the commission was enthusiastic about the resulting mosaic: "It gives an exact representation of the country which can be used for the revision of existing maps and the preparation of new ones."[22] Burpee's comments attracted widespread attention and his 8 February 1922 letter of appreciation also noted the practical value of the photographs that "in many cases have thrown a new and important light on the engineering problems which the Commission has had to deal with."[23] The photographs were the basis of engineering work that saw the St. Lawrence Seaway open to ocean traffic 38 years later.

In co-operation with Ottawa, "refresher courses" in aerial photography were introduced at Camp Borden. Although aerial photography was in its infancy, returning airmen had brought home a considerable body of information. As far back as 1914, enthusiastic pilots used their personal cameras to take vertical front-line pictures through a hole cut in the cockpit floor of their flimsy machines, or "obliques" taken with hand-held cameras while peering through wings and bracing wires. By 1918, thousands of photographs were taken daily of enemy fortifications and troop movements in France, and as far afield as the Balkans, East Africa and around the Indian Ocean. Cameras, film and photo development, and printing techniques were all improved to satisfy the perpetual need for "intelligence," which made a major contribution to Allied victory.

Operations from Roberval were handicapped by smoke from forest fires during the early part of the flying season, followed by unusually wet weather, which quenched the fires but did not allow much photography. Nevertheless, 102 flights were made for preliminary surveys and fuel caches established as far north as Lake Mistassini in preparation for the following year's work. Quebec

19 F.H. Hitchins, Photographic Operations (Ottawa: Directorate of History, Department of National Defense).
20 F.H. Hitchins, Air Board, Canadian Air Force, Royal Canadian Air Force, 58.
21 Ibid.
22 F.H. Hitchins, Photographic Operations.
23 Ibid.

Establishing a fuel cache at Stacker Lake, Quebec, 1921. K.M. Guthrie carrying fuel, T.A. Lawrence standing on the HS-2L flying boat.
NAC PA114524

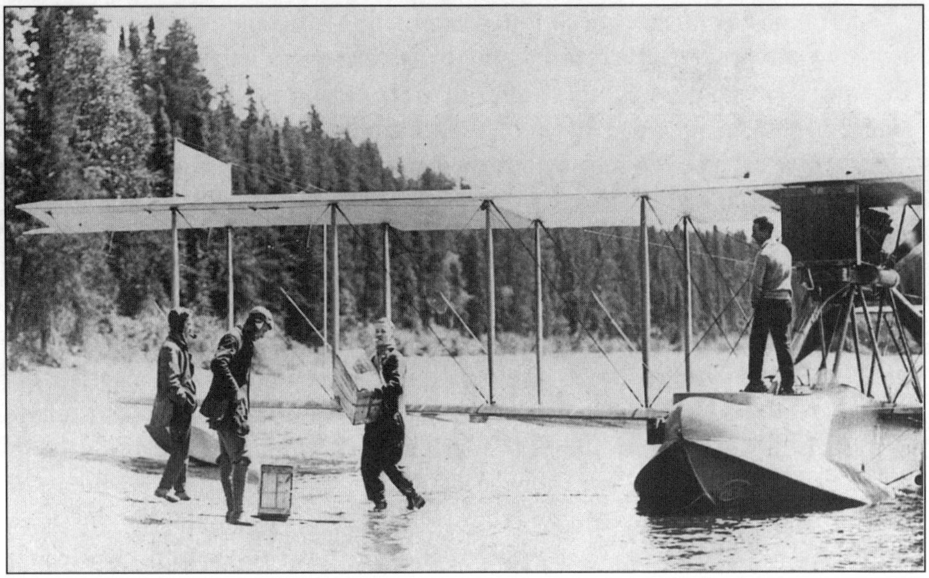

provided $20,000 towards base maintenance costs and the province invested in the erection of permanent buildings.

Despite the success of basic triangulation surveys in British Columbia, the CAF reluctantly turned down a similar request for support of the Geodetic Survey Branch in the Magdalen Islands. Reasons given included lack of landing and maintenance facilities and, "the questionable suitability of the only available machine."[24]

Dartmouth was used primarily for the repair and overhaul of flying boats, but the station CO, S/L A.B. Shearer, was asked to undertake a photo mission in his HS-2L to resolve a dispute by the International Boundary Commission. Determination of the main course of the St. Croix River would establish the nationality of an island near St. Stephen, New Brunswick. Shearer could only struggle to 3,000' because of engine overheating and, deafened by the roaring Liberty engine, could only communicate with the camera operator by handwritten notes. Nevertheless, a successful run was completed from St. Stephen to Milltown, contributing to award of the disputed island to Canada.

About 8,000 aerial photographs were produced during the 1921 flying season, a fivefold increase over the previous year.

By 31 March 1922, 1,335 officers had been commissioned and 1,384 airmen had enlisted in the CAF. More than half the officers, however, were members of the service in name only, not even being summoned for training. About 500 officers and most of the airmen served for only one training month before being sent on unpaid leave. Of the remaining officers, many were civil servants who donned uniform only as required. Only about 50 officers could be considered on permanent duty. The non-permanent, non-professional nature of the force inevitably resulted in constant change of personnel. Even

24 Ibid.

at headquarters, there were three changes of command during 1921. Lt.-Col. Stedman added "Supply" to his technical responsibilities, and was assigned an aide, Capt. S.G. Tackaberry, who held the title of assistant director of stores for the next 18 years. Much of the correspondence from "Ottawa" during this period was signed by Tackaberry.

To replace the departed Stuart Graham, Laurentide hired charismatic and flamboyant William Roy Maxwell, who had served as an instructor with the RAF and CAF, barnstormed in a Curtiss JN-4 Jenny, and established the Canadian Aero Film Company. A second pilot, Allan Gordon McLerie, with a similar background, was also recruited. Two mechanics, Roméo Vachon and Irenée (Pete) Vachon, transferred from the CAF. J.F. Hyde and J. Sherbourne completed the team. Four days after operations started on 1 May, the Seagull crashed with pilot Maxwell blamed for "improper flying." He and Roméo Vachon were only slightly hurt, but passenger W.M. Bowden, an executive with Laurentide, was seriously injured. The first photographic expedition photographed 140 square miles of timber about 100 miles northwest of Grand Mère, providing the required information at a significantly lower cost than by ground survey. At the end of the season, Laurentide, however, decided that it was too expensive to carry the whole cost of aerial operations themselves. Maxwell was offered the job if he could establish a company able to attract enough other work to be self-sustaining.

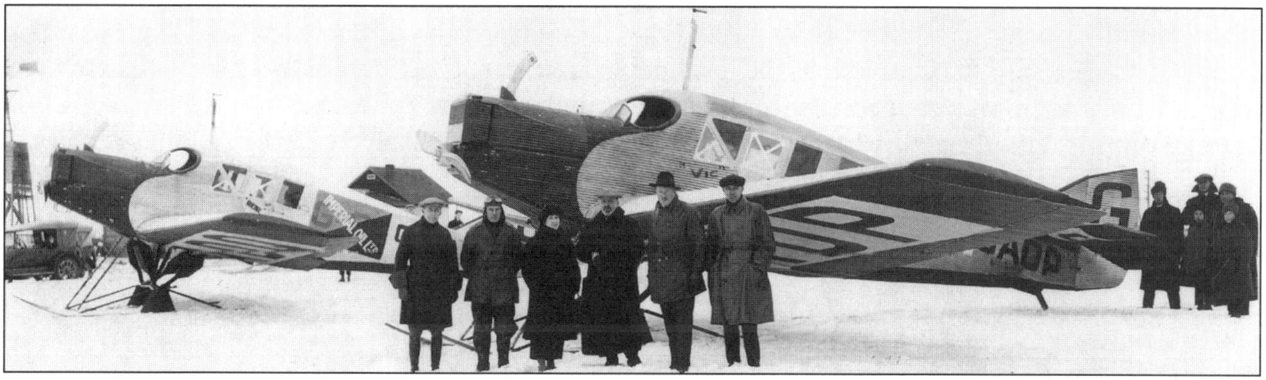

Junker-Larsen JL-6s René and Vic at Edmonton prior to departure for Great Slave Lake, March 1921. Elmer Garfield Fullerton, *Vic* pilot, on the left; George Gorman, *René* pilot, second from the right. Both were World War I pilots, Fullerton with the RNAS and Gorman in the RAF. *GLENBOW ARCHIVES NA2301-1/MRS E.G. FULLERTON*

Imperial Oil Limited was the first company in Canada to exploit the potential of aerial survey in the search for oil. Two Junkers JL-6 monoplanes were sent in 1921 from "end of rail" in the Peace River country over the trails blazed by 12 ground teams in 1919 through the foothills of Western Canada, during which oil had been discovered near Fort Norman on the Mackenzie River. The two single-BMW engine aircraft, named *Vic* and *René*, made the first flight into the Northwest Territories. They took supplies and transported geologist W. Waddell, DLS and RCMP Sgt. Hubert Thorne as far as Fort Simpson on the Mackenzie River where the ski undercarriage and propeller of *René* were damaged when it broke through the snow crust. As *Vic*'s engine was giving trouble, its skis and propeller were transferred to *René*, but this aircraft stalled and crashed soon after takeoff, fortunately not injuring the crew. Enough parts were available between the two wrecks to salvage *Vic*, but the propeller could not be repaired.

Having no contact with "the outside," the crews seemed sentenced to a long wait at Fort Simpson until the spring breakup allowed a request for a new propeller to be dispatched by canoe. One of the first entries into the annals of bush plane heritage was made, however, when mechanic William R. Hill and Hudson's Bay carpenter Walter Johnson combined their talents. Fullerton recalled: "There was enough of the broken propeller available that could serve as a pattern from which templates could be made. We learned that the Catholic Mission had a few oak sleigh boards 10 feet long and 7 inches wide, which was just about the size we needed to make a laminated propeller, except for being about one inch too narrow. The difficulty was solved by the eventual realization that by 'fanning' the boards we could get the additional width needed. We further learned that moose-hide glue was available at the Catholic Mission."[25]

After a search that was not concluded until 1945, both huge, hand-crafted propellers, 115 inches in length, are now at the National Aviation Museum. Arguments have been made that they are more deserving of a place in the National Art Gallery than some of the acquisitions made in recent years.

Meanwhile, *Vic*'s engine was given an overhaul by Fullerton and the other mechanic, S. "Pete" Derbyshire. Two propellers were completed and balanced in less than three weeks. *Vic* was the only airworthy aircraft and returned safely to the Peace River base 500 miles south, fortunately just one day before the snow melted and landing on skis would have been impossible. Derbyshire stayed with *René* and the repaired aircraft was flown out the following spring to reach its intended destination of Fort Norman. *René* transported men and supplies to "bring in" the landmark first producing well on the Mackenzie River at Norman Wells. The use of aircraft for mineral exploration and the support of isolated resource exploitation soon became common practice with all the larger oil companies.

Dr. Theodore A. Link, Imperial Oil's chief geologist and an articulate advocate for aerial photography, explained: "Truly, aerial photographs are to the geologist what x-ray photographs are to the medical doctors." He noted that a geologist may see detail of which he was previously entirely ignorant in spite of the fact that he had already covered the area on foot. He believed that aerial photography was a science, dependent on the greatest co-operation between several independent departments:

> Namely, the surveying branch, the aviation branch, the photographic department, and the map-making department. If one of these falls down on the job, unsatisfactory results are to be expected. If the pilot of the plane is unable to fly his machine along a true course at a reasonably constant altitude the best photographic work will be of little use. On the other hand, if the pilot has completed his work in a perfect manner, poor photography will ruin his painstaking effort.

25 Elmer J. Fullerton, "Pioneering Flying in the Sub-Arctic," *Canadian Aviation*, 7, no. 6 (1934): 12.

> There are two kinds of aerial photographs which may be employed in map making—oblique and vertical. The latter are the most commonly used and give the best results from a scientific and practical standpoint. If the camera is pointed through a hole in the floor of the plane straight down at the landscape, so that the focal plane (the film or plate) lies parallel to the surface of the earth, a vertical photograph will result. This amounts to nothing more than a map of the area in question. An oblique photograph is one in which the film or plate of the camera lies at an angle with respect to the surface of the earth. In a picture of this kind, the various objects will necessarily be distorted and the result is not a true map of the area photographed. Oblique photographs bring out the relief of an area, i.e. mountains, valleys, trees, houses, etc., much more satisfactorily, but not in true perspective.[26]

Link continued, explaining that accurate measurements of relief could be obtained from overlapping vertical photographs with the aid of a stereoscope. He cautioned that the techniques used varied widely depending on the local topography, atmospheric conditions, use to be made of the photographs and the overall economics of the project—"It costs plenty of money to send a high-powered plane up to a height of 16,000 to 18,000 feet and keep it up there for from four to six hours."

While the basic idea of taking photographs from high altitude and converting them into detailed and accurate topographical maps sounds simple, many problems had to be solved. Cameras of the day were large, cumbersome,

26 T.A. Link, "Aerial Photography Applied to Geology," *Imperial Oil Review*, 14, no. 4 (1930): 42.

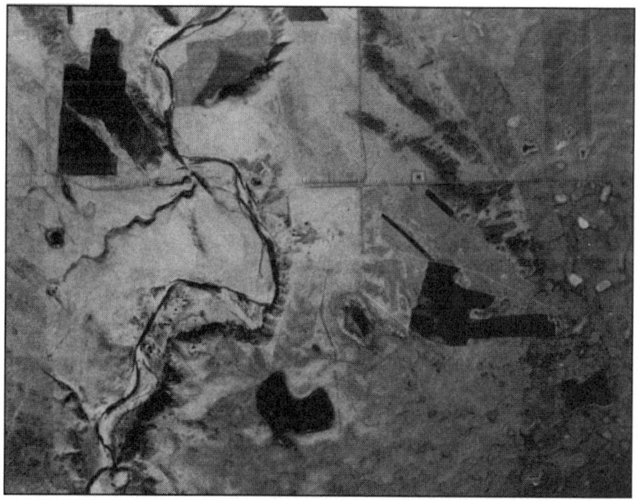

PHOTOGRAPH BY WESTERN CANADA AIRWAYS LIMITED

FIGURE 1

A vertical aeroplane photograph taken from an altitude of over 16,000 feet. This is essentially a map of the area covered on the picture. For an explanation of the photograph see Figure 2.

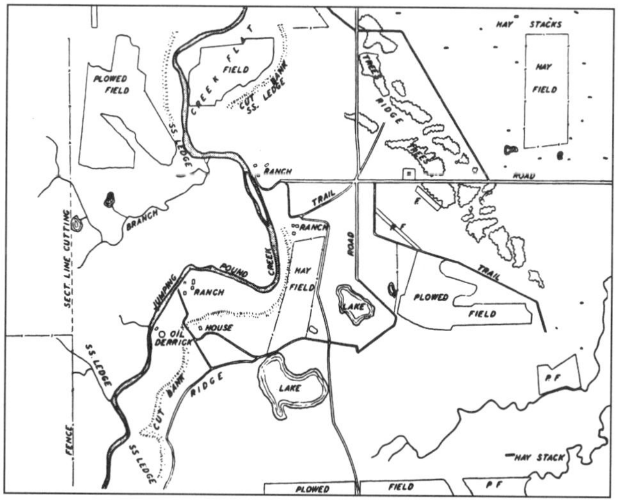

FIGURE 2

Map constructed on data taken from photograph in Figure 1. Only the most outstanding features are shown on this map. Note the hay stacks, cow trails, gullies, ploughed and unploughed fields, ranch houses, barns, cut banks, hills, bed rock outcrops, etc.

Reproduced from *Imperial Oil Review*, Aug/Sept 1930, illustrating the early use of vertical photographs to prepare detailed topographical maps.
BY PERMISSION OF IMPERIAL OIL LIMITED

temperamental and difficult to handle in cramped, open cockpits. Film plates and strips were only in the early stages of development and required long exposure to obtain acceptable results. Flying straight and parallel "lines" was an extremely taxing operation requiring close co-operation between crew members. The navigator would establish the initial line by sun or compass, preferably at between 45 and 135 degrees to the sun in order to minimize glare. While flying along this line, he would sketch prominent details visible along the parallel line several miles distant to be flown on the return trip. This information would assist him to calculate drift and determine the precise course for the second line. This routine would be repeated until the operation was complete, or fuel or weather called a halt. He would also have to determine ground speed, compare it with altitude and calculate the frequency of exposures for a given overlap. Until the introduction of automatic cameras, he would advise the photographer to trigger the shutter by means of a string tied to his wrist. The photographer, meanwhile, had to juggle the camera and change magazines while standing in his open cockpit. The pilot concentrated not merely on holding the line, but also on

Illustration reproduced from Report on Civil Aviation 1925, showing vertical and oblique methods of photography for mapping purposes. A vertical view covers a relatively small area and calls for extreme accuracy in flying parallel "lines." A set of three oblique photographs covers a much wider area, but perspective distortions have to be corrected.

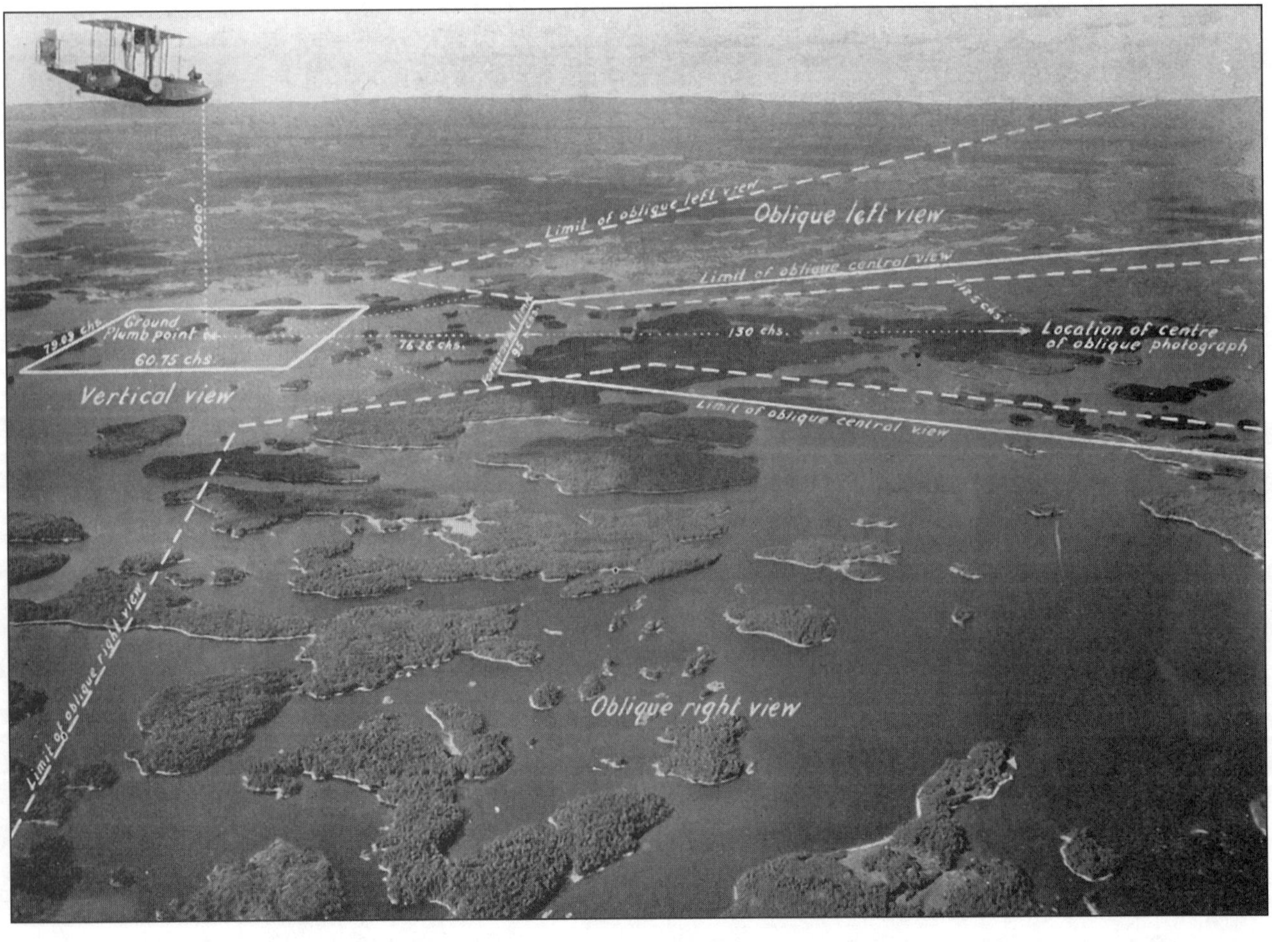

keeping the aircraft on a horizontal plane at a constant altitude and speed, no mean task when buffeted by turbulence and winds that could match the aircraft's speed. A variety of drift indicators, sun compasses, perspective plotting grids, depression indicators, etc., were improvised by the crews and hung on the aircraft. To such an extent that one Viking with its registration letters EZ obscured by navigation aids was known as "Earnest Zebra."

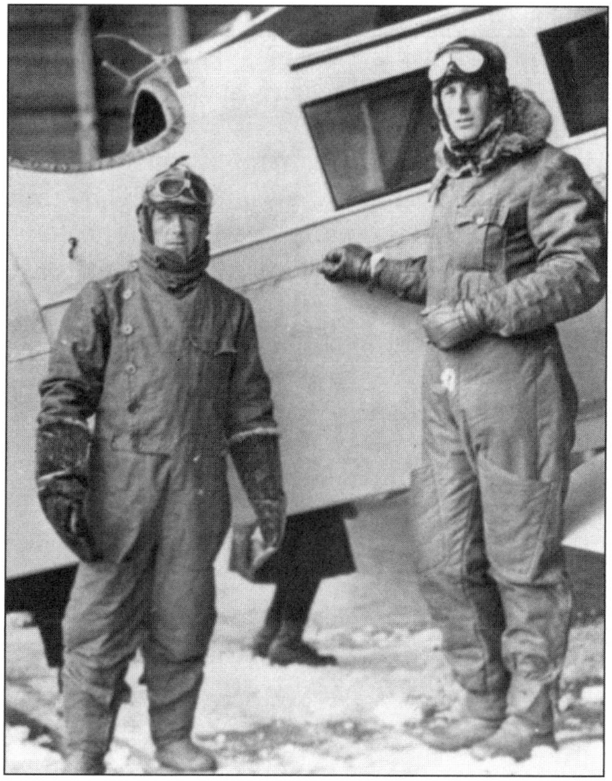

F.S. Cotton, right, and H. Stanward with a Martinsyde Type A Mk 1 aircraft of the Aerial Survey Company (Newfoundland) Ltd. Both wearing Sidcot Suits. Probably at Botwood, Newfoundland, c. March 1922. *NAC PA140793*

Frederick Sidney Cotton had many inventions to his credit, one of the earliest resulting from service with the RNAS during World War I. He found the issue flying suit woefully inadequate and commissioned his London tailors, Robinson & Cleaver, to make one to his design—a thin fur lining, waterproof silk, an outer layer of light Burberry material and fur at neck and cuffs, with capacious pockets below the knees. The "Sidcot Suit" was adopted by many other fliers and Cotton generously waived any royalty payments. Versions of it were used for many years by Canadian photographic aircrews in their open cockpits.

Difficult though it was, taking the photographs was possibly the easiest part of the job. Contact prints of the negatives comprising a mosaic usually did not join to make a perfect match. Changes in altitude and attitude of the aircraft and variations in ground level would change the scale of adjacent photographs. This was overcome by enlarging or reducing each print until they did fit. Any tilt of the aircraft—an almost continuous situation as its flight path was corrected manually by the pilot trying to fly a straight line—produced distortions that frustrated attempts to bring all the area covered by a photograph to the same scale. This was remedied by mounting both enlarging camera and easel on gimbals, enabling them to be individually manoeuvred until a match was achieved. Light flashes on the film caused by static electric discharge plagued the photographers. Even the bromide paper on which the prints were made caused its share of problems by contracting and expanding during development and fixing. A better understanding of the mathematics involved and the development of ingenious mechanisms enabled maps to be produced with great accuracy, but it was still a tedious process. The accuracy of a topographic map, however, ultimately depended entirely on ground surveyors determining the exact location of the triangulation points on which it was based.

Frederick Sidney Cotton was one of the most colourful characters in the early days of aerial photography. With Alan S. Butler and V. Sidney Bennett, he formed the Aerial Survey Company (Newfoundland) in January 1921, pioneered airmail service in Newfoundland, and made the first flights to Labrador and to the French islands of St. Pierre and Miquelon. He became a leading figure in the RAF aerial photography unit during the Second World War with a reputation for getting results in unorthodox ways. He was fond of quoting his Eleventh Commandment, "Thou shalt not be found out." One of his most audacious ventures took place just before the war. With hidden Leica cameras whirring, he flew senior German officials in his "Kolossal Lockheed," a new Model 12A purchased with the rumoured connivance of British Intelligence, over military installations. His was the last civil aircraft to leave Germany before the invasion of Poland, photographing the fortified Siegfried Line along the Franco-German border on the way out.

1922-1923

1922 WAS A CRITICAL year for the CAF. The Conservative-Unionist government of Arthur Meighen had been displaced the previous December by Mackenzie King's conservative-minded Liberals. The CAF estimate for 1922–23 was reduced to an even $1 million from the previous year's $1.6 million, but revenues from British Columbia, Ontario and Quebec did provide an additional $64,000. In a general reorganization of the air force, the previous practice of appointing semi-permanent staff for six- to twelve-month periods was discontinued with the temporary establishment approval for 45 flying officers, 24 ground officers and 238 other ranks. Refresher training for ex-RAF flying personnel was discontinued in favour of infusing "new blood" from Canadian universities. The senior levels of the CAB experienced important changes. Col O.M. Biggar, vice-chairman, and Sir Willoughby Gwatkin, inspector general, who had been largely responsible for the administration of the organization during its formative period, resigned. Lt.-Col. Robert Leckie, director of Flying Operations, completed his three-year secondment and returned to the RAF, as did several other officers.

Aircraft on strength during 1922, almost all survivors from the Imperial Gift and the US Navy contribution, included: Avro 504K (33), D.H.9A (9), SE5A (11), HS-2L (17), Bristol F.2B (2), Curtiss JN-4 (8), D.H.4 and 4B(9), Sopwith Snipe (2), Felixstowe F.3 (5), Curtiss H-16 (1), and Martinsyde F-6 (1). Several remained in their packing cases and others were not in flying condition. Plans to replace the obsolescent aircraft were stopped in their tracks by the reduced funding.

The CAB was very conscious that aviation had to be "sold" to a skeptical and parsimonious government, and to an even more suspicious taxpayer whose impressions of aircraft had been formed during their development to assist the slaughter in France. The 1922 annual report identified work undertaken for a variety of other government departments and underlined the improvements achieved in operating efficiency and the resultant financial savings. Tangible results during 1922–23, boosted by continuing favourable comments on the

1921 St. Lawrence River photographs, resulted in an encouraging report on aerial survey. Benefits were claimed for improved maps of the prairies, which, the report grumbled, would have been greater "in better financial conditions." The aerial surveys provided basic information for water conservation and irrigation projects in central and southern Alberta that eventually converted dry prairie into fertile farms. Proponents continued to drive home the message that surveys by air could be completed at less cost and much faster than by ground parties.

Note was made in the annual report of the insistence by "practically all branches" on the importance of establishing a base for air operations in the MacKenzie River Basin, perhaps with Imperial Oil's pioneering work in mind: "Communications are poor and distances are great. The length of time taken to reach the scene of operations unduly shortens the period during which field work can be undertaken each year." However, "financial considerations" did not permit any operation along the MacKenzie during 1922, and the thought was expressed that this region should be left to the private sector.

The Pacific coast presence at Jericho Beach was strengthened during 1922. A superintendent, two pilot navigators, twelve mechanics, one storekeeper, one photographer and three miscellaneous employees brought the establishment to 20. Temporary Besonneau hangars deteriorated in the damp climate of the Pacific coast and a permanent steel and wood hangar was built to replace them, along with a 5,000 gallon fuel tank and concrete apron. Equipment included two Felixstowe F.3 twin-engine flying boats and three HS-2L single-engine flying boats. The F.3s were used for the longer-range work and the HS-2Ls for "shorter patrols, photography and work where a useful load not exceeding 800 pounds is required."[27] A long drought initiated many forest fires, 57 starting in one week during June. Effective support was provided to the firefighters, but smoke frustrated photography until late in August when a mosaic of the Fraser River Valley in the Nicomen Island area was completed in five hours and ten minutes' flying time to show the status of the Sumas reclamation work.

Maj. Claire MacLaurin, one of the first ten graduates of the Curtiss Aviation School in Toronto on 20 June 1915, had an impressive war record with the RNAS and RAF. After the war, he could have taken over the family's prosperous timber business at Lachine, Quebec, but decided to make the air force his career. He was a leading pioneer in the development of aviation in Canada, playing a key role in establishing the CAB.

Jericho Beach superintendent Maj. Claire MacLaurin died in the crash of his HS-2L G-CYEA off Point Grey on 11 September. Shortly after takeoff, a fuel leak was detected and, rather than risk a fire in the air, MacLaurin decided to make an emergency landing, but stalled and hit the water in a nose-down attitude about 200 yards from shore. Passenger John R. Duncan, engineer in charge of the Sumas Lake reclamation scheme they were planning to inspect, died of his injuries, but Engine Fitter A.L. Hartridge escaped with only shock and a sprained ankle. This was the first flying boat fatality in three years of operation. S/L A.E. Godfrey, MC, AFC, took command at Jericho Beach on 18 October 1922.

Maj. George M. Croil, AFC, CO at High River, Alberta, had a staff of 24 in 1922, including four pilots and one photographer. Their aircraft were

27 Canadian Air Board, Report of the Air Board for the Year Ending March 31, 1922 (Ottawa: Dominion of Canada, 1923), 33.

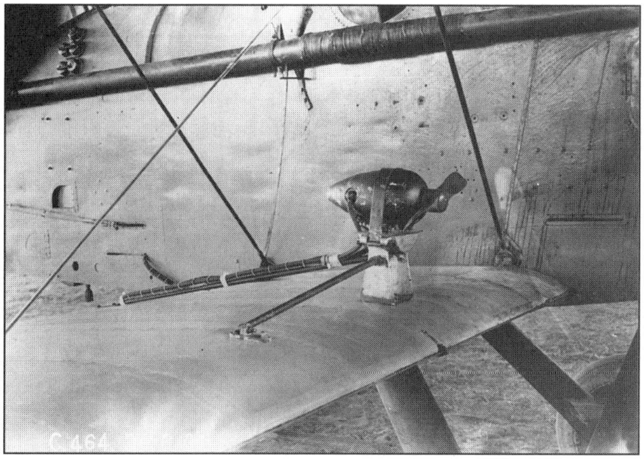

This wind-driven generator powered the wireless set in a High River D.H.4 patrol aircraft. Similar generators were subsequently installed in photographic aircraft to drive automatic cameras. *NAC PA52137*

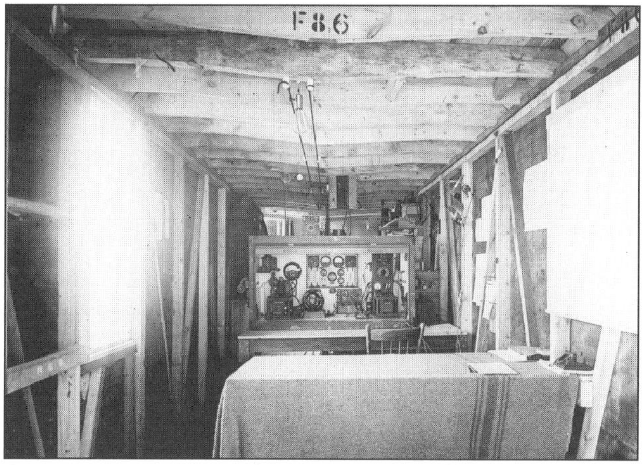

Communications were the responsibility of the Royal Canadian Corps of Signals. The Wireless Room at Victoria Beach, 8 August 1922. *NAC PA53255*

somewhat-weary D.H.4s and D.H.4Bs powered by Rolls-Royce Eagle VIII engines. These aircraft were reported to be very reliable: "practically every patrol"[28] undertaken was completed, and no forced landing occurred during three years of operation. Over 906 flying hours were carried out for the Forestry Branch, including vertical and oblique photography in Dutch Creek, Ptolemy, North Kootenay and Middle Kootenay Passes, and vertical photographs of Dutch Creek Valley and between Dutch and Racehorse Creeks. A mosaic of Banff townsite was completed, oblique photographs were made of the principal valleys in Rocky Mountain Park (renamed Banff National Park in 1930) for the National Parks Branch, and photographs of

28 Ibid., p. 37.

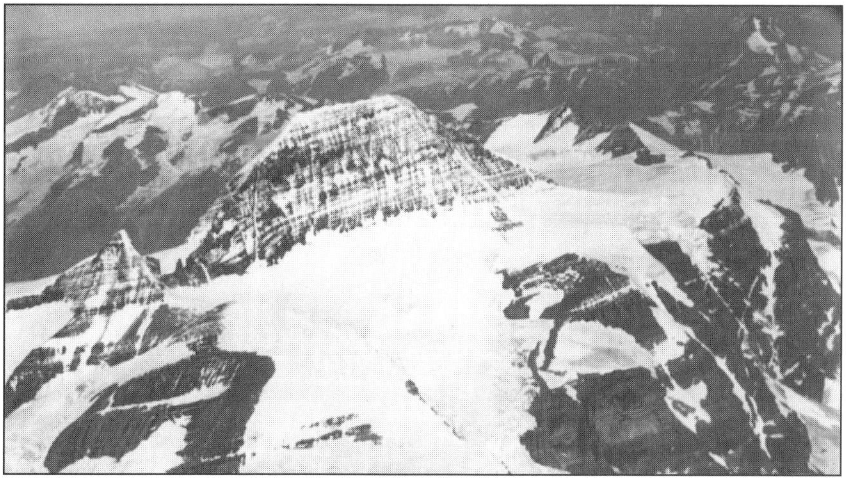

Mount Sir Alexander Mackenzie from an altitude of about 12,000 feet. One of the 24 photographs taken by H.F. Lambart, DLS, from a D.H.4B on 16 July 1922 before a jammed magazine prevented further exposures. *NAC PA 202783/EVELYN LAMBERT COLL.*

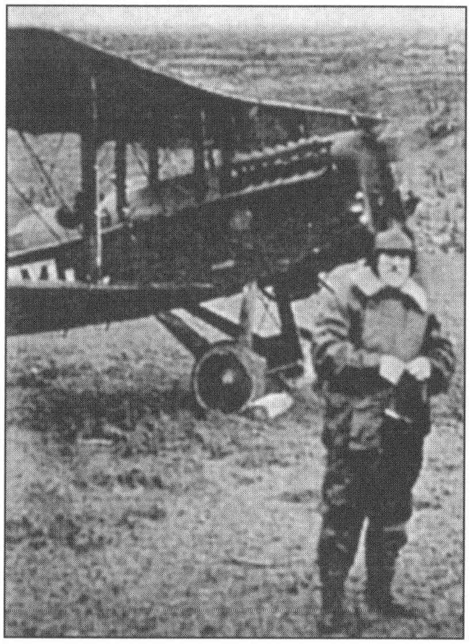

Photograph of A.O Wheeler "Ready To Start" by H.F. Lambart. *COURTESY OF THE ALPINE CLUB OF CANADA*

Jasper Park were made available to the Geodetic Survey. For the Water Powers Branch, flights totalling 17 1/2 hours were made on photographic missions over the Bow River drainage basin, Devil's Gap in the Lake Minnewanka District, Spray River Valley, and the headwaters of the Ghost and Red Deer Rivers. The temporary Besonneau hangars were replaced by permanent, wooden structures with concrete foundations, and workshops and storehouses were constructed.

Impressed with his experience in using aircraft to establish basic triangulation points in the Fraser River Valley, H.F. Lambart requested similar assistance along the line of the continental watershed from Yellowhead Pass northwest to the intersection of the 120th meridian and Jarvis Pass. Lambart and A.O. Wheeler, who had extensive experience of mountain phototopography on the British Columbia-Alberta Boundary Commission, met Capt. J.H. Tudhope and a "mechanician" at a level stretch of prairie at Henry House by the Athabaska River on 12 July 1922.

Tudhope made five flights in his two-seater D.H.4B, totalling about 20 hours, taking the surveyors in turn over unmapped territory in Jasper Park and the provincial boundary north of Mount Robson. With Wheeler aboard he was forced down by low cloud at McBride, 125 miles down the Fraser River, and had to purchase 12 gallons of gasoline for the return flight. Tudhope recalled problems in changing magazines at Henry House where fine dust managed to infiltrate everywhere, compounded by the fact that his hands began to perspire as soon as he put them in the changing bag. Both surveyors were favourably impressed by their adventure and Lambart telegraphed J.A. Wilson in Ottawa: "Weather magnificent . . . Mount Robson clear. . . . Information obtained far reaching." Wheeler's impressions of the 1922 flight were quoted in the CAB annual report:

> As a means of mountain reconnaissance the aeroplane offers exceptional facilities. Given a clear day and the ability to keep to known landmarks it is to a topographer a study of a living map, the most accurate that can possibly be had. I was enabled to get a clear conception of the country my future surveys would cover and the nature of the access to them, and in one case was able to obtain information that will prevent a considerable loss of time. In conclusion, I may say that the aeroplane service can be applied to distant and difficult surveying with a great advantage and I should think could be used as a means of saving expense.[29]

Lambart also was an articulate advocate for aerial survey, noting that in less than four hours' flying time on 16 July he had covered over 300 miles on the round trip to Jarvis Pass. He compared this with 14 days by pack train to just reach this destination.

29 Ibid., p. 38.

For the Reclamation Service, Department of the Interior, 43 hours and 30 minutes of aerial photography were undertaken from High River over the streams rising in the eastern slopes of the Rocky Mountains. Location of reservoir sites was the key objective: "A complete photographic record of the courses of the more important streams in this district has now been obtained. This cannot but be of great value to the Reclamation Service in its work of preparing plans for the conservation of water in this district and schemes for irrigation."[30] Photographs were also taken of the Waterton, St. Mary's, Belly, Saskatchewan, Red Deer and Clearwater Rivers, Sherbourne Reservoir, Swift Current Creek, and proposed dam sites on Battle River and Tail Creek.

For the Topographic Survey Branch, Department of the Interior, mosaics of the Bow River from Kananaskis to Calgary, Township Number 28, and Moose Mountain Forest Reserve were made to provide topographical details for the ground survey of these areas. Completing a busy program, the mountain areas at the head of the Ghost River and Warparous Creek, The Gap, and Red Deer River were also photographed to assist surveyors for a total of 43 hours and 10 minutes' flying time. High River was by far the busiest station in 1922 with 267 departures, accumulating a total of 1,075 flying hours.

Vertical photographs were taken during flights totalling 32 hours 49 minutes to assist with the survey of the Ontario–Manitoba boundary and along the river from Winnipeg to Lac du Bonnet to assist the Water Powers Branch

30 Ibid.

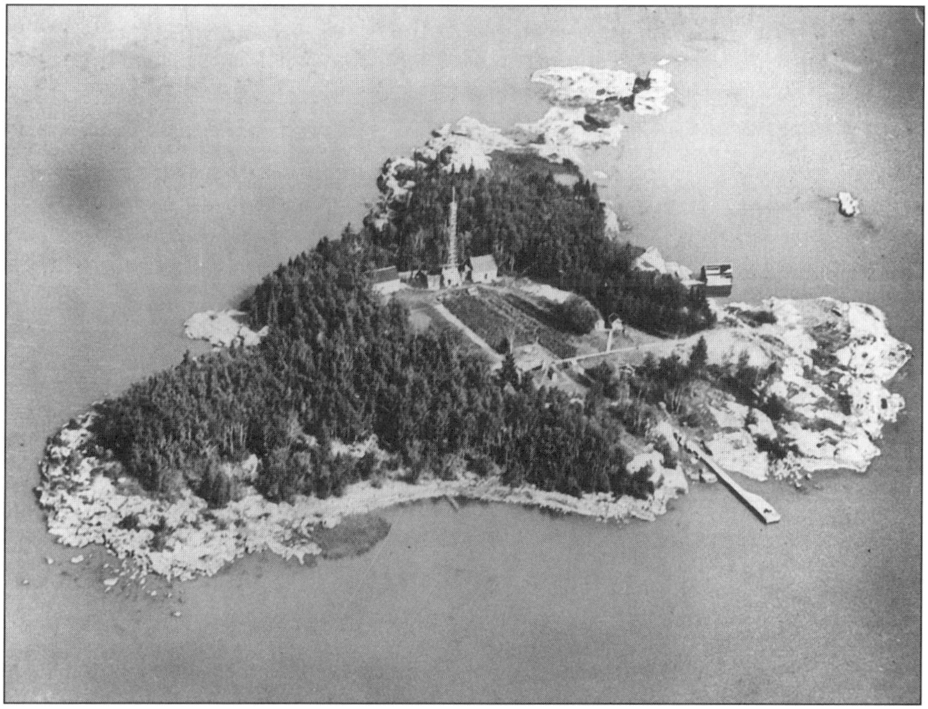

Victoria Beach's sphere of operations was extended again in 1922 with a sub-base at the District Forester's Headquarters on Forestry Island in the Nelson River near Norway House, about 40 miles north of Lake Winnipeg. The Pas sub-base at the junction of The Pas and Saskatchewan Rivers was also activated. Under the command of S/L B.D. Hobbs, DSO, DSC, 307 hours and 40 minutes of flying were carried out from the three bases by four pilots supported by one photographer, one storekeeper, 24 mechanics and seven other miscellaneous employees.

The substation at Forestry Island, 8 September 1921. *NAC PA52355*

select hydroelectric power generation sites. For the Geological Survey, a depot of 1,100 pounds of supplies was established for a survey of Beaver Dam Lake, and the officer in charge taken for an extended observation patrol. Additional supplies were to be taken in and the survey party transported back to civilization, but this was not undertaken due to the unavailability of aircraft. For the Department of Public Works, four and one-half hours of flying enabled the district engineer to check various water level gauges.

While most of the season's work from the Winnipeg Station was devoted to forestry operations and the transportation of treaty-paying parties, "a considerable program of aerial photography" was undertaken. Athol Maxwell

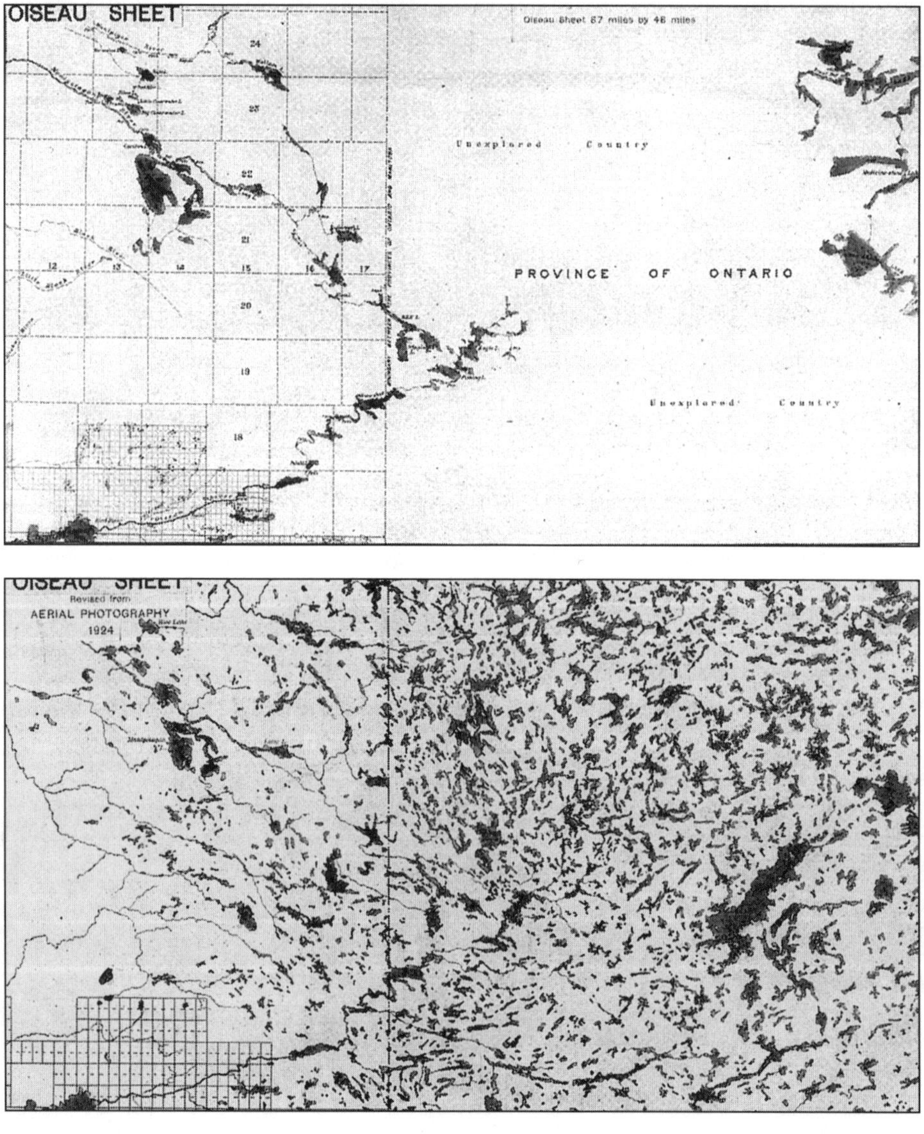

Comparison of the Oiseau topographical map, on the Manitoba–Ontario border before and after aerial survey. Oblique photographs of "Unexplored Country" were "tied in" to the border, which had already been accurately surveyed, without the necessity of any additional groundwork. *REPORT ON CIVIL AVIATION, 1924*

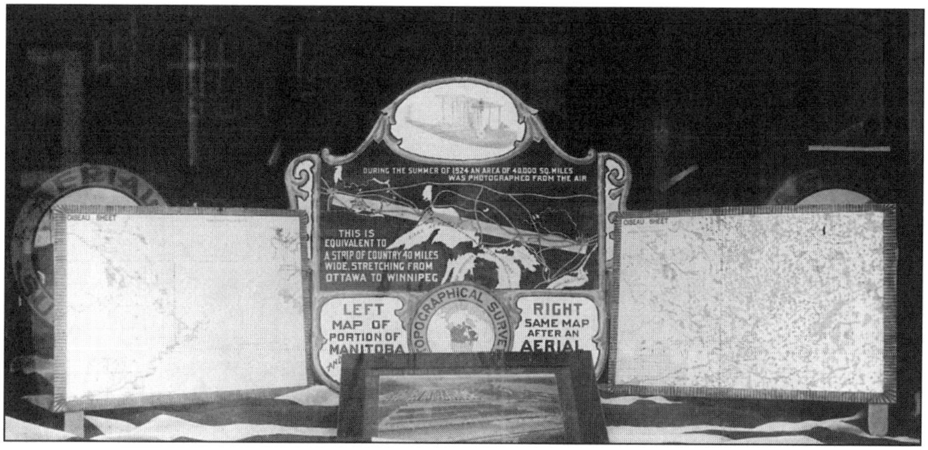

The same two maps formed part of a window display at the Topographical Survey offices in the Labelle Building, 105 George Street, Ottawa, in 1924. *NAC PA45658*

Narraway, controller of Surveys, Topographical Survey Branch, Department of the Interior, was flown on forestry patrols for over 14 hours during which he experimented with oblique photography. He returned from a 1,100-mile flight from Victoria Beach an enthusiastic and vocal proponent of aircraft for both aerial photography and for the supply of ground parties. The first leg of his flight was 150 miles to a land surveyors' camp on the Ontario–Manitoba border—a two-hour flight that would have taken three weeks by ground travel. His journey continued to Norway House, The Pas and Reindeer Lake, mostly unmapped territory. Narraway commented: "The knowledge of the country to be gained on such a trip by canoe would be confined to within the shore lines of streams and lakes followed. . . . A clear unobstructed view was obtained of the country twenty miles on each side of the route followed. . . . Many lakes and water courses previously unmapped are being noted."[31]

The CAB's annual report emphasized the economic value of preliminary exploration by air: "By ordinary methods of travel this inspection work would have taken practically the whole season and would have been very expensive." Narraway was also quoted:

> Perhaps the most interesting and useful part of the trip was north from Pas to the Churchill River and to the camp of the surveyor mapping the waterways of the district; the most interesting because of the wonderful views obtained of this country of lakes; the most useful because of the experimental photographic work which we were able to do. Flying over the course of the survey, oblique photographs were taken with the Eastman K-2 camera at intervals of about two minutes. In some of the photographs, the survey pickets are clearly discernable and in all of them the prominent land features surveyed can be accurately located. Using the survey of these main features as a control, the details of the shoreline and the intricate mass of islands can be filled in from the photographs

31 A.M. Narraway, Report, Natural Resources Canada 1, no. 9 (October 1922): 1.

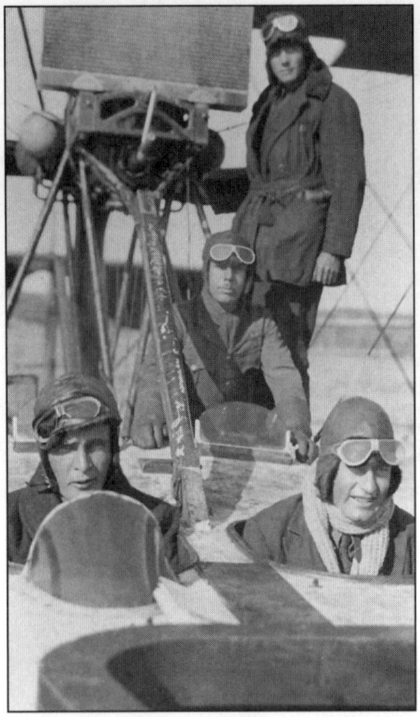

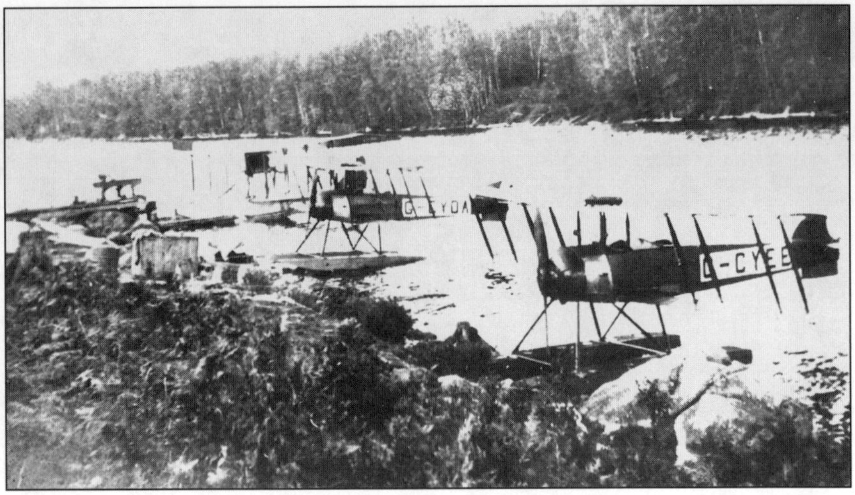

Avro 504Ls, two in the foreground at Whitney, were employed to assess the ability of the smaller and cheaper aircraft for survey in isolated areas. Two seats and a meagre payload severely limited their usefulness, and most of the work was done by the larger flying boats. The Avros also had a series of minor accidents necessitating all five float-equipped versions being brought into service at some time in order to keep two in the air. The HS-2Ls, one in the background, fared better, with only one "spare" having to be brought in while repairs were effected.

A detachment of three pilots, two storekeepers, seven mechanics and one labourer under the command of F/L C.M. McEwen was dispatched from Rockcliffe as soon as the ice broke up to temporary bases at Whitney and Parry Sound. Two Curtiss HS-2L flying boats and two Avro 504Ls on floats were assigned and the Ontario Forestry Branch provided two observers. Overall control was by District Forester W.A. Delahaye, on right, with aircrew members, left to right, Jack O'Gorman, George Brookes and Frank Watson in their HS-2L.

with accuracy consistent with the requirements. The surveyors were obliged to spend considerable time mapping the shoreline details and islands and even then could only obtain what was confined between the two shores. With the plane, the time required to get in the details can be estimated in minutes instead of hours and in some cases days, and the scope of the camera widens the survey to several miles on either side of the waterway and includes lakes and topographical features which otherwise would be left unmapped.

As a result of this trip I am of the opinion that invaluable service to surveying in unexplored and unmapped territories will be rendered in the future by the seaplane; in transporting supplies and men, thereby saving for the work many days of the working season which is already short enough; in permitting reconnaissance and exploratory flights for purposes of planning surveys; in keeping in communication with the surveyors during the season; and in taking photographs along the course of the survey for filling in detail. To those who have a thorough knowledge of the ground conditions and experience in sketching and mapping such as surveyors have, the value of flights over the districts to be surveyed can scarcely be overestimated.[32]

Narraway, a friend of CAB secretary J.A. Wilson, spoke at many public meetings in favour of aerial survey and exerted his considerable influence on his government colleagues. Narraway's superior, Surveyor General Dr. E.G.D.

32 Report of the Air Board for the Year Ending March 31, 1922, op. cit., p. 42.

Deville, reported enthusiastically on the progress made by R.B. (Bob) McKay, a mining engineer, experienced land surveyor and mathematician, in co-operation with Professor Cooke to develop a system for transferring photographs to topographical maps. Vertical photographs from 5,000' to 11,000' and obliques from lower altitudes at various locations were studied and experiments undertaken with various cameras and lenses over the Gatineau Hills, north of the Rockcliffe base, which led to a practical solution the following year.

Following successful aerial forest surveys and firefighting support during 1921 in northeastern Ontario by the Northern Ontario Mobile Unit at Sioux Lookout, the provincial government requested another similar operation over 10,000 square miles south of the Ottawa River and Lake Nipissing, reaching from Pembroke to Georgian Bay.

The Whitney and Parry Sound aircraft completed 242 patrols totalling 575 hours and 41 minutes between 23 May and 4 October 1922. Delahaye reported enthusiastically, but not statistically in agreement with the CAB, on 166 patrols averaging 165 miles per flight— "a total considerably in excess of any previously attained by a flying station in the Dominion."[33] Evidently protective of their own responsibilities, the CAB reported that the work was carried out in the greatest co-operation with Delahaye whereas he claimed direct supervision of the whole operation. Ontario established its own air service in 1924.

Photography was only a small part of the operation with Ontario in 1922, but the experience gained by the air force detachment on independent operations led directly to the formation of the specialized detachments that photographed large expanses of Canada during the 1920s and 1930s. It also generated a welcome $38,909.23 for the CAF coffers. Also establishing a pattern followed for several years, the HS-2Ls were flown to Rockcliffe for overhaul in Ottawa during the winter while the Avro 504Ls went to Camp Borden via Allandale on Lake Simcoe. Most of the photo detachment (PD) personnel wintered at Ottawa where they were kept busy on maintenance and repair, and preparing for the next season's missions.

In 1920, the Quebec government demonstrated an early appreciation of the possibilities offered by aviation in providing the site at Roberval on Lac Saint-Jean, north of Quebec City, and the facilities were continuously improved during the next few years. The CAB reported that the base was not officially operational until 30 June 1922, "owing to delays in reaching an agreement [with the province] as to terms of operation."[34] Quebec agreed to contribute a maximum of $20,000 in 1922 towards operational costs. The superintendent was, for the second year, F/L W.R. Kennedy, DFC, with a staff of two pilots, photographer, storekeeper, eight mechanics and five other employees. One hundred and forty-one flights totalling 186 hours 30 minutes were made before closing down on 22 October as ice formed on the lakes. As the main objective was exploration, survey and fire protection of the largely

33 Ibid., p. 44.
34 Ibid., p. 45.

uncharted region north and west of Lac St. John, a sub-base was established at Stacker Lake, about 100 miles farther north. One of the three HS-2L aircraft was permanently assigned to hauling supplies to the sub-base, otherwise accessible only by a five-day canoe journey.

Detached operations were also carried out by a lone HS-2L, G-CYEK, along the north shore of the St. Lawrence River to survey the forest resources in the basins of the St. Marguerite and Natasquan Rivers. While attempting to take off from rough water in the Gulf of St. Lawrence on 21 September, 'EK was damaged and sank off Anticosti Island before it could be beached. Pilot F/L W.R. Kennedy and his crew were rescued, along with some of their equipment.

Alphonse Landry, chief of Quebec's forestry department, reported very favourably on the 1922 program from Roberval, "exploring 7,145 square miles of land about which we had little or no information from a forestry point of view."[35] Bad weather restricted photography from Roberval to only 15 days from the middle of July to the middle of September, but Landry reported 2,225 vertical photographs taken, "which, in an unbroken length, would represent 600 miles, and cover a surface of 325 (square) miles."[36] Only 75 obliques were possible, however, before the equipment went with G-CYEK to a watery grave in the Gulf of St. Lawrence.

Although it grew considerably over the following two decades, in 1922, Ottawa Air Station at Rockcliffe consisted of only one small storehouse and a Besonneau hangar. The hangar was sheathed with wood that spring, but workshop facilities were non-existent. Wheeled aircraft were sent to Camp Borden for overhaul and winter storage until the winter of 1921–22, when temporary quarters were leased at 362 Sparks Street, Ottawa, for the repair and overhaul of some engines and airframes. In the summer of 1922, "admirable"[37] winter quarters were leased in the government shipyards on Victoria Island, located in the Ottawa River between Ottawa and Hull. Waterborne aircraft could be flown in, drawn up a slipway, and worked on "under excellent conditions."[38] For many years, all flying boats operating in Eastern Canada followed this procedure, flying directly to their assignments in the spring.

Rockcliffe, now the site of the National Aviation Museum, was the only station able to accommodate both water- and land-borne aircraft. The 107 flights made there in 1922 were primarily concerned with experiments and testing related to the emerging technologies of aerial photography and wireless telephony. Oblique photographs of the Lièvre and Gatineau Valleys and the intervening area were taken for the National Parks Branch in parallel with Professor Cooke's experimental work on aerial survey and mapping for the Topographical Survey Branch. In September, with the co-operation of the Royal Canadian Corps of Signals, a successful demonstration was made of air-to-ground voice communication.

34 Ibid., p. 45.
35 Ibid., p. 48.
36 Ibid., p. 47.
37 Ibid., p. 49.
38 Ibid.

The death of Maj. MacLaurin at Vancouver in September 1922 triggered the transfer of S/L A.E. Godfrey, MC, AFC, Ottawa CO, to command at Vancouver in October. Godfrey was succeeded at Ottawa by F/L C.M. McEwen when he returned with his unit from Whitney and Parry Sound on completion of forest surveys and fire patrols in the Algonquin Park area. The permanent staff at Rockcliffe during the flying season consisted of four officers and fifteen airmen. The designation "Ottawa Air Station" was changed to "CAF Unit, Ottawa" on 25 November 1922.

Dartmouth, on the north shore of the outer harbour at Halifax, was the only year-round, ice-free base on the Canadian east coast, but was falling into disrepair. It was used mainly as an erection and repair base for flying boats with only 16 flights taking place in 1922. Less "air-minded" than some provinces, Nova Scotia made few demands on the services offered by Superintendent S/L A.B. Shearer, who was supported by one pilot, one storekeeper, five mechanics and one labourer. Aerial photography was started during the fall for the Water Power Branch, Department of the Interior, over the watersheds of the Roseway, Bear and East Rivers, Sheet Harbour, and the hydroelectric developments at Musquash and St. Margaret's Bay. Poor light late in the season and uncertain weather prevented completion of the task.

Thoughts were already turning to survey and mapping in the Arctic by air, and the CAF recognized the need to establish an information base for the operation of aircraft in this harsh environment. An opportunity offered itself when the Department of the Interior, conscious of an increasing foreign presence in the Eastern Arctic, mounted an expedition to establish police posts and customs houses. The necessary funds were not available to send an aircraft and pilot, as originally planned, but it was possible to include a CAF representative. Maj. Robert Archibald Logan, a pilot and qualified surveyor,

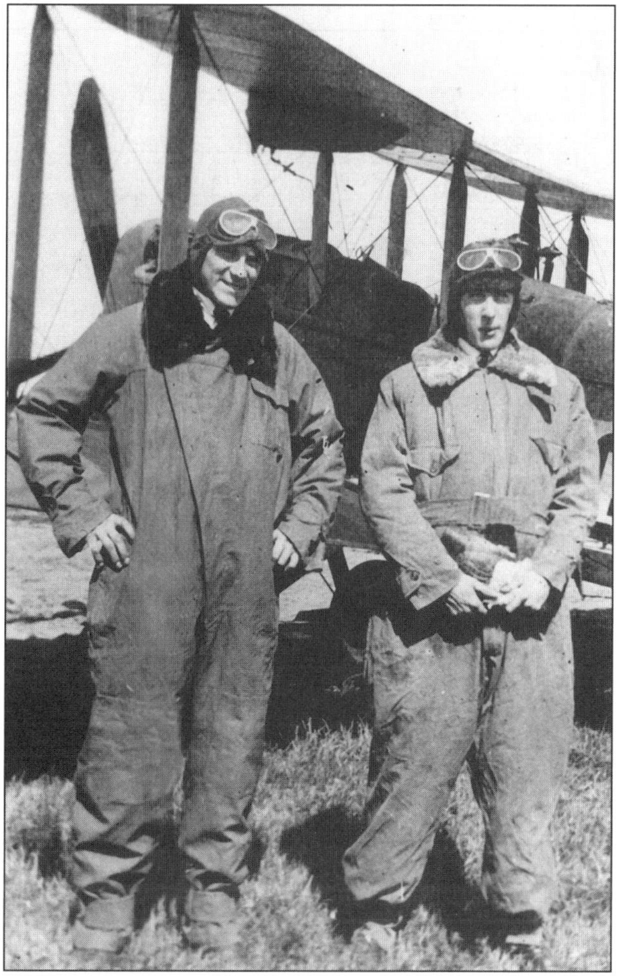

On 23 April 1922 Capt. Hubert L. Holland, CO of the Ottawa Station, and Lieut. Edward R. Owen were en route from Camp Borden to complete photographic coverage of London started the previous year, when their D.H.4B crashed near Brantford, killing Holland and seriously injuring Owen. Here, Holland is seen on the left with photographer A.E.E. Church and a D.H.4 at Rockcliffe in 1921.
NAC PA117841/JOHN V. MULVEY COLLECTION

The task was completed in November by F/L R.S. Grandy and Sgt. A.J. Le Sueur, exposing five rolls of film from a Camp Borden Bristol Fighter. The director of the geodetic survey was evidently pleased with the results as he commented that the photographs, "constitute perfect work by the officers concerned . . . they are to be congratulated on the accuracy of their work and skill with which it was carried out."[39]

39 F.H. Hitchins, *Air Board, Canadian Air Force*, Royal Canadian Air Force, 78.

Captain Joseph Bernier was the right man to lead an expedition into dangerous and largely unknown waters. He was born in 1852 at L'Îslet, Quebec, went to sea as a cabin boy at 14 years, and was captain of a ship hauling lumber to Britain by the time he was 17. Exploring the Arctic Archipelago from 1904 to 1911 with the CGS *Arctic*, he established Canadian claims to the region and became an expert polar navigator. After a period as a trader on Baffin Island, he commanded merchant ships crossing the Atlantic throughout World War I. He retired in 1925 and died in Lévis, Quebec, in 1934.

Nova Scotian Robert Logan, born in 1892, was a dominion land surveyor. He qualified as a pilot at his own expense and enlisted in the RFC where he developed aerial photography techniques before being shot down and captured in 1917. After his Arctic journey, he played a pivotal role in the formation of the CAF and RCAF before pioneering aerial survey operations in Central Africa and route development across the Atlantic. He returned to Canada from the post of general manager of the Irish Government Airlines at the beginning of World War II and was sent to survey airfield sites in Greenland. After transfer to the US Army Air Corps, he was injured in 1944 while on a

experienced in meteorology and radio communication, with previous experience in the North, was detached from his instruction duties at Camp Borden.

Under the command of northern explorer Capt. J.E. Bernier, the CGS *Arctic* sailed from Quebec City on 18 July 1922 with 43 members of the expedition. Posts were established at Craig Harbour[40] on Ellesmere Island and Pond Inlet on Baffin Island where Logan identified and surveyed "aerodrome" sites. Photo-topographic surveys were quickly completed from high ground at both sites and two other promising locations. Dr. J.D. Craig, DLS, MEIC, leader of the expedition, noted the abundance of aerodrome sites, "thus obviating the necessity of using hydroplanes in the north, an exceedingly dangerous operation over waters apparently clean and open, but necessarily containing many small pieces of ice, difficult to 'spot' from a rapidly moving plane."[41]

The two police posts established in the Arctic Archipelago administered an area extending about 1,500 miles north from Hudson Strait and 500 miles wide, containing islands with a total area exceeding that of Alberta and Saskatchewan. Radio communication made it possible to obtain accurate time checks and the expedition navigators corrected many points that were found to have been previously plotted "seriously in error." Logan pointed out that the interior of the islands was practically unknown and the coastlines only roughly sketched. He emphasized the value of aircraft for topographical survey, exploration, transport of men and materiel, and assisting marine navigation by locating ice fields and open water. He felt that, as in the rest of Canada, flying operations would be feasible during five months of the year. He recommended

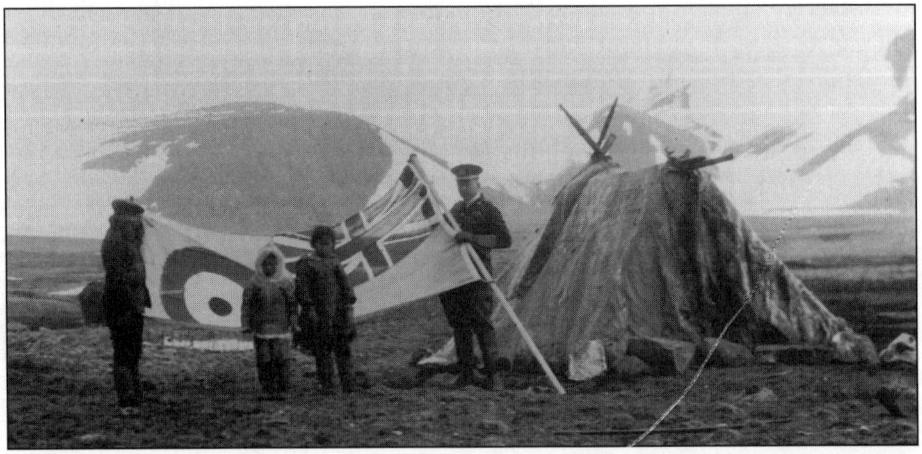

S/L R.A. Logan, with Eskimo guide Katko and two of his children, displaying the RCAF flag at Craig Harbour, Ellesmere Island, 29 August 1922. The latitude of 76° 12′ was the farthest north reached by the RCAF for many years. A bronze tablet "Canada NWT" marked the spot. On his return to Ottawa, Logan was asked to return the flag to "stores," but he had left it flying at Craig Harbour. His response is not on record. *DNDPU*

40 Named after J.D. Craig, DLS, officer-in-charge of the expedition.

41 J.D. Craig, *Canada's Arctic Islands: Log of Canadian Expedition 1922* (Ottawa: Department of the Interior, 1923), 24.

that two pilots and two mechanics with two specially equipped small aircraft travel north with the next expedition and stay a whole year, "making flights each day when conditions permit."[43] While cold weather operations were an important consideration for the fledgling air force and continuing cold weather experiments were undertaken at the established bases, little further was done, except on an *ad hoc* basis, to develop an Arctic capability until dictated by wartime pressures some two decades later.

While economic and operational advantages of aviation were emphasized, the 1922–23 report gently grouses about shortcomings in the type and reliability of equipment. The large, unwieldy, expensive to operate, and unpopular Felixstowe F.3 flying boat was the only one available at Victoria Beach for long-range patrols and gave rise to a typical complaint: "With the small crews available for handling such a large machine and the few facilities in the way of slipways and protected anchorages for refuelling, the task of maintaining the patrol has thrown a great burden on the staff of this district. Only by the most strenuous effort have the machines been kept in the air. Until more modern types of machines, smaller and easier to handle, but at the same time capable of giving the same service are available, operations in this district cannot be completed with complete success."[44]

The 1922–23 report expresses satisfaction with the CAF's contribution to aerial photography at the end of what is probably the most formative year of military flying in Canada until the buildup for World War II:

> The developments, during the past year, in the use of aerial photography for exploration and preliminary survey work show great promise. The Survey services are alive to the importance of the problems. The development of methods of mapping from aerial pictures rests with them. There is little doubt that the lines on which they are

special mission with Admiral Byrd in the South Pacific. A man of many talents, Logan was also an authority on the Cree language and a prolific writer on many subjects. Robert Logan laid claim to being the first dominion land surveyor to take a vertical aerial photograph. On 12 March 1916 he piloted his RFC B.E.2 on a photo assignment at 8,000′ behind the German lines in France. An enemy pilot took exception to this intrusion and Logan was wounded in the exchange of machine-gun fire. Nevertheless, he kept at it and exposed all his 18 glass plate negatives before leaving the scene. In 1930, a remarkable sequel to this event occurred at a diplomatic reception at the Mexican Embassy in Rio de Janeiro, Brazil. Logan met a German businessman who turned out to be the pilot he met in very different circumstances on 12 March 1916.[42]

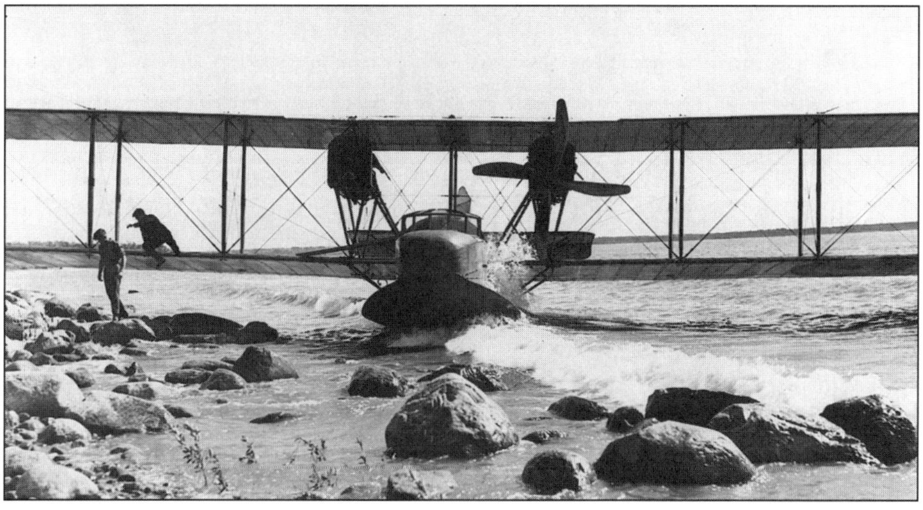

Felixstowe F.3 near Victoria Beach, Manitoba, 5 September 1921. *NAC PA53251*

42 R.A. Logan, Letter to Don. W. Thomson, The Canadian Surveyor xviii (December 1964), 427.

43 *Report of the Air Board for the Year ending 1922*, op. cit., p. 51.

44 Ibid., p. 39.

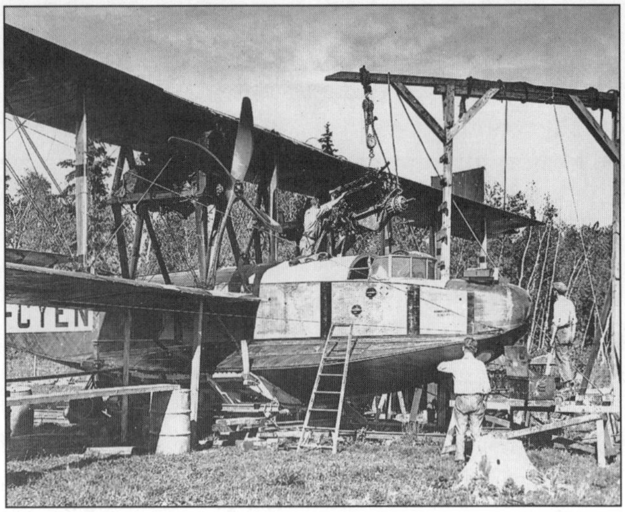

Improvisation was standard practice. Installing an engine in a Felixstowe F.3 at Victoria Beach, 8 August 1922. *NAC PA53249*

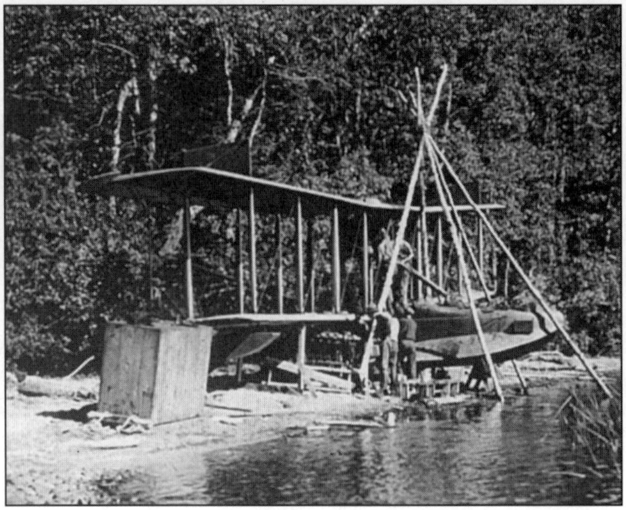

When engines failed away from home base, they had to be changed, as with this Curtiss HS-2L, at lakeside. *ALEX MILNE*

> now working will be further developed with far-reaching results. Even today it is not too much to say that, in unsurveyed country, where there are no great differences of elevation, given geodetic control, that is, points accurately determined at intervals of from thirty to fifty miles apart, the topographical details of the whole district may be filled in from aerial pictures with a speed and certainty unobtainable by any but the most minute ground surveys. The assistance aircraft can be to surveyors, geologists and others, working in the remoter parts of the country, by providing a rapid, economical and easy method of transportation is proved beyond all reasonable doubt.[45]

Perhaps bearing the mark of another hand, sentences become concise and direct when the future is considered:

> What is now required is the provision of aircraft specially built to meet the conditions which obtain in Northern Canada. Without such machines little further progress can be made. The types now in use are obsolescent war machines. Without the aircraft so generously given by the Imperial Government at the close of the war, the work of the past three years would have been impossible. They have given wonderful service. Many of them, after five seasons' constant use, are still fit for much further work, but to obtain the full benefit of the work done, more efficient types are essential. If such machines are available continued progress can be made, without them little further can be accomplished.[46]

Emerging by this time was the idea that ALL of Canada could be accurately mapped from aerial photography. Some idea of the immensity and difficulty of the task is demonstrated by the fact that Prince Charles Island and

45 Ibid., p. 51.
46 Ibid., pp. 51-52.

the adjacent Air Force and Foley Islands in the Foxe Basin, east of Baffin Island, were only "discovered" in 1948 by chance through a break in the cloud cover by an RCAF Lancaster on a photographic mission from Rockcliffe. Presumably, the islands were already known to the native people, but they have a total area twice the size of Prince Edward Island, and were unknown to later arrivals in Canada and appeared on no maps.

Prince Charles was born 14 November 1948 and the island named for him was the second-last major island identified in the Canadian Arctic, the last being in 1952, Stefansson Island in the Western Arctic. Foley Island was named in memory of John Hilliard Foley, RCAF navigator on the 1948 photographic flight, who was killed in a flying accident near Maxville, Ontario, in 1949.

The steady increase in requests for aerial photography led to the establishment of the Topographical Survey Branch within the Department of the Interior, the primary customer, to act in an advisory and consultative capacity for all federal departments. The CAF retained the responsibility for operational aspects while the new branch advised on technical aspects, such as what should be photographed, altitude, flight path, focal length, types of film and, for oblique photography, camera inclination.

After the successful forest inventory and fire control program over Algonquin Park and adjacent areas in 1922, Ontario was moving towards establishing its own provincial air service. Quebec followed suit, but contracted the work to the Dominion Aerial Exploration Company, formed from the flying operations of forestry company Price Bros. & Co. of Chicoutimi. Dominion was operating from Roberval by September 1922 with three Martynside Type A Mk 1s and two HS-2Ls. Dominion's president, Capt. H.S. Quigley, MC, DCM, ex-RAF and ex-CAB, had been chief pilot for Price Bros for the previous three years and continued to conduct aerial surveys and fire detection patrols on behalf of the forestry company.

The defence department had a vested interest in a domestic aircraft industry, which it did its best to promote, introducing the capability that became such an important factor during World War II and a leading contributor to the gross national product today. The 1922–23 budget included $175,000 for eight flying boats to replace the tired HS-2Ls. Under the auspices of senior university professors comprising the Associate Air Research Committee, a variety of projects concentrating on winter flying were under way at Canadian universities. Indicating the problems being experienced, the projects included the operation of aero engines at low temperatures, barograph diaphragms, anti-freeze mixtures, friction of runners on snow, wind tunnel tests, storage of oil at low temperatures, and low temperature effects on rigging.

Two pioneering companies registered in 1922 made a lasting impression on aerial photography: Fairchild Aerial Surveys Company (of Canada) Limited, a subsidiary of the Fairchild Aerial Camera Corporation of New York, and Laurentide Air Service Limited. The latter had, effectively, been operating since 1919, under the banner of the St. Maurice Forestry Protective Association. Both companies owed their birth largely to the initiative of Ellwood Wilson, chief forester for Laurentide, headquartered at Grand Mère, where both companies were registered.

Sherman Fairchild, possibly in reaction to rejection by Eastman Kodak of suggestions he made during the war for improving the K-1 aerial camera, had decided to put his own ideas into practice. With financial support from his father, chairman of one of the companies that later became IBM, he founded Fairchild Aerial Camera Co. in New York. By 1922, he had indeed produced a superior camera that incorporated a between-lens shutter, hitherto considered impossible, and other refinements. Fairchild also had plans to improve photographic film, but a tacit agreement left this to Kodak while he concentrated on cameras.

All aerial cameras to this time had focal plane shutters—a slotted curtain drawn across the lens—with the inherent disadvantage that, as the aircraft moved, the image seen across the width of the film would vary, distorting the picture to the extent that it could be inadequate for accurate map-making. To compensate for aircraft movement, fast shutter speeds of at least 1/150 second were necessary. As the lenses used were about three inches in diameter, a between-lens shutter exposing all the film simultaneously would have to move so rapidly that it was believed it would self-destruct. Fairchild succeeded in designing and manufacturing a between-lens shutter that would open and close a large aperture at 1/150 second. This K-3 camera was one of the most significant milestones in the development of aerial photography for precision mapping purposes.

The K-3 design permitted film changes by exchanging an exposed magazine for a new one without interference with the photographic sequence. A clever mechanism held the film flat at the moment of exposure, preventing buckling and significantly improving the resulting photograph. An electric motor could be set to automatically trigger the shutter at predetermined intervals with a warning light to the pilot a few seconds in advance so he could hold the aircraft steady at the vital moment. This innovative system was gradually adopted until it, and its successors, had a virtual monopoly in North America. The occasional breakdown could usually be repaired by the cameraman while the pilot circled to maintain his hard-won altitude.

In parallel with camera developments, improved film played a large role in making aerial photography both economically and technically practicable. Most wartime cameras used 4"x5" glass plates giving good results, but with obvious disadvantages in terms of weight, size, convenience and cost. The introduction of nitrate-based roll film solved most of these problems but, if ignited, generated its own oxygen leading to a conflagration. Also, it could not be held as flat as a glass plate, thereby introducing an element of distortion. Early emulsions were slow, allowing little leeway in exposure times, and development of the negative suffered from the same handicap. W/C Harold Pearce, MBE, CD, FRPS outlined some of the problems inherent in the processing procedure:

> For many years the old Eastman apron system was employed. This consisted of a massive celluloid, corrugated-edged apron upon which the film was wound and upon which it remained throughout its processing. The only drawback to this system was its bulk, but from a photographic or sensitometric viewpoint it was much better than the modern small tank which winds film from one roller to another back and forth and in so doing lowers the resolving power of the first few exposures on either end of the roll by differential development.[47]

Pearce noted that contrast could be enhanced by adding a stain developer to the developing solution contained in 25-gallon tanks, changed for each roll of film. During the 1930s panchromatic film was introduced, but it took the demands of World War II to motivate radical improvements in photo processing.

Ellwood Wilson and Sherman Fairchild shared the same belief in the potential of aerial photography and became good friends. While it is generally believed that Wilson left his Laurentide position to manage Fairchild Aerial Surveys Company (of Canada) Limited, he probably worked for both companies with the full blessing of Laurentide's manager George Chahoon, who even loaned the infant company a building for use as an office and laboratory. The friendly arrangement had Fairchild both leasing aircraft from and undertaking forestry surveys for Laurentide. The available aircraft, however, were inadequate for Fairchild's photographic requirements and the company started to acquire its own aircraft. Fairchild's expansion into manufacture influenced an eventual move from Grand Mère to better facilities at Longueuil, circumstances that influenced Wilson to sever his contact with Fairchild in 1929 to concentrate on his post as chief forester with Laurentide.

Laurentide withdrew from direct involvement in aerial operations at the end of the 1921 season and Roy Maxwell accepted the company's offer to take over the work. With the help of two partners, he formed Laurentide Air Service Limited, utilizing facilities at Lac-à-la-Tortue, generously transferred without charge. The two borrowed CAB HS-2Ls were adopted by the new company, and a new Vickers Viking IV, G-CAEB, was purchased from Britain.

Laurentide Aviation completed a variety of tasks in its first year for the governments of Canada, Ontario and Quebec. The largest was a forestry survey for the Ontario government of the vast area lying north of the Canadian National Railway as far as James Bay, bounded by Pembroke in the east and Lake of the Woods on the Manitoba border. The objective of the survey was to identify what was thought to be valuable timber resources. The extensive mapping, largely by sketching, however, determined that timber only existed in economic quantities along riverbanks where there was adequate drainage, and soon petered out into muskeg. The survey was based at Sudbury with seven sub-bases, and HS-2L G-CAAC was lost en route to one of them, Remi

47 H.P. Pearce, "Development of Photography, Cameras, Aircraft and Camera Mounts," *The Canadian Surveyor* 13, no. 10 (October 1957): 637.

Lake, on 2 September 1922. Pilot Don Foss and Air Engineer Jack Caldwell made an emergency landing on a small, unnamed lake to wait out a heavy downpour. Despite jettisoning unnecessary items, the left wing tip float struck the water on takeoff and the aircraft crashed. The two crewmen were not hurt and walked out. The engine was recovered right away, but the remains of the fuselage were not salvaged until 1969 and are now displayed at the National Aviation Museum, alongside a beautiful HS-2L in Laurentide colours constructed to original Curtiss drawings.

Individual entrepreneurial pilots also entered the stage in 1922. In August, John M. Miller, pilot of the *Chicago Tribune*'s Curtiss Seagull flew the company's local manager to survey the newspaper's Franquelin River and Shelter Bay timber properties on the north shore of the St. Lawrence. One of the first "mercy flights" was also written into the records when Miller transported a doctor from Shelter Bay to collect some much-needed medicine from Clarke City.[48]

While technology, knowledge and enthusiasm were advancing, the Canadian government's attitude was not keeping pace. Mackenzie King's Liberal government came to power in December 1921, introducing economies that included amalgamation of the army, navy and air force into the Department of National Defence. The air arm became officially known as the Canadian Air Force (CAF). This was the midpoint in two years of organizational changes that did not effectively settle down until the formation of the RCAF in 1924. It was gradually realized that conducting the air force as a part-time refresher-training organization was not cost effective. Constantly changing and aging personnel made for a very inefficient operation. Furthermore, if the service was not to "fade away," it must train new pilots to take over from those who won their wings during the war. Accordingly, after much discussion, it was decided to establish a permanent nucleus and to recruit pilots from the engineering and science schools at Canadian universities. A great plan—but it was deferred for one year due to lack of funds, and the refresher courses were also cancelled.

Every cloud has a silver lining, however, and the consolidation of all federal government air operations enabled proponents to present their arguments on a common front, resulting in a $250,000 allocation for new aircraft and the eventual training of permanent air and ground crews. Another detail sorted out was elimination of the CAF's duality of rank. Daily Routine Orders issued by the Training Depot Station at Camp Borden on 28 November 1922 read: "The use of Military Equivalent of Canadian Air Force ranks will be discontinued throughout the Service forthwith and only the ranks hereunder quoted will be used both in correspondence and conversation." RAF ranks from pilot officer to group captain were listed.

48 John M. Miller, "A Timber Survey for the Chicago Tribune," *Aviation* 8, no. 11 (11 September 1922): 319.

1923-1924

A SIGNIFICANT event in the newly organized CAF was activation of the pilot training scheme deferred the previous year, albeit to a slow start. "Unavoidable delays"[49] limited the first intake to only nine cadets, given the rank of provisional pilot officers. Of the four graduates who obtained their wings and a permanent commission, two died in flying accidents in 1927 and one resigned in the same year, leaving only Roy Slemon, who went on to become an air marshal. The door had been opened, however, and a steady flow of university graduates entered in the following years, supplemented by "gentlemen cadets" from the Royal Military College—the men destined to lead the RCAF in World War II.

The estimate for air services remained at $1 million for 1923, but $250,000 was added for the purchase of new aircraft.

About 3,000 aircraft of 13 different types were built in Canada from 1917 to 1919, but production ceased until 1923 when industry responded to the federal government's invitation to supply eight new single-engine flying boats, on condition that they be manufactured in Canada. The Air Force was anxious to establish a domestic manufacturing industry to support its own activities and emphasized the demonstrated economic benefits of aircraft for forest management and aerial mapping. Specifications called for a single-engine seaplane capable of carrying a pilot and four passengers at a cruising speed of 80 miles per hour for six hours and with a ceiling of not less than 12,000'. The successful tenderer, at $175,000 for eight aircraft, was Canadian Vickers Limited, established by the British shipbuilder Vickers Sons & Maxim in 1911 to build vessels for the Royal Canadian Navy.

Canadian Vickers was primarily a shipbuilding company, but had some experience with aircraft. A Felixstowe F.3 and a Fairy III Atlantic were assembled in 1920, and Curtiss HS-2L flying boats were being repaired and overhauled for Laurentide Air Services Ltd. Vickers offered the British-designed Viking flying boat, but had problems in starting up production at its new aircraft production facility, so the first two Viking IVs were shipped from

Located at the foot of the Lachine Rapids on the St. Lawrence River at Notre Dame and Viau Streets in the east Montreal suburb of Maisonneuve, the Canadian Vickers plant was built on historic ground. Lachine, starting point for many explorers, fur traders and missionaries, was named in 1669 in mockery of its owner, Robert Cavalier de la Salle, who dreamed of a route westward to China. Also, Ethan Allan, famous for leading his Green Mountain Boys to the capture of Fort Ticonderoga from the British in 1775, is believed to have been captured at Lachine four months later, remaining a prisoner until exchanged in 1778.

During World War I, Canadian Vickers built nearly 300 vessels of various types for the navies of Britain and the USA.

49 F.H. Hitchins, *Air Board, Canadian Air Force, Royal Canadian Air Force*, 100.

England. The remaining six were assembled in Montreal, albeit with some delays.

The CAF gave itself credit for re-establishing a domestic aircraft manufacturing capability. It would, at the time, have needed a vivid imagination, however, to forecast that this initiative would evolve through Canadair and de Havilland Canada into Bombardier Aerospace, the third-largest manufacturer of civil aircraft in the world by the 1990s with 25,000 employees and annual sales approaching $5 billion. The CAF was then more concerned regarding the supply of "aeroplane timber" and recommended that the supply of Sitka spruce, in particular, should be conserved for the benefit of aircraft industries throughout the Empire. A modest industry was forecast for Canada, based on licensed airframe construction, but not large enough to justify engine manufacture or the supply of instruments and accessories for many years.

Sitka spruce, *Picea sitchensis*, grows only in the coastal rain forests along the west coast of North America. Exploding demand by the aircraft manufacturing industry during World War I for this straight-grained wood with a high strength-to-weight ratio initiated one of the largest forestry operations in the history of British Columbia, the only source of supply in the British Empire. Under the direction of BC's chief forester, H.R. MacMillan (of later MacMillan Bloedel fame), more than 3,000 men dispatched in excess of 150 million board feet of "Aeroplane Spruce" to the war machine in the closing 13 months of the war. The spruce was shipped east on "spruce trains," receiving priority over all other goods, much as did the silk trains in the 1930s. There was considerable waste in the desperate harvesting of Sitka spruce. It was estimated that only 15 per cent of the superior quality trees felled met the standards of the Imperial Munitions Board; the remainder was left in the forest to rot. Of this select harvest, many logs escaped the rafts en route to sawmills. It is not surprising that fears were expressed that the extractable Sitka spruce would be exhausted within one more year.

While forest inventories and fire protection continued to be the prime concerns of the CAF, experimental photography over the previous four years was beginning to show practical results. Methods of transferring information from aerial photographs to topographical maps had been developed by the Topographical Survey Branch, Department of the Interior. By the end of the 1923 season, 4,421 photographs had been taken.

As the demand for aerial photographic services grew, the CAF turned to the surveying profession to fill a serious deficiency in its navigation capability over unmapped northern regions. The first dominion land surveyor loaned by the Topographical Survey Branch was Robert Douglas Davidson, whose navigation feats became legendary. He was soon joined by surveyors John Carroll, Eric Stanley Fry, Frederick Hay Wrong, Cecil Barnwell Coslette Donnelly and, later, by Colin Stone Macdonald, among others. A/C R.C. Gordon, a photo pilot in the 1930s, noted in *The Roundel*, February 1949, that they were rugged individuals: "They enjoyed their work, seemed to have a great capacity for living among mosquitoes and ate and slept comparatively little."

With the exception of Donnelly, all the surveyors had served in Britain or France during World War I. Wrong was killed when his photo aircraft crashed at Hilbre, Manitoba, in July 1927. By the mid-1930s the RCAF had an adequate roster of its own navigators and no longer needed the civilian surveyors. Davidson worked for a brief period on the St. Lawrence Ship Canal before becoming a technical adviser on the General Staff in 1932. The others returned to the Topographical Survey Branch, although the RCAF called on Donnelly's services again to help chart two northern air routes from North America to Britain and Russia during World War II. Davidson received the Kodak Award in 1959 for his outstanding contribution to aerial photography on the unanimous recommendation of the Canadian Institute of Surveying.

Until 1923, the CAF had been undertaking work for some of the provinces on a partial repayment basis. With amalgamation of the CAF into

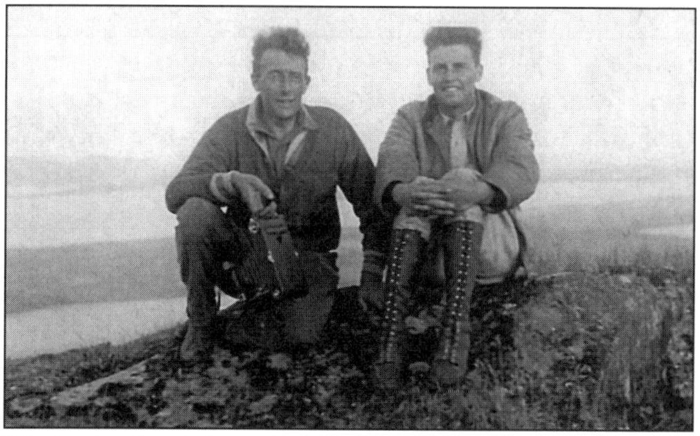

Eric Stanley Fry, DLS, on the left, photographed with F/L Ralph McBurney overlooking Burnside Harbour at the south end of Bathurst Inlet, Northwest Territories, 14 July 1931. *R. E. MCBURNEY*

Fry was a sniper in France during World War I. Wounded, he returned to Canada to qualify as a dominion land surveyor in 1924. From 1926 to 1938, he was on loan to the RCAF from the Topographical Survey Branch as navigator and surveyor on photographic flights. Borrowed by the RCAF again in 1941 because of his knowledge of Labrador, he located the site for a new airfield, now known as Goose Bay Airport. He spent the remainder of the wartime years mapping with the RCAF and US Army Air Force in Labrador and Baffin Island. In 1946 he was awarded the OBE in recognition of his wartime efforts. He went on to map vast areas around Great Slave and Great Bear Lakes. After "retirement" in 1955 he played a major role in selecting Distant Early Warning Line station sites across the Arctic and was senior Canadian observer on a US Navy task force controlling mapping in the high Arctic.

Fry Inlet, the southwest arm of Contwoyto Lake, southwest of Baffin Inlet, is named for Eric Fry.

the Department of National Defence, it was mutually agreed to gradually withdraw this service and leave the work to commercial aviation companies.

The CAF's top priority was to justify support of its existence by demonstrating that it could perform "civil" tasks for other government departments with greater efficiency and at lower cost than by conventional means. Nevertheless, a detailed plea from the Geodetic Survey Branch to support the establishment of some 35 triangulation stations along a 150-mile stretch of the Rocky Mountains between Yellowhead Pass and Jarvis Pass was rejected:

> We are most reluctantly forced to the conclusion . . . that the objectives to be gained do not justify the risk entailed to the flying personnel of the Air Force or the members of your staff who would be carried. As you are aware, the country is of a very rough nature and suitable landing places are few. The only two pilots who have flown in the district are both of the opinion that the risks are considerable and, provided that other means of transportation are available to carry out the work, it should not be undertaken.[50]

The CAF was careful to include a nudge to the government by noting that refusal was only because suitable types of machines were not available in the country.

Construction of a wireless station was started in 1923 at Jericho Beach, complementing the permanent hangars, workshop, concrete platform and launching ramp completed in 1922. The CO was S/L A.E. Godfrey, with a staff of two pilots, one photographer, three NCOs, 20 mechanics and one civilian storekeeper. Total 1923–24 flying time was 271 hours by three HS-2L flying boats and one Viking. Most of the flying was on "preventive patrols" for

50 NAC RG24, Vol. 4889.

the suppression of smuggling and illegal fishing, but various aerial photographs were taken, including water power sites, the new dock at Esquimalt, Burrard Inlet waterfront, and the southern branch of the Fraser River.

High River, Alberta, similarly took on an air of permanence with the completion of three hangars, a workshop, offices, storerooms and a wireless station. S/L G.M. Croil commanded the station from its formation until he was posted to Ottawa and relieved by S/L A.L. Cuffe in October. The High River staff in 1923 included six officers, seven NCOs and 15 mechanics. All government communications were then the responsibility of the Royal Canadian Corps of Signals, which co-operated actively in developing lightweight equipment for the aircraft, although for several years weight considerations limited aircraft to transmission-only capability, leaving the pilot in ignorance of whether his message had been received. Substations at Pincher Creek and Eckville were activated during the flying season to extend the range of the unit's forest patrol D.H.4 aircraft.

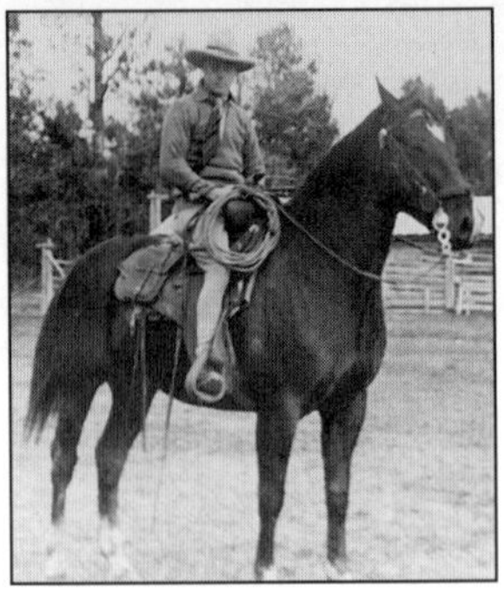

A diversion for the photo and patrol pilots occurred when spring breakup made dirt roads impassable. At the request of Edward, Prince of Wales, photographed on *Midnight*, mail was dropped at his EP Ranch at Pekisko, 17 miles southwest of High River.
GLENBOW ARCHIVES, CALGARY NA2626-11

One High River D.H.4B had a busy photographic season despite wet and snowy weather in the early months. From a temporary base at Wetaskiwin, now the home of Canada's Aviation Hall of Fame, 800 negatives were exposed in 16 flying hours to make a complete mosaic of the Cooking Lake Forest Reserve east of Edmonton. A similar amount of flying was done for the Reclamation Service to locate suitable reservoir sites on the headwaters of streams flowing towards the prairies. Other photographic assignments were carried out for the Topographical Survey, Forestry, and National Parks

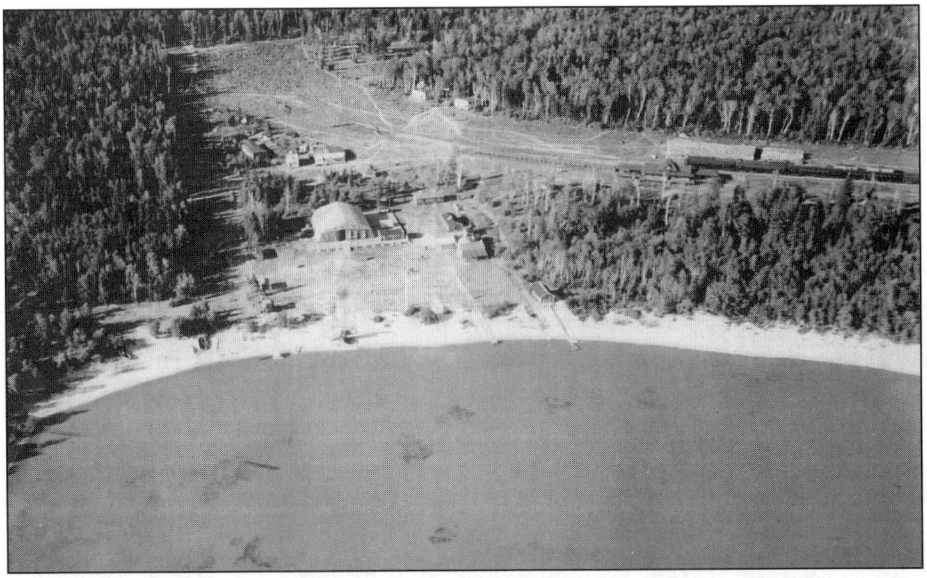

Victoria Beach at the terminus of Canadian National Railway on the east shore of Lake Winnipeg, 5 October 1923. The area is now a cottage community known as Main Beach. *NAPL FA50-20*

Branches of the Department of the Interior for a grand total of 55 flying hours, ten percent of the station's total flying time in the 1923 season.

At the Winnipeg station, patrol ranges were again extended from Victoria Beach with sub-bases at Norway House and The Pas. The 1923 report complains of the late arrival of new Vickers Vikings to replace four aging HS-2L flying boats and a fifth that was blown from its moorings and destroyed early in the season. A venerable Curtiss H-16 flying boat was made airworthy and sent from Ottawa in July to help fill the gap, but was of little use and the result was hardly worth the effort, as it was beaten to Victoria Beach by the first Viking in an impromptu race. The three new Vikings eventually arrived, one each month from July on.

The new aircraft helped, but did nothing to solve a problem endemic throughout the CAF at the time, as noted in the annual report: "A number of the older mechanics on the station had preferred to remain in civil occupations rather than enlist in the Air Force, with the consequence that the personnel on the stations, however willing, were inexperienced. The operations suffered accordingly." The Viking's Rolls-Royce Eagle engines were a constant source of trouble and needed experienced nursing to keep them operating.

One airman who did sign on in 1923 was Thomas (Tommy) Frederick Cooper, ex-RAF, who wrote of his experiences at that time in *Roundel*, Jan–Feb 1959, after he retired as a wing commander with an OBE for his wartime services in command of No.8 Repair Depot at Winnipeg. The Winnipeg base's equipment in 1923 consisted of one Leyland truck, one Model T Ford "staff car," and a Model T canvas-covered tender with two rows of seats. At Victoria Beach was a mobile workshop with two gasoline engine-powered generators, a 40 hp Clayton tractor, a Ford Model T truck, and a Cadillac electric winch for hauling out flying boats. Firefighting equipment was a Model T engine hooked up to a geared pump and a cart carrying about 100' of hose. Cooper recalled that opening the Victoria Beach seaplane base in the spring was quite a chore, and mechanics had to be versatile. Power plants had to be made serviceable, latrines dug, water systems assembled, and aircraft readied for flight as soon as the lakes were free of ice. Issue tools were scarce and most mechanics relied on their personal set to carry out the daily tasks. An indication of the type of repairs carried out is demonstrated by the contents of tool kits carried by aircrew: "lots of copper wire, blow torch, solder, friction tape, chewing gum, fabric, dope [to tighten fabric], a piece of thin plywood, Amberoid glue, spark plugs, engine tool kit and in some cases spare magnetos." On photographic flights, a few extra tools would be carried to attend to the temperamental cameras. Chewing gum sounds like an unlikely spare part, but Cooper said it was far from unusual for a crewman to get out of his seat in mid-air and transfer a wad of gum from his mouth to plug a leaking radiator or hose.

Mechanic Alex Milne recalled the difficulty of navigating in northern Manitoba when his Viking was assigned to transport a Department of Indian

Affairs representative to make treaty payments to Indian bands. "Now, I'm telling you, finding those places, with the maps we had of that country at that time was a marvel." Heading for Island Lake, they had to quarter the countryside, trying to deduce the water route the trappers and traders would have followed. When they eventually found Island Lake, they established that it was 60 miles farther east than shown on their map.[51]

Shortage of serviceable machines prevented a planned survey of Lake of the Woods and other districts, but an important survey was completed for the Topographical Survey Branch in the fall of 1923 along a 180-mile-long stretch of the Manitoba–Ontario boundary north of the Winnipeg River. With the help of a civilian surveyor as navigator and camera operator, about 2,500 square miles comprising the Kississing Sheet was photographed during a series of flights from The Pas. The resulting maps proved the feasibility—and economy—of the process and laid the groundwork for a more ambitious venture the following year.

One of the new Vikings crashed during a load test on 16 September, injuring the pilot and completing a rough year for the Victoria Beach commanding officer S/L Basil D. Hobbs and his establishment of seven officers and thirty airmen. Due to late delivery and mechanical problems, by the time the three Vikings were dismantled for 1923–24 winter storage at the Winnipeg depot they had accumulated only 68 flying hours.

The Ottawa station, under command of F/L C.M. McEwen, with a staff of three officers and 15 airmen, operated in the summers from Rockcliffe. During the winters, they moved to Victoria Island where a slipway and

51 Alex J. Milne, "Pioneer Survey Flights in Northern Manitoba and Sakatchewan," *CAHS Journal 21*, no. 4 (winter 1983): 102.

In the early days of operations at Winnipeg, flying boats were hauled out of the Red River at the foot of Brandon Avenue for winter overhaul and storage. The Legislative Buildings can be seen in the background. A pair of Felixstowe F.3s are shown, 28 October 1921. The small bungalows were removed shortly afterwards and a new, wider ramp and workshops built for both service and commercial operators.
ALEX MILNE

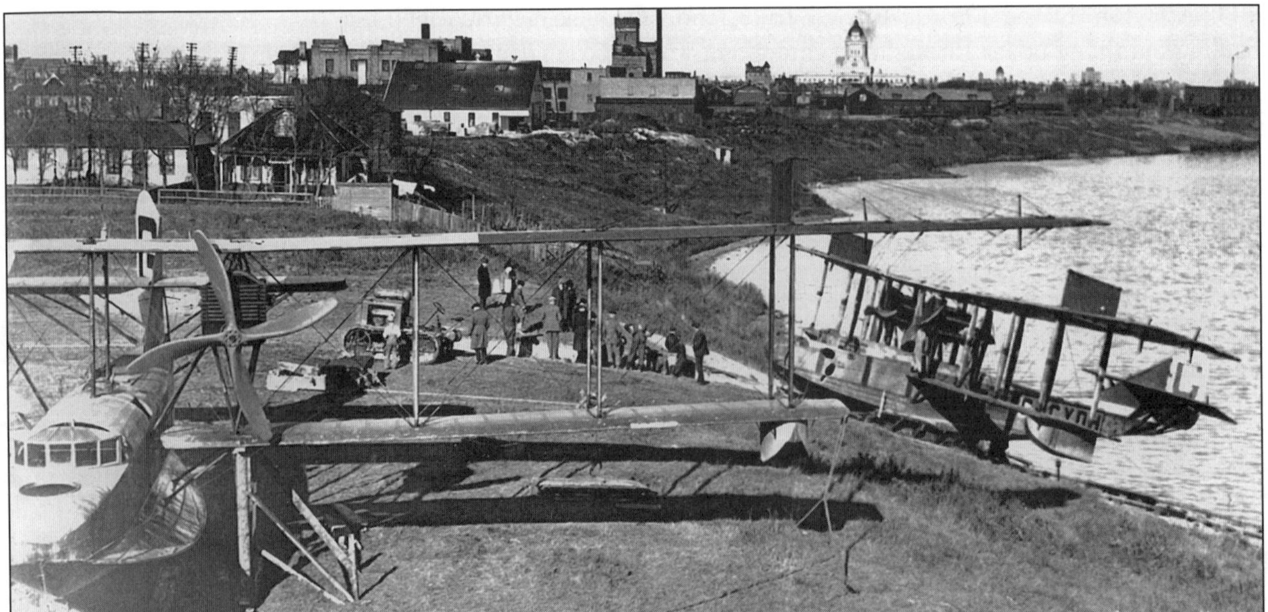

permanent buildings were made available by Public Works. The unit was primarily involved in flight-testing eight new Vikings and their delivery to other stations. Considerable effort also continued in co-operation with Prof. Cooke to develop aerial photographic techniques. Photographs were taken of the canal systems and principal water-power sites and harbours in eastern Ontario and western Quebec. A series of photographs showing manufacturing plants, water-power sites, and pulp and paper mills were taken for the Government Exhibition Commission to display at the British Empire

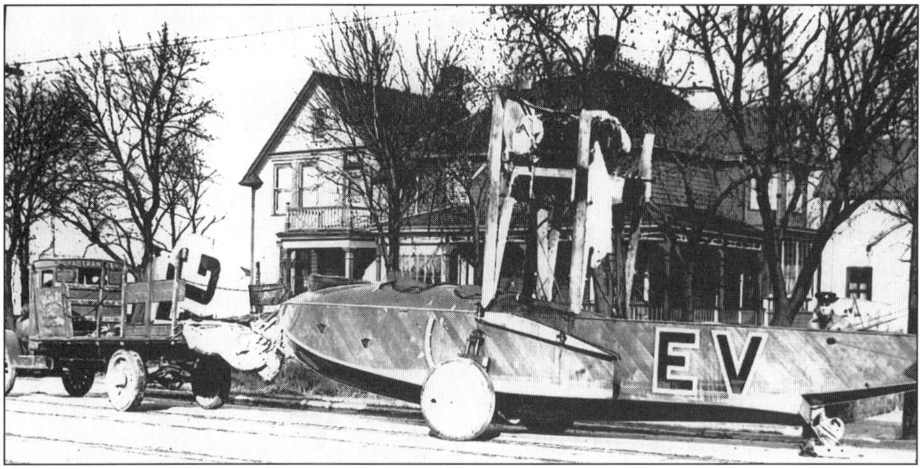

Aircraft were towed, with a police escort, along Main Street and Portage Avenue to a rented garage for overhaul and storage, 19 September 1923.
NAC PA53362

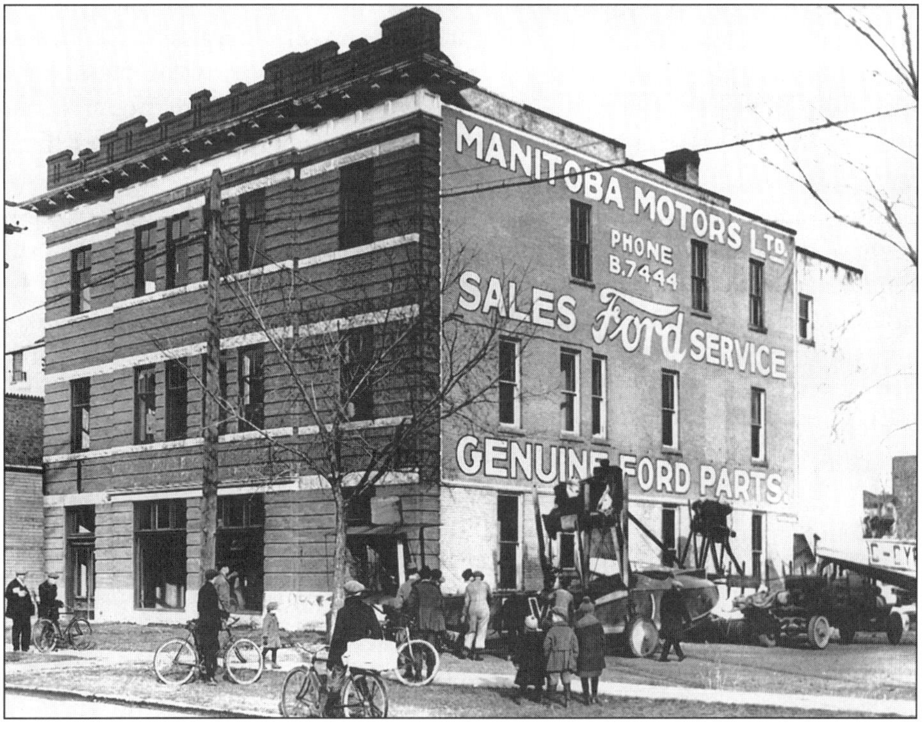

Arrival at winter quarters was an entertaining event. Wings were hoisted to the second floor and engines to the third for winter overhaul, 19 September 1923.
NAC PA53363

By 1927, winter overhauls were conducted in a warehouse at 797 Notre Dame Street, Winnipeg Air Station HQ.

Exhibition at Wembley, England, later that year. Experimental photography in co-operation with the Topographical Survey Branch, Department of the Interior, accounted for 17 of the total 63 hours' flying time. Ottawa also co-ordinated the experimental aerial survey work of the other stations.

Under contract to the Quebec government, the Dominion Aerial Exploration Company assumed management of the Roberval station in July 1923. One HS-2L, G-CYEJ, operating from Roberval on a photographic mission was lost in rough water off the north shore of the St. Lawrence in September of that year. The crew were rescued and most of the equipment salvaged. The three crewmen of another HS-2L, C-GYAG, which, for unknown reasons, stalled and went into a spin on approaching Roberval the same month were not so lucky. The aircraft dived into the lake and pilot B. deSalaberry with crew A. Landry and C.M. Guise were killed. A third HS-2L, G-CYAE, was loaned by the CAF to Dominion Aerial, which employed three pilots, three air engineers and a dozen mechanics in 1923. Their aircraft flew 19,410 miles in 304 hours' flying time, over half concerned with forest survey and reconnaissance, covering 24,300 square miles.

Nova Scotia awoke to the potential of aerial photography in 1923 and the Dartmouth station, under the command of S/L A.B. Shearer with two officers and ten airmen, undertook a variety of photographic assignments with two HS-2Ls. For the Department of the Interior, photographs were taken of hydroelectric power developments on the Jacques Cartier, Montmorency, Ste-Anne-de-Beaupre, Nipisquit, Aroostock, Musquash and St. Margaret Rivers, and of the watersheds of the Musquash River, St. Margarets Bay, Sheet Harbour, and Bear, Economy and Rosewater Rivers to determine their limits. A 64-square mile mosaic of the New Glasgow area and a strip mosaic of the coast from Brule Harbour to Knoydart Point were completed. A survey flight from Charlottetown to Grindstone Harbour for the post office identified landing and ice conditions. Oblique photographs were taken during a six-hour flight of the Annapolis Valley from Windsor to Digby, following the line of the Dominion Atlantic Railway, to assist with the survey of land cleared under the Soldier's Settlement Act: "This information is not shown on the existing maps and to get it by ground methods would require much time and no little expense."[52] Some relatively minor photographic operations were also undertaken for the Halifax Town Planning Board and other local municipalities.

While significant progress was made during 1923–24, difficulties continued to dog the experimental photographic work of all the units. A fundamental problem with both vertical and oblique photography was the inability of the HS-2L flying boats to reach the desired altitude of 10,000'. Eight new Vickers Vikings were delivered that summer, but the annual report grumbles about late- or non-delivery of the new flying boats and their cameras. The point was made that by the time the new equipment arrived, the year was

52 *Report on Civil Aviation for the Year 1923* (Ottawa: Dominion of Canada, 1923): 36.

so advanced that failing light detracted from the value of what was accomplished. Vertical techniques were by now well established, but required great accuracy in maintaining direction, altitude and stability along very narrow, parallel flight paths. Dr. Deville took a great personal interest in the work and commented:

. . . the effect of tilt and change of altitude shown in the prints becomes very serious and the training of pilots should be directed to reducing these errors as much as possible and this reduces greatly the ultimate cost of the survey. If not, the rectification of individual prints and additional groundwork necessary to determine the tilt by means of a tilt finder will introduce considerable expense and be prohibitive in any but exceedingly rough and inaccessible important areas.[53]

Oblique photography with its much larger coverage showed great promise of satisfying the increasing demand for topographical maps for settlement, mineral exploration and forest inventories. Dr. Deville reported positively on the results of oblique photographs taken of the Churchill River north of The Pas, although the HS-2L struggled to only 2,600' altitude. This experiment gave adequate confidence to establish fuel caches during the winter in preparation for S/L Basil Hobbs's survey of Reindeer Lake in 1924.

The concept of oblique photography had some very practical attractions. The immediate need was for better maps—indeed, in many cases, for any map—of vast areas of the Precambrian Shield and the wooded areas of the central plain. R.D. Davidson, DLS, summarized the advantages of oblique photography:

> Although local relief abounds, changes in elevation were small and large sections of the two areas did in reality approximate a plain. For example, the extreme variations in lake elevations over 180 miles of the

53 Ibid., p. 43.

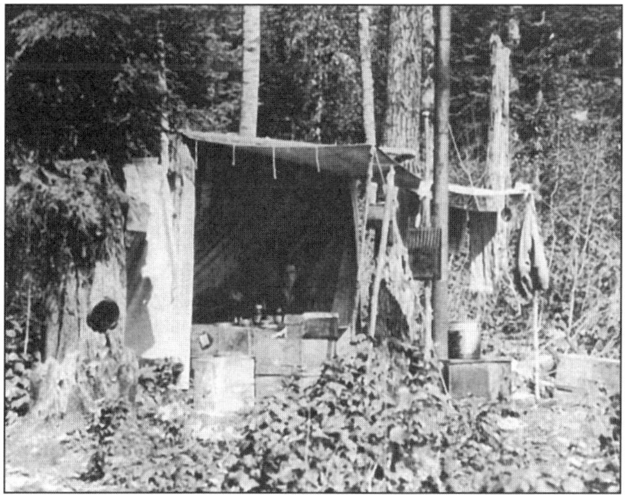

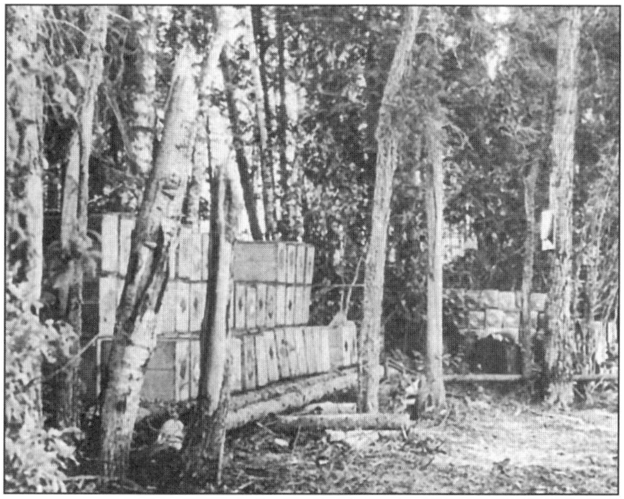

Typical cook shack and fuel storage used by detached flights, c.1923.
ALEX MILNE

> Manitoba–Ontario boundary, which crosses some fairly rugged country, is only 170 feet, and if we omit two lakes out of hundreds it is only 135 feet. Such conditions were almost ideal for oblique plotting. Also, there were numerous lakes and rivers on which to land, but no airfields and this meant using flying boats, which limited the ceiling to between 2,500' and 5,000'. Thus, to cover a worthwhile area with vertical photography both flying time and number of exposures would be excessive, and, lacking maps and precise navigational instruments, gaps would abound. The photograph for photo oblique coverage was not only much larger, but also more elastic, so much so that it could, in a pinch, be stretched to cover almost any gap.[54]

R.B. McKay and Prof. Cooke perfected their procedure for transferring oblique photographs to topographical format. Their system—known internationally as The Canadian (High Oblique) Method—was used for many years instead of the time-consuming procedure of drawing individual perspective diagrams on each photograph. A series of .020-inch-thick transparent celluloid sheets was engraved with "grids" comprised of lines converging towards the horizon and parallel lines at right angles to the camera. Over 2,000 grids were made for combinations of focal lengths at .05-inch intervals from 8.10 to 8.35 inches; altitudes at 25-foot or 50-foot intervals from 3,400 feet to 5,600 feet; and depression angles, measured as distance from the top of the photograph to the horizon, at 0.1-inch intervals from 0.1 to 1.2 inches. An appropriate grid was selected and superimposed over the photograph. Features shown at the intersection points were then transferred to a plotting sheet divided into squares, usually on a scale of one inch = forty chains.

In relatively settled areas, where initial ground surveys had already been completed, grids could be established from the visible township and section lines, greatly simplifying plotting. All aerial photographs required knowledge of fixed ground points in order to be converted into accurate maps, but the wide-angle oblique photograph vastly increased the coverage compared with the thin strip of overlapping images from vertical photography. The CAF confidently predicted that these techniques would be perfected with far-reaching results in the exploration and mapping of large areas throughout Canada. The surveying fraternity soon realized that aircraft could obviate most of their traditional drudgery of climbing mountains, cutting trails through forest and swamp, or paddling and portaging for weeks at a time while enduring the torments of insect hordes and inclement weather.

Camera experiments continued with encouraging results. The camera in general use until the advent of wide-angle lenses in the early 1930s was the Fairchild K-3. An improved version, the F-3, incorporated a focal plane shutter in the camera rather than in the magazine. Both cameras automatically

54 R.D. Davidson, "Development of Mapping from Air Photography to 1930," *The Canadian Surveyor* 13, no. 8 (July 1957): 502.

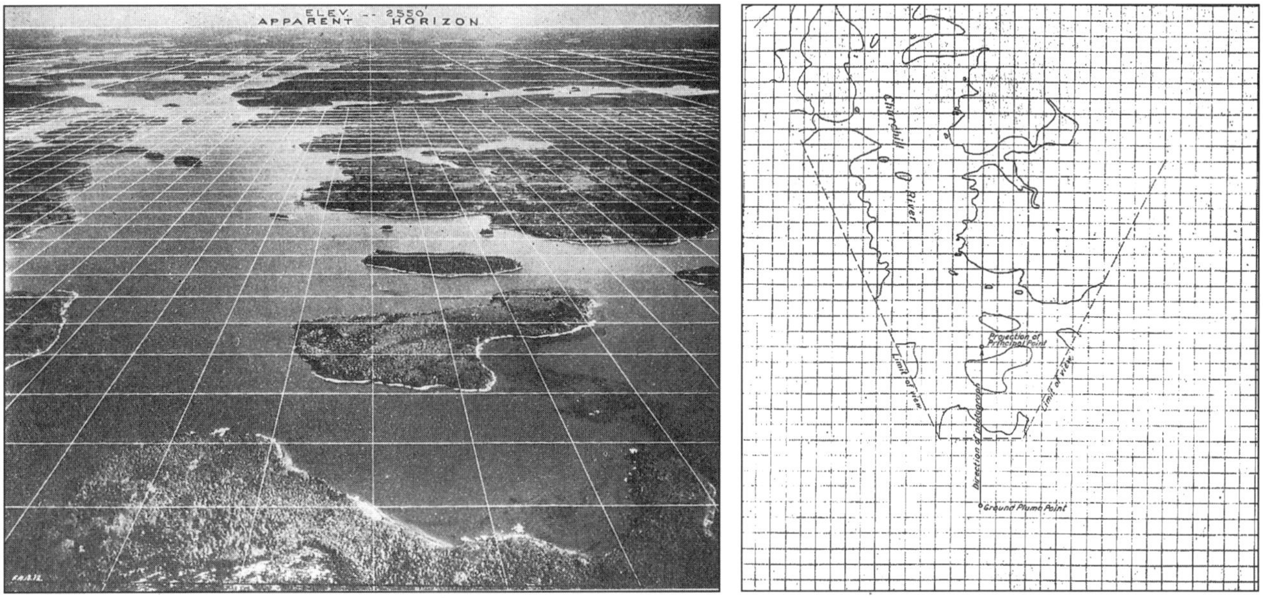

Oblique photograph of the Churchill River taken from 2,550′ and transferred to a square grid. A transparent overlay etched with appropriate converging lines was used to locate points then transferred to a square grid that became the basis of an accurate topographical map. *REPORT ON CIVIL AVIATION*, 1923

advanced the film, set and released the shutter, and could be adjusted for aircraft altitude and speed, and the focal length of the lens installed. Seventy-five-foot-long rolls of film were contained in interchangeable lightproof magazines.

Experimental oblique stereoscopic photographs were taken during 1923 in the Ottawa area and vertical stereoscopic photographs over Sheep Creek, Elbow River and Beaver Creek, Alberta. These showed potential for identifying contours and providing extremely detailed information.

The private sector had scented the potential commercial benefits of aviation. The CAF, with no aircraft or crews to spare, emphasized that its role was limited to the support of federal government, not commercial, operations. Anxious to cultivate a viable aviation industry, the CAF enthusiastically, but not exclusively, referred enquiries to the new commercial enterprises. Fairchild and Laurentide, in particular, really got underway during the 1923 flying season.

Fairchild acquired a Curtiss Experimental flying boat with an experienced crew, pilot Kenneth F. Saunders and engineer Bill Kahre. Six mosaics were completed ranging from the town of Nominingue, covering two square miles, to a map of the McLaren Lumber Company limits, covering 370 square miles. For these mosaics, a total of 510 square miles was covered during 98 hours' flying time, all by vertical photography requiring parallel flight lines one-quarter of a mile apart; not an easy task. Oblique photographs of water power plants and industrial plants were taken for advertising purposes. The largest contract was for a complete mosaic of 320 square miles of forest, for which no map existed, in the Rouge River Valley. Strip mosaics of the Batiscan River identified hydroelectric power generation sites and the location of power lines. Mosaics of Three Rivers and Grand Mère were completed for town planning purposes. Fairchild's new, gyroscopically stabilized camera gave significant improvements in vertical photography for the town mosaics that formed an increasing proportion of the company's business. Picture postcard and illustrative views generated a profitable sideline. These tasks fully occupied the company's Curtiss Experimental and its successor, a Curtiss J-1 Standard, that operated throughout the 1923–24 winter on skis, probably the first winter flying undertaken in Canada on a regular basis.

Laurentide's growth exceeded expectations, increasing from six aircraft flying 688 hours in 1922 to 12 aircraft and 1,480 flying hours in 1923. It was reported that Roy Maxwell went to the United States and purchased 14 surplus HS-2Ls in their original packing crates. A Curtiss JN-4 was purchased for pilot training and its standard 90 hp Curtiss OX-5 engine replaced with a 150 hp Hispano-Suiza engine to allow the "Jennie" to tow a long banner. Pilot staff increased to nine, including engineers Jack Caldwell, who obtained his pilot's licence in 1922, and Pete and Romèo Vachon, who qualified in 1923.

Most of Laurentide's efforts were expended on fire patrol, transport and continuing the previous season's forestry survey for the Ontario government, a total of 1,045 flying hours. Three thousand square miles of timber limits east of Sault Ste. Marie were surveyed for Spanish River Pulp and Paper Mills Limited. Other tasks included route survey of a planned extension of the Temiskaming and Northern Ontario Railway towards Moose Factory. The machines employed included nine HS-2Ls, one Vickers Viking, one Curtiss JN.4 and one Loening M-23 Air Yacht.

A unique concept by Grover Cleveland Loening, the first North American to earn an aeronautics degree (Columbia University, 1910), the Air Yacht, G-CADV, was damaged beyond repair at Lac-à-la-Tortue on 23 June when a fire extinguisher fell and jammed the rudder bar on takeoff. The aircraft hit the water and the hull collapsed. Pilot Jack Caldwell and passenger S. Moir were shaken, but uninjured.

On 17 July, Laurentide had another setback when HS-2L G-CACV was badly damaged at Biscoe, Ontario. The facilities at Grand Mère were too small for the greatly expanded operation, and the water base at Lac-à-la-Tortue was ice-bound late in the spring season. In the fall of 1923, the company, therefore, moved to the old plant of the Three Rivers Shipbuilding Company at Three Rivers on the relatively ice-free St. Lawrence River. Possibly due to rapid expansion in the grand Roy Maxwell manner, the company was losing money and towards the end of the season made representations to the federal government for financial support.

Largely as a result of John Miller's flight to Shelter Bay the previous year, the Ontario Pulp and Paper Company purchased an HS-2L flying boat in 1923 to explore their logging limits along the Franquelin River on the north shore of the St. Lawrence. Reconnaissance flights covered more than 6,000 square miles and aerial photographs were taken of the principal waterpower sites in the region.

The Dominion Aerial Exploration Company continued forest patrol, aerial sketching and photography for the Quebec government from Roberval with its two HS-2Ls. Forestry engineers sketched some 4,300 square miles of forest in the Saguenay area, supplemented by aerial photography as necessary to obtain finer detail. Some aerial photographs were taken of fires and their aftermath for Price Bros., but the aircraft used, a Martinsyde Type A Mk II floatplane, stalled during an unauthorized flight and crashed at Chicoutimi on 11 July. J. Oldham, a wartime pilot, but unlicenced, and his passenger, L. Evans, were killed.

"Aerial sketching" was essentially the hilltop sketch used by surveyors for many years to fill in detail where the position of base points had already been established by ground survey. As early as 1886, photographs taken by cameras laboriously hauled up mountainsides in Western Canada gave more accurate results than hand sketches. Aircraft offered a higher observation point disclosing more detail, and, if required, from directly above the target. In addition, the observer was delivered speedily to the area of interest, reducing transportation costs. The 1923 *Report on Civil Aviation* included a statement by R.N. Johnson of the Ontario Forestry Branch emphasizing the probability of greater accuracy, as more detail could be seen. He also noted the relative ease of transferring to a two-dimensional map surface what is seen from the air as a flat plain compared with a three-dimensional view from the ground. Sketches, however, shared a common need with aerial photography for accurately established reference points. Johnson concluded: "Base maps, whose accuracy must be assumed in order to proceed, are unfortunately very often inaccurate, and for a great deal of Canada's timbered country are not even in existence, except for the barest of outlines."

The need for aerial photography was obvious and the technology in place. Time to get on with the job.

1924-1925

1924 WAS THE YEAR of majority for aviation: 21 years after the Wright brothers made their first flight at Kitty Hawk, North Carolina. In Canada, the government year started with all military aviation placed under the flag of the Royal Canadian Air Force (RCAF).

The RCAF was formally established on 1 April 1924 with a staff of 66 commissioned officers, 43 of whom were pilots. All 263 airmen in the CAF had been discharged on 31 March and 66 declined the invitation to re-enlist in the RCAF, but 65 new recruits enlisted. After three formative years and two years of continuous reorganization, government flying operations settled down into a formal organization that promised direction and stability. RAF-style grey-blue uniforms were specified, although the dark-blue CAF outfit continued in use until 1926. The RAF motto, *Per Ardua Ad Astra* (Through Toil To the Stars), appropriately replaced the CAF's *Sic Itur Ad Astra* (This Is the Way To the Stars).

The basic appropriation for the RCAF remained at $1 million for 1924, but the budget for new equipment was increased by $50,000 to $300,000. Priorities were indicated by a supplementary vote of $261,000 for civil operations. All in all, an increase of almost 25 percent over the previous year.

Photographic operations up to 1923 had been *ad hoc*, largely learning what was possible—and what was not—by experiments with whatever equipment was available, or could be improvised. However, cameras and aerial survey techniques were now developed to the stage of commercial viability. New aircraft and cameras, much better suited for the task than their ex-World War I predecessors, were being designed and manufactured by Canadian and US companies. Air and ground crews had gained experience—often the hard way—in "photo ops" and were keen to get on with the job. Private industry was taking an active role. The federal and provincial governments recognized the value of mapping Canada and appeared willing to provide necessary funds. The federal Topographical Survey Branch was feeling the pressure for better maps (indeed, in some cases, any maps) for resource exploration and exploitation purposes, as well as for settling virgin land, identifying hydroelectric power generation potential, and building roads, railways and canals. The evident feasibility of quick, accurate and economical aerial surveys led to the RCAF and the Topographical Survey Branch being charged with systematically photographing Canada. One unfortunate event in this exciting time was the death of one of aerial photography's leading pioneers, Dr. Édouard Deville. He was succeeded as surveyor general by H.F. Peters.

A historic expedition set out from Victoria Beach on 18 July 1924, charged with photographing sections of the Churchill and Reindeer Rivers and the whole of Reindeer Lake. Trading posts of the Hudson's Bay and Révillon Frères Companies offered some supplies and shelter, otherwise most of the region to be visited was inhabited only by Indians and a few white trappers and traders. Emergency rations, weight 56 pounds, headed the list of supplies to be carried. Cooking utensils and tents were 39 pounds, and engine spares 80 pounds. A collapsible canoe, shotgun, lantern, ropes and refuelling hoses added another 52 pounds. The Fairchild camera, 25 rolls of film, magazine and a Graflex hand-held camera weighed 206 pounds. Four crewmen and their kit at a mere 180 pounds each came to 720 pounds. No changes of clothing were included: the crew washed their underwear in the evening, when they could, and hung it out to dry for the next day. Gasoline (112 gallons = 840 pounds) and oil (8 gallons = 70 pounds) plus aircraft weight (3,800 pounds) brought the total to 5,863 pounds. The allowable all-up weight of the Viking was 5,800

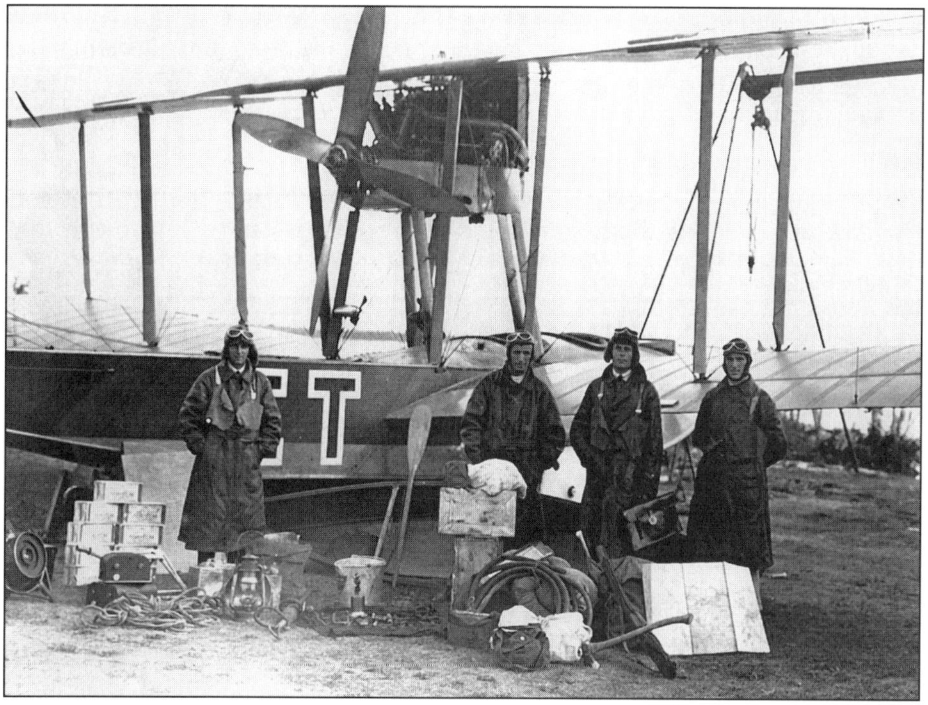

Setting the pace for the fledgling RCAF was a highly successful expedition to northern Manitoba by a lone Canadian Vickers Viking, G-CYET. The crew consisted of, left to right, P/O James R. Cairns (photographer), Mr. Robert D. Davidson, DLS (navigator), S/L Basil D. Hobbs, DSO, DSC (pilot), and Cpl. Alexander J. Milne (mechanic). Shown is all their gear laid out for inspection. Note the folding canoe, and the two wind-driven "dynamos" (generators) on the engine supports. Wheels and tail skid were removed before departure to save weight. Alex Milne recalled that an enthusiastic Hobbs said, "We're going to make the air force or break it—one of the two!"[55]
NAC PA-53239

55 Alex J. Milne, "Pioneer Survey Flights in Northern Manitoba and Saskatchewan," *CAHS Journal 21*, no. 4 (1983): 104.

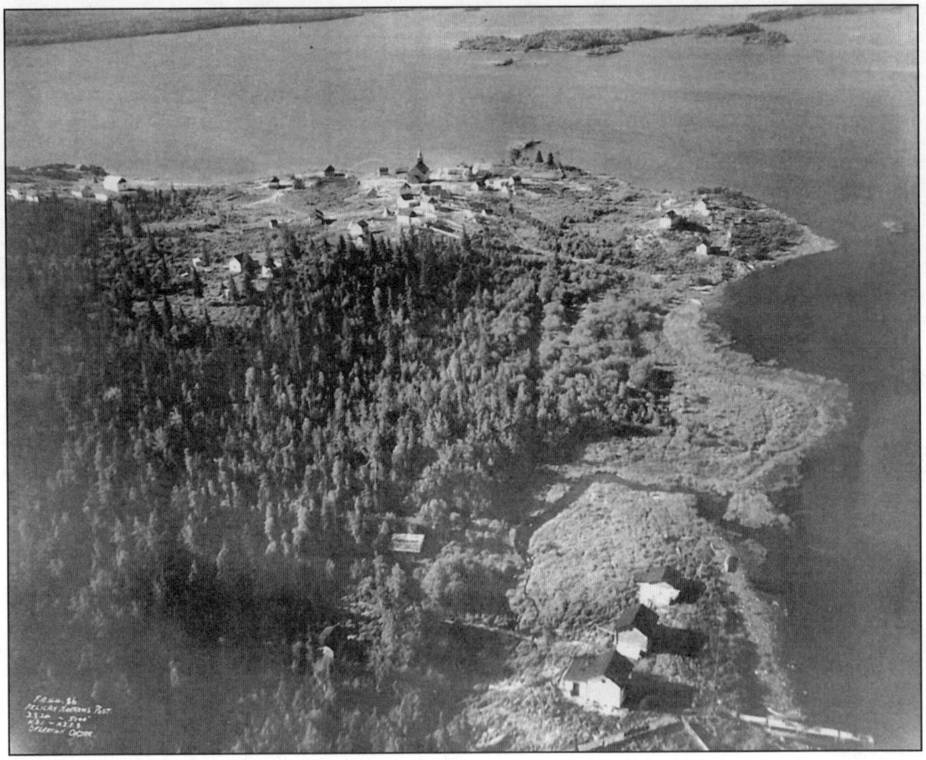

Pelican Narrows Post, Saskatchewan, was strategically situated on the primary route between Eastern Canada and Winnipeg, Edmonton, Fort McMurray and the McKenzie River system. This photograph was taken by P/O Cairns from Viking G-CYET at 5,000′ in 1924. *NAPL FA44-86*

pounds and it was felt that such a heavily loaded aircraft would not even be able to take off in rough or very calm waters, much less reach its assigned altitude of 5,000'. So 200 pounds of wheel assembly, useful for beaching the Viking, but unlikely to be required on this trip, were removed.

Cameraman Jimmy Cairns could traverse the Fairchild camera fitted on a machine-gun mounting in the front cockpit through 180 degrees. A wind-driven dynamo supplied power for the camera's automatic drive, but he was responsible for aiming the camera to bring the horizon one inch below the top of the picture. The automatic drive worked well flying upwind, but could not wind the film fast enough to take photographs downwind, so Cairns had to spin the wheels by hand. Hobbs had to fly straight and level at 5,000'. At a ground speed of 60 mph, a set of three pictures—one straight ahead, one to the left and one to the right—had to be taken with a 50 percent overlap. Many years later, Milne remembered with admiration Hobbs's skill in flying a steady line: ". . . we would be flying almost sideways, depending on how the wind was . . . sometimes we thought the old Viking was skidding to a degree that made us wonder which end the elevators ought to be on." It took one hour to climb to 5,000' altitude.

The route of Hobbs's survey covered the watercourses from The Pas to Pukatawagan, then west and north along the Churchill and Reindeer Rivers to cover the entire Reindeer Lake on the northern Saskatchewan–Manitoba

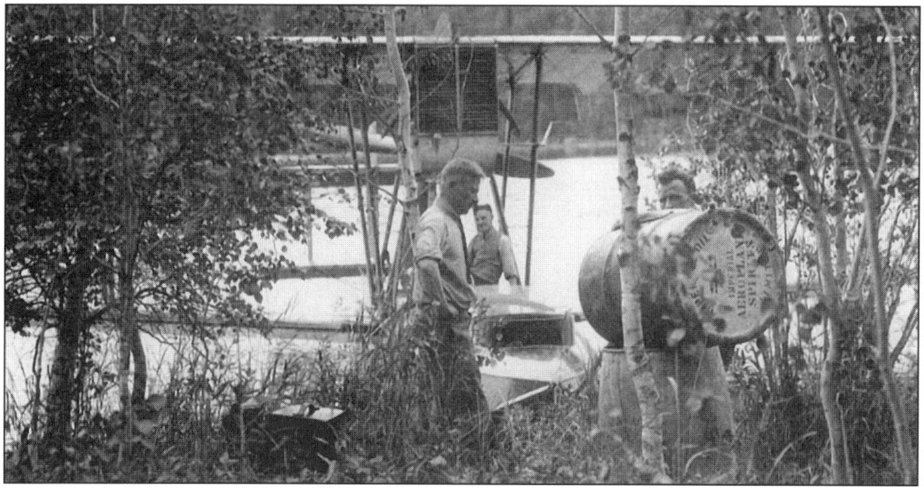

Fuel had been dispatched to strategic locations during the previous winter, but few assurances had been received of its arrival. The aircraft started on many "legs" of the journey, some over 200 miles long, without assurance that there would be fuel available for the next stage. Another unknown was the number of drums that would still be full after the locals found that a nail could give them ready access to a free fuel supply. Here, 'ET is being refuelled at Rabbit River, Saskatchewan. Hobbs is on the left, Milne on the aircraft and Davidson behind the drum of "Aeroplane Spirits." *NAC REINDEER LAKE REPORT*

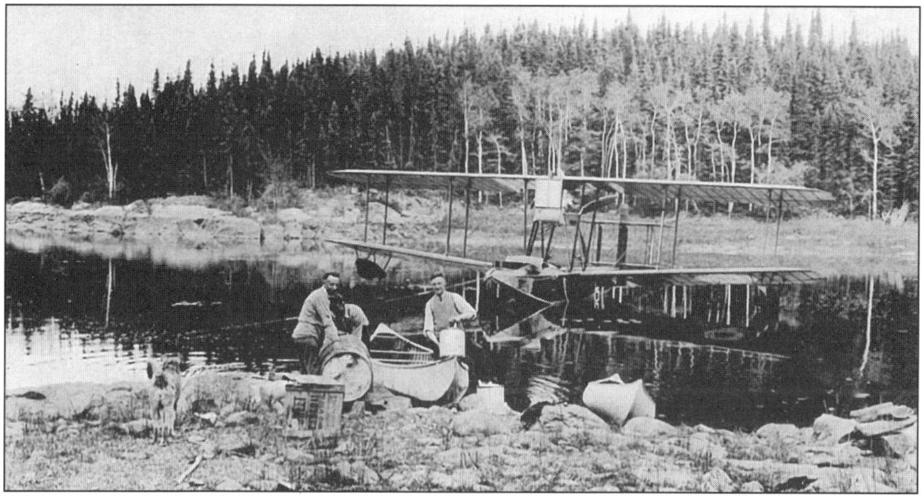

Refuelling was usually a task for all hands on the Reindeer Lake expedition. Gasoline in 45-gallon drums delivered by land and water transport to Pukatawagan, Manitoba, had to be transferred to the Viking in five-gallon cans by canoe borrowed from the local Révillon Frères post because the water was too shallow and rocky near the shore to risk the fragile Viking. "All right, get your pants off," Hobbs commanded himself before swimming to the far shore with a rope to hold the Viking safe in midstream.[56] Left to right, Cairns, Davidson, Milne. *NAC REINDEER LAKE REPORT*

border. The return route took them to the Churchill River, westerly to Stanley Mission and Isle-à-la-Crosse, then directly to Prince Albert and along the Saskatchewan River to The Pas. Brochet, at the head of Reindeer Lake, the most northerly point reached, was 1,000 miles from the base at Victoria Beach, and hundreds of miles past the limits of railway and steamboat communication.

Hobbs and his crew performed valiantly, despite severe discomfort and unco-operative weather. Takeoff was always in a shower of spray. The aircraft sat so low that the pilot could dangle his hand from the cockpit and touch the water (some wags suggested that he should be issued a paddle to assist the aircraft to become airborne). Cairns, in the forward cockpit, had to stand when photographing and was exposed to the 60-mph wind, protected by only goggles, gloves, helmet and Sidcot suit (a fleece-lined cover-all, similar to a snowmobile suit). The mechanic would have to fix whatever went wrong in the most primitive conditions. And the noise of the engine was, literally, deafening.

56 Alex J. Milne, "Pioneer Survey Flights in Northern Manitoba and Saskatchewan," *CAHS Journal 22*, no. 1 (spring 1984): 4.

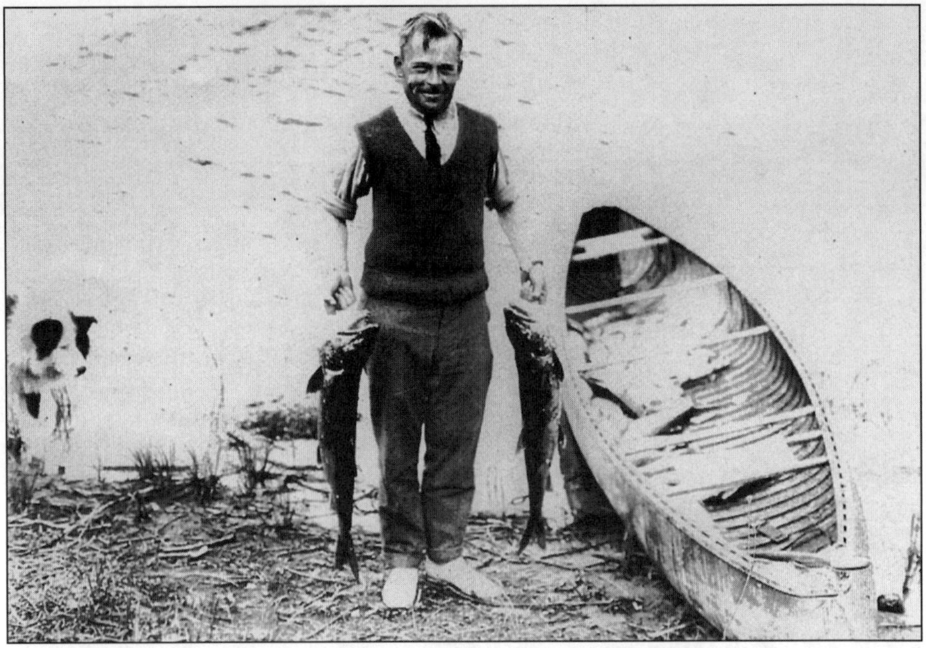

The limited ability of 'ET to carry supplies meant that the crew had to "live off the land" whenever possible. Basil Hobbs is evidently looking forward to a fish supper—if he can keep his catch out of the jaws of the advancing dog. Rabbit River, Saskatchewan. *ALEX MILNE*

Designed for a 400 hp Napier Lion engine, Hobbs's Viking had only a 360 hp Rolls-Royce IX Eagle engine. When takeoff, cruise and landing were all performed at close to 60 mph, the missing 40 hp was important. On calm water, Milne would have to clamber over the windshield and crouch with Cairns in the forward cockpit to move the centre of gravity far enough for Hobbs to haul the flying boat onto the step, then Milne would return to his station for takeoff. With a full load, the Viking often had to run three or four miles until it gained flying speed, resulting in the radiator boiling over and spitting hot water at the crew, who had to protect themselves with engine covers. Once in the air at cruising speed, there was adequate airflow to cool the radiator.

Except for a few problems expertly fixed by Milne, the Rolls-Royce Eagle engine "ran like a motor car." Not so with the camera. Its electrical system failed completely and had to be hand-cranked until this became impossible. It was discovered that the teeth of one gear were completely destroyed. At the north end of Reindeer Lake, near the hamlet of Brochet, Milne hand-filed teeth on a piece of metal from his tool box hoard to make a new gear, and re-calibrated the camera, which then worked perfectly. However, not until many years later was a footnote to this story divulged. It appears that a few unofficial bottles of whisky were carried in the Viking to share with hospitable trading post managers. One was tapped at Brochet while Milne was repairing the camera, but Hobbs believed in equal sharing and assured Milne that a shot would be put aside for him with every round. Milne said he enjoyed his accumulated share with great satisfaction after successfully repairing the camera essential to the success of their mission.

While at Brochet, the expedition had two sets of visitors, each a prelude to future RCAF undertakings. An agent of the Department of Indian Affairs arrived with treaty payments for the Indians and compared his months of travel time by canoe with the few days taken by air. In future years, most of this work was undertaken by the RCAF, its officers even being deputised on occasion as Indian agents. C.S. Macdonald, DLS, had been in the area for some months running ground traverses against which the aerial photographs would be plotted by the Department of the Interior. Hearing the Viking pass over his camp, he quickly made his way to Brochet and was very pleased to receive a bag of mail from Hobbs. Davidson and Macdonald spent a long time comparing notes that resulted in significant modifications to their survey plans, and probably influenced MacDonald's future career as he subsequently served as a civilian navigator with the RCAF.

Viking 'ET returned to Victoria Beach on 14 August, completing 2,810 miles in 44 hours and 10 minutes' air time. Cameraman Cairns had taken 1,700 photographs, mostly by winding the camera by hand after the electrically driven "automatic" mechanism failed. The area surveyed totalled about 15,000 square miles and, discounting preparation time, took four weeks. In a rare abandonment of military modesty, in its annual report the RCAF hailed this expedition as "certainly the greatest ever undertaken for aerial survey; and, when considered in the light of practical results, one of the most brilliant achievements in the history of flying."[57]

Surveyor Robert Davidson was equally enthusiastic, as quoted in Hobbs's report:

> The distance travelled and the air photographs taken demonstrate clearly that in no other way can the surveyor hope to cover unmapped and remote territory with such ease and rapidity. The visibility on a clear day is so good that well-defined points, such as prominent headlands or fairly large bodies of open water, can be easily distinguished at a distance of forty miles or more from an elevation of

57 *Report on Civil Aviation for the Year 1924* (Ottawa: Dominion of Canada, 1925): 73.

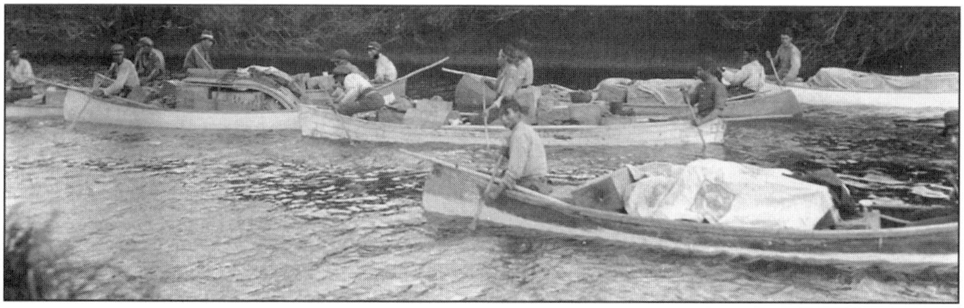

Basil Hobbs and Bob Davidson introduced Colin Stone Macdonald, DLS, to aerial survey at Brochet in 1924. Macdonald was impressed and went on to navigate the epic flight by pilots Mawdesley and Winny to the Northwest Territories in 1930. He took this photograph of canoes transporting "aeroplane gas" to a cache at Lac la Ronge, Saskatchewan, in 1926. *NAC PA2007*

Colin Macdonald gave due warning to any violators of the gasoline caches. Near Lac la Ronge, Saskatchewan, 1930. *R.E. McBURNEY*

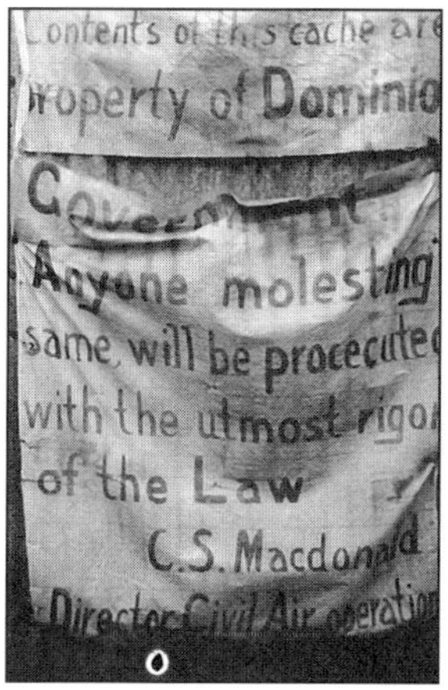

> five thousand feet. The sketch maps now in existence show these points, together with a vague outline of shore, and very little more. It is just here that the air traveller gets one of his greatest advantages over the man on the ground. The latter cannot jump from one easily identified point to the next, often many miles distant, but must follow the watercourse in all its intricate details.[58]

Hobbs correctly believed he had demonstrated that photographic survey flights could be extended much farther north, "provided the necessary fuel depots can be made." He gave glowing reports on the performance of the team members. Robert Davidson, the civilian surveyor who acted as navigator, for example, "seemed to have an instinctive sense of direction and to be able to follow his course unfailingly, without reference to known landmarks and often across country absolutely unknown. Without such skill and knowledge, this flight could not have been successfully undertaken."[59]

With no threats of conflict on the horizon in 1924, the RCAF concentrated on its civil missions, primarily in support of federal government departments. A series of informal conferences were held every winter with representatives from all interested branches with the objective of satisfying as many clients as possible with the minimum number of flights. Coordination of an increasing number of requests for aerial photographs became a problem, and this task was assigned to the Topographical Survey Branch of the Department of the Interior. As usual, the RCAF's annual report did not miss an opportunity to underline the value of aerial photography to other departments. A photograph of a river taken during an investigation of water power possibilities was offered as an example that could also contain information as to the topographical features of value in map revision, show the nature of forest cover of interest to the federal and provincial forest services, and identify geological features of greatest interest to the Department of Mines.

Before 1924, less than 4,000 square miles, less than one-tenth per cent of the Dominion's land mass, had been covered by aerial photographs. In 1924 alone, over 40,000 square miles were covered by vertical and oblique photography: 27,000 in Manitoba, 10,000 in Alberta, 3,000 in Nova Scotia and about 200 in the vicinity of Ottawa. Most of this work was accomplished by a Viking in Manitoba, a D.H.4B in Alberta, and an HS-2L in Nova Scotia. Photographs taken during 1924–25 totalled 14,075 and about 30,700 contact prints were received from the RCAF by the Topographical Survey Branch. The photographs were used by government departments for a wide range of applications, including topographical mapping, forest and mineral resource identification, locating roads and railways, identifying hydroelectric power sources, improving harbours and navigation channels, planning towns, administering national parks, developing federal lands, and advertising historic sights and scenery in national parks, "the playground of our tourists." The two

58 Ibid, p. 101.
59 Ibid., p. 103.

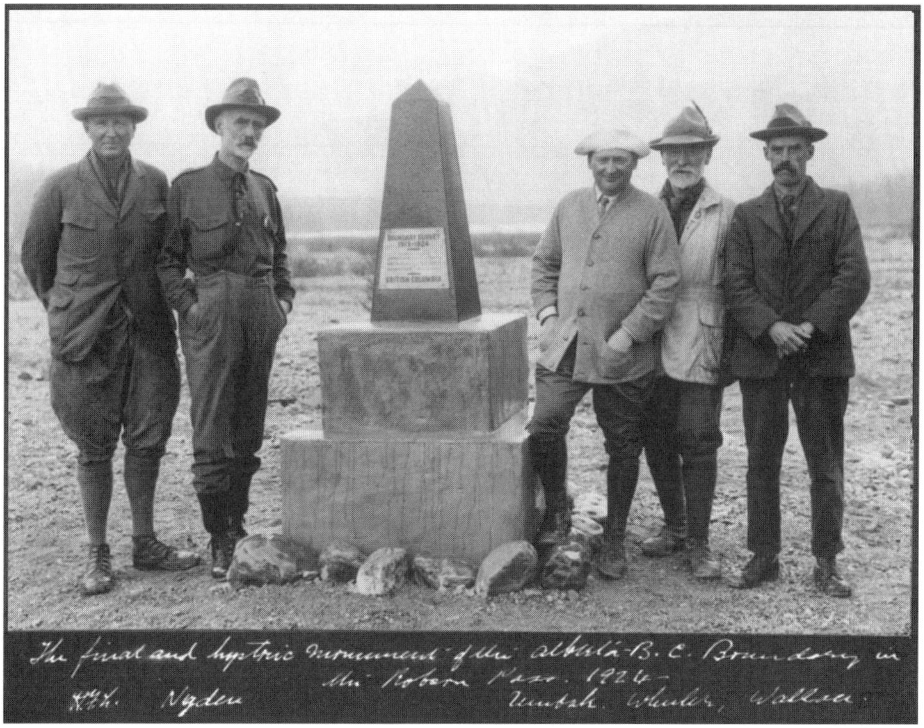

The final and historic monument of the Alberta–BC Boundary in Mt. Robson Pass, 1924. Aerial photography was vital to accurately establishing boundaries through mountainous territory. Howard Frederick Lambart on left, Arthur Oliver Wheeler second from right.
NAC PA 202781/EVELYN LAMBART COLLECTION

HS-2Ls and single Viking at Jericho Beach were primarily engaged in fisheries protection operations during 1924. Photographic missions were limited to oblique pictures of canneries and harbours along the west coast for the Department of Marine and Fisheries, and 186 photographs of the Point Grey area to make a mosaic for town planning purposes. Wireless systems were installed in the Jericho Beach HS-2Ls on fishery protection patrols off the west coast, identifying an unexpected problem. Due to frequent fog and low-level cloud, aircraft often had to fly so low that they could not extend the aerial necessary for transmission. Successful communication by both radiotelephone and Morse code was, however, achieved in more clement weather.

A series of photographs illustrating mountain scenery in Alberta, from Lake Louise to Canmore, were taken for the Parks Branch by the High River aircraft. Some 25,000 square miles were covered in the Vermilion District by oblique photography. A 520-square-mile mosaic was made of the Edmonton area, later extended to a complete map sheet covering 3,224 square miles. Photography was hampered by the usual problems: bad weather at Wainwright and lack of ground-surveyed landmarks at Brazeau. Vertical photographs were taken in the Calgary and Edmonton areas, and 1,263 photographs covering about 2,500 square miles south of Chip Lake, 80 miles west of Edmonton, were photographed in connection with preliminary land classification field examination for the Topographical Survey Branch. In total, aircraft from High

D.H.4B G-CYCW over the Rockies.
DNDPU PL143835

During 1924, High River's D.H.4Bs flew 226 hours on aerial survey, work that S/L A.L. Cuffe, the station commander, with military understatement noted, "was not without hazard as the country in many parts of western Alberta is not particularly suited to flying. Landing grounds in the forest reserves are few and far between; and the operations often entail long flights of as much as 60 to 100 miles before the scene of actual work can be reached." To emphasize his point, he waxes poetic regarding flying through the heart of the mountains, "where the scene presents nothing on all sides but tumbled masses of rock, snow and ice, with deep forested valleys lying between the ranges."[60]

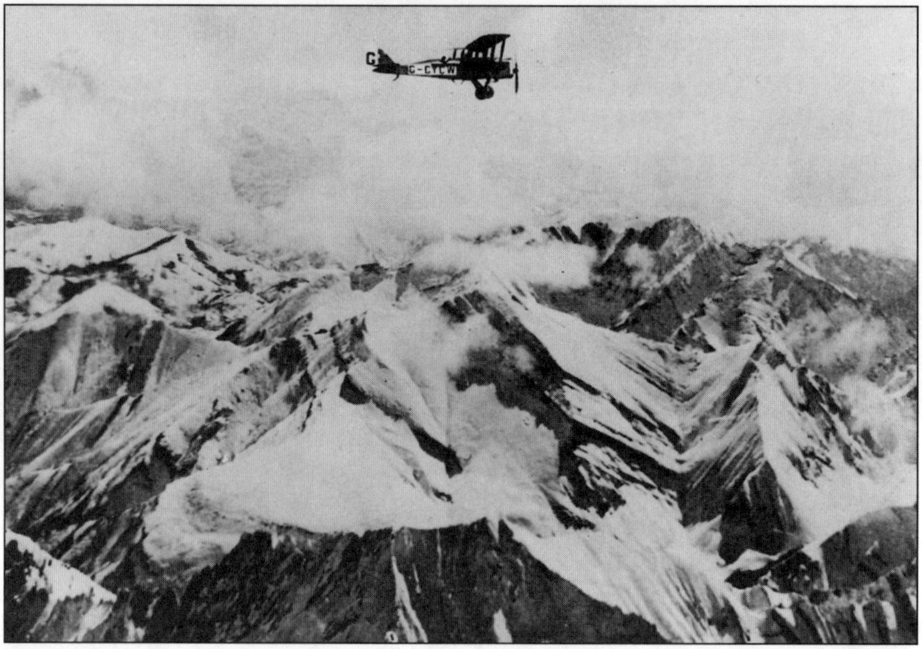

River flew 687 hours during 1924, 382 on fire patrol and 251 on photographic missions.

An unexpected by-product of High River's busy season was the reported hasty dismantling of several illicit liquor stills, as the operators thought the aerial surveillance was on their account. In another area, the RCAF was given credit for a sorely needed heavy rain that fell on the heels of a photo detachment's arrival![61]

By 1924, the World War I D.H.4Bs were getting tired and suffering badly from moisture absorption. S/L Cuffe commented on several other aspects of flying them on photographic missions in western Alberta, which illuminated some of the difficulties faced. When photographing near Brazeau, west of Red Deer, he reported great difficulty in identifying the very few established reference points on which to base his flight lines, and had to use points on larger rivers or natural features. "This area is chiefly muskeg, possible landing grounds are few; and, had the engine not been running perfectly, it would not have been possible to have undertaken it. The lines of flight were sometimes 120 miles from the base and photography often commenced after a flight of fifty miles."[62] During these flights he observed that north-south section lines made better marks on which to fly than those running east-west. This was because the railway ran east-west and all the roads leading to it were more travelled than those running parallel and were, consequently, easier to see from the air.

Farm fields were generally utilized as temporary bases on photo missions. These were convenient in many ways, but most were wheat stubble, which,

60 Ibid., p. 61.
61 Ibid., p. 65.
62 Ibid., p. 62.

Cuffe maintained, cut the wooden propeller to pieces, soon rendering it useless. He recommended the supply of metal propellers. All the D.H.4 variants lived in the open during the summer and stood up well to weather, although the varnish applied over the doped wings turned to powder and rain infiltrated both fuselage and wings. Cutting drain holes helped, but the insides of the wing were not free of moisture for several weeks after being exposed to a rainstorm.

Responding to complaints from Ottawa that his pilots failed to maintain a constant altitude during photo flights, so varying the scale of photographs, S/L Cuffe stated that constant attention was given to the altimeter where the fault must lie. The instruments were all scheduled to be tested before the next season, but he noted that the problem may have been caused by changes in barometric pressure over the long distances flown, and added, "On one occasion it was found on landing after a long flight that the altimeter registered 400 feet though it had been set at zero on taking off."

Cuffe was not happy using the D.H.4Bs for photographic duties and detailed its shortcomings:

> As a photographic machine it is unsatisfactory. The cockpit is too small, making so small an operation as changing magazines an exhausting process. The observer's feet are wedged under the seat and as temperatures of five or ten degrees were not uncommon at 12,000' (14,000' above sea level) the discomfort was acute. The blind spot caused by the lower wing is so large that directional errors are easily made and difficult of correction when found. In the oblique work the position of the camera diverts the slipstream into the cockpit, which adds to the discomfort.[63]

Today's aircrew usually turn on the oxygen before reaching 10,000', but lack of oxygen was not even mentioned as a discomfort in 1924. As late as 1932, F/L R.C. Gordon and LAC J. Fortey were sent on Operation Cosmic, which involved flying a supercharged Fairchild at 18,000' to 23,000' for several hours to investigate the effects of cosmic rays. Enquiring about the provision of oxygen, they were advised to "chew gum and swallow frequently." Both became ill, but recovered after returning to *terra firma*.

The new Fairchild camera introduced in 1924 was viewed favourably, although the High River crews had initial problems until they fully understood its operation. The "intervalometer" used to automatically expose the film at predetermined intervals was discarded "owing to functional failures."[64] Instead, the camera was hand-triggered, with the added advantage that it could be adjusted prior to each exposure to make it absolutely vertical, so removing any problems due to aircraft tilt. In this procedure, the photographer gave directional instructions while the pilot concentrated on his altimeter and keeping the aircraft level.

63 Ibid., p. 63.
64 Ibid.

Wireless sets were installed in five new 200 hp Wolseley Viper-powered Avro 552A single-seat aircraft, ordered from Canadian Vickers to replace the tired Rolls-Royce Eagle-powered D.H.4s employed the previous four years on fire patrol at High River. The Avro Vipers, built by Canadian Vickers in Montreal, were not popular with the crews, who endured severe vibration until the Vipers were replaced by D.H.60 Moths in 1927.

In addition to the historic flight to Reindeer Lake, Winnipeg crews operating from Victoria Beach and Norway House also took 3,904 oblique photographs covering 16,000 square miles near The Pas and several areas east of the Manitoba–Ontario boundary for mapping purposes. Verticals were taken of the Cooking Lake forest reserve near Edmonton. Late in the season, yet another photo mission was undertaken in conjunction with forest patrols along the east shore of Lake Winnipeg from Fort Alexander north to the Berens River. Using a refuelling base at Matheson Island, the Viking crew completed the assignment in 35 flying hours, including one day of 8 1/2 hours of photography in exceptionally good weather. Other flights of relatively short duration were made to photograph hydroelectric developments and potential sites on behalf of the Water Powers and Reclamation Service. A diphtheria epidemic broke out at Norway House in October after the aircraft had been withdrawn for the winter, but a quick rescue flight was made from Victoria Beach to deliver antitoxin just before freeze-up.

Rockcliffe continued to be the centre for test and experimentation. CO F/L C.M. McEwen, with a staff of three officers and 26 airmen, also undertook photographic missions with his two aircraft, a Viking and an HS-2L. The principal task was a mosaic of islands in Georgian Bay for the National Parks Branch, completed from a base at Parry Sound despite unco-operative weather. Photographs were taken for the Forestry Department of the same forest area using different altitudes and focal lengths to establish the best combination for forest type identification. Twenty-nine obliques were taken along the Ottawa River east of Hawkesbury for the Water Power and Reclamation Service. Eighteen photographs of Kingston, "illustrative of industrial ordnance areas," and 12 of Charleston Lake east of Kingston were taken on behalf of the Ordnance, Admiralty and Railway Lands Branch for mapping purposes.

Only one HS-2L, G-CYEL, was available for operations at Dartmouth in 1924. It was fitted with a high-compression Liberty engine that enabled it to reach the unheard-of, for an HS-2L, altitude of 10,000' preferred for photography. Four hundred and fifty-seven vertical photographs were taken in the New Glasgow area to update maps. Oblique photographs for topographical mapping were also taken of about 2,830 square miles of Nova Scotia lying west of a line joining the Sable River and the Annapolis Basin. On 3 September 1924, Dartmouth CO S/L J.H. Tudhope flew G-CYEL to meet the US Army's round-the-world flight from Seattle and escort it to Pictou, Nova Scotia.

Canadian Vickers embarked on the design of a new, all-Canadian flying boat and the prototype was flown at Montreal on 4 November 1924. Superficially similar to the Viking, the three-place Vedette was in service by 1925 and nine were still operational in 1939. The Vedette was popular with pilots as, after some initial modifications, it had good handling characteristics. Performance deteriorated during the season as the wooden hull soaked up water, but when dry it could haul a full load to 10,000' in thirty minutes. The hull was constructed of cedar strips retained by copper rivets and the wings were fabric-covered wood. The RCAF encouraged the development of a domestic aircraft design and manufacturing capability, also purchasing the Avro 552A Viper single-seat, single-float planes from Canadian Vickers. A support industry also evolved, building floats and other aircraft parts, but reliance continued on Britain and the USA for engines and propellers.

In 1924, Dominion Aerial Exploration was still operating the two HS-2Ls. In addition, three ex-RN Norman Thomson NT2B flying boats, each powered by a single Hispano-Suiza 210 hp engine, were purchased. Parts from one, G-CAEM, were used to refurbish G-CACG and G-CAEL, which commenced operations from Roberval in June. Advance bases were established an hour or two out of Lac Saint-Jean and supplied by HS-2Ls while the Norman Thomson aircraft undertook surveys in the area. One of the bases was at Lac Édouard for photography and survey in the Laurentide National Park and Game Preserve. An indication is given of the continuous concern regarding engine reliability by a comment in the RCAF's annual report: "The Park is rugged and mountainous and the better performance of the Norman Thomson boats was found to be of great use as it was at times necessary to gain a height of 9,000 feet to ensure that a good water surface could be reached in case of engine trouble during flight."[65] None of the flying boat pilots liked to be out of gliding distance to a large enough body of water.

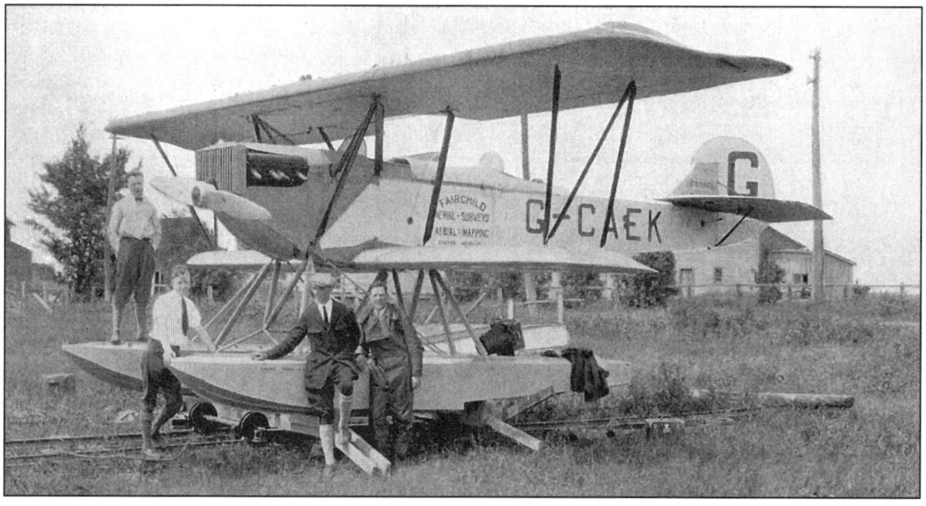

Fairchild Aerial Surveys' Curtiss J-1 was returned to the manufacturer in April 1924 and replaced by this Hispano-powered D-19 Petrel V from Ogdensburg, NY, manufacturer Huffman-Daland. Left to right, Edward R. Olley, Ellwood Wilson, Kenneth F. Saunders and Hubert M. Pasmore.
REPORT ON CIVIL AVIATION 1925

65 Ibid., p. 31.

Fairchild Canada continued to expand and purchased a Curtiss MF Seagull flying boat. A second pilot, Hubert Martyn Pasmore, was hired and he subsequently played the leading role in the company's development. Fairchild accumulated 232 hours' flying time in 138 flights, but could not satisfy the demands for its services and had to contract with Laurentide Air Service for another 45 hours. At $30 per hour for photographic mapping, Fairchild was beginning to show a profit, but preferred first to make fast and economical sketches of unexplored forests, following up with photographs if the area showed promise.

A stable platform, the two-seater was similar to one modified by the US Army for aerial photography. With the money-making business of town mosaics in mind, the Huff-Daland was fitted with wheels, floats or skis according to the season and task in hand. All-season aerial photography was offered and oblique photographs were taken over an area of 400 square miles near Lake Shawinigan, Quebec, in December 1924's freezing temperatures. The Huff-Daland continued this type of work until August 1928.]

Fairchild assisted the Southern Abatibi Railroad with the preliminary survey of a proposed line from Amos, Quebec, on the Canadian National Railway, to Maniwaki and Mont-Laurier. These sites were the termini of the Canadian Pacific branch lines to the upper waters of the Gatineau and Lièvre Rivers (approximately today's Highway 117). Consulting engineer Henry K. Wicksteed, MEIC, FRGS, used this "reconnaissance" to document the value of aerial photography in railway location and his paper was attached to the RCAF's 1924–25 annual report.

Wicksteed criticized early methods of estimating railway survey costs at so much per mile of length: "The examination is one of not merely of a narrow ribbon of country with a view to laying down a practical and reasonably inexpensive line topographically. It is this and more, much more." He emphasized the need for a railway to thrive and expand—for which it must have an adequate traffic load. Outlining what we today would call an economic feasibility study, in addition to topography, he listed for consideration commercial resources, agricultural lands, mineral deposits, waterpower, timberlands, and summer and winter playgrounds. Further, he argued, this evaluation must cover a wide swath on each side of the line, lune-shaped and roughly half as wide as it was long. For his 200-mile-long example, the area to be considered was 8,000 square miles. By ground survey, this would incur a great expenditure of time and money. By air, however, the task could be accomplished quickly and relatively easily, and with much greater accuracy. He saw vertical photography, requiring great accuracy and clear weather, as useful only in the precise stages of line location. Obliques, however, for his reconnaissance purposes, were "far and away superior to the rough sketches which are all that is now available over much of our northern wilderness, and they are the result of direct instrumental vision, not merely what a man fancied he saw through the trees or what an Indian drew with a stick on a sandy beach, and in at least nine cases out of ten, they are sufficient for the purposes of the reconnaissance engineer."

Vertical photographs were dismissed by Wicksteed as failing completely to provide information on gradients, a subject of great importance to railway engineers, although he did admit that the location of lakes and streams had value in identifying the lower levels. While stereoscopic photographs could provide contour information, he felt the cost was exorbitant and unreasonably delayed by dependence on an office several hundred miles away performing rather abstruse mathematical analysis, "a matter of interesting 'mathematical gymnastics' rather than utility." The accuracy of such contours was, he maintained, nowhere near that required to establish gradients of less than 1:100 (53 feet per mile) required by the railways. Oblique photographs had value in identifying gradients. Wicksteed, however, contrary to most other users who desired high-altitude photography for maximum coverage in mapping, wanted them taken at the lowest possible altitude to show the comparative heights of geographic features along the route. Wicksteed concluded by stating that, from the overall perspective, it was difficult to overestimate the value of aerial photography in comprehensive railway survey.

The double blows of losing money in 1923, and Ontario, its major customer, establishing its own air service in 1924, forced a critical change in Laurentide's operation. Thirteen Liberty-engined HS-2L flying boats together with company bases and equipment in Ontario were sold to Ontario in the spring of 1924. Roy Maxwell went with them to head the Ontario Provincial Air Service (OPAS) and many of the Laurentide pilots and mechanics joined him. Reconditioning these aircraft before delivery to OPAS, at a reported $5,500 each, gave Laurentide the basis of a manufacturing capability at Three Rivers that was never fulfilled in competition with Canadian Vickers. Three HS-2Ls and one Napier Lion-powered Viking were retained by Laurentide. They were used primarily for passenger, mail and freight services to the new gold fields at Rouyn, Quebec, from Angliers on the Quebec–Ontario border and terminus of a Canadian Pacific Railway branch line, the Interprovincial & James Bay Railway. Flights were arranged to connect with CPR's trains, and a surprising lack of fanfare marked the initiation of Canada's first scheduled air service on 23 May 1924. The shallow water at Angliers was full of logs and deadheads, so the main operating base was moved to better conditions at Haileybury on Lake Temiskaming, Ontario, in July. A six-seat Westland Limousine III and a D.H.9 three-seat aircraft were purchased and equipped with skis to continue this service during the winter. The Limousine was the one brought to Canada by Sidney Cotton in 1921 for service in Newfoundland and Labrador; on inspection the longerons were found to be rotten and the aircraft was condemned. Pete Vachon enclosed the rear cockpit of the D.H.9, which made a belated start to passenger service on

21 January 1926, but crashed three days later with Jack Caldwell at the controls, bringing an end to the Rouyn winter service.

Laurentide continued aerial survey and photography, although amounting to only 182 hours out of 933 flown during 1924. In addition to the 45-hour Fairchild contract, 2,000 square miles of forest were surveyed in co-operation with James D. Lacey & Co in the Algoma District for the Spanish River Pulp and Paper Company; similar work was undertaken northeast of Quebec City and in the Gatineau region north of Ottawa. One HS-2L was shipped by rail to Prince Rupert, British Columbia, from where pilots Scott Williams and Jack Caldwell spent six weeks in Alaska, the Yukon and the Northwest Territories prospecting for a US mining syndicate. The aircraft was flown to Vancouver and shipped to Sault Ste. Marie where it remained in storage until sold in 1926. All remaining Laurentide aircraft were scrapped, except for HS-2L C-CAGT, sold to Bill Broach, who started Northern Air Services Ltd. Laurentide would cease operations in 1925, but not before entering two more Canadian "firsts" into the record book—scheduled airmail stamps and, a noteworthy advance for the time, a company operating manual. Another "first" in 1924 was an aerial stowaway—an unidentified homesick mineworker stealing a free ride from Rouyn in Jack Caldwell's dangerously tail-heavy HS-2L.

Ontario went into aviation in a big way. The 13 ex-Laurentide HS-2Ls were supplemented with a 14th and a Loening flying boat later in the year. The Sioux Lookout station was taken over from the RCAF and a second station established at Sudbury together with many sub-bases and refuelling depots utilized for extended fire patrol and forest mapping. All the aircraft were flown to a permanent facility at Sault Ste. Marie for the winter. Indicating the main area of interest, Ontario air operations were assigned to the Forestry Branch. Mapping was also carried out over 1,500 square miles west of Lake Nipigon, north of Port Arthur (now Thunder Bay), and south of the Canadian National Railway. Oblique photographs of some 750 square miles were taken for comparison with existing maps prepared from aerial sketches and ground survey.

1925-1926

CONSIDERABLE ATTENTION was given in the 1925–26 *Report of Civil Aviation* to progress made abroad, highlighting financial support by foreign governments to passenger, mail and freight services, construction of airfields, establishment of flying clubs, formulation of regulations, and the development of many new aircraft types. In contrast, it was emphasized, "In Canada . . . flying was used principally as an improved method of observation. Straight transportation work has been subsidiary to this, and has been undertaken only when other modern facilities were lacking."[66] Some satisfaction was, nevertheless, expressed with the improvement of existing organizations, both civil and military. Despite the comparison with investment by other governments, the basic vote for aviation in 1925–26 was reduced by $121,000 to $1.14 million. Funds for new equipment, however, were increased by $440,850 to $740,850, a net increase of $319,850.

At Jericho Beach, no specific photo missions were flown, but the station's four HS-2L flying boats were in continuous demand for forestry, fishery and customs patrols, and pilot training. A reconnaissance of Graham Island for the Soldier's Settlement Board was appreciated by the chairman as being "of the greatest value" in eliminating areas unsuitable for settlement and permitting him to concentrate on areas that would repay further examination.

The brace of D.H.4Bs at High River covered 8,600 square miles in 161 hours' flying time. Ambitious plans to photograph 70,000 square miles of Western Canada were foiled by smoke from forest fires in British Columbia. The largest task completed was vertical photography of a 4,350 square mile area in the Edmonton district, using an eight-inch lens from 12,000'. Other operations covered the Saskatchewan River near Rocky Mountain House, the Buffalo Park Reserve at Wainwright, and the Fort Assiniboine area. The latter mission was over largely unsurveyed territory and proved that existing maps were inaccurate; one hill feature located six miles out of its mapped position. On this flight, covering 4,350 square miles of ground with considerable differences in elevation, the aircraft followed the ground profile, keeping the

66 Report on Civil Aviation for the Year 1925 (Ottawa: Dominion of Canada, 1926): 12.

scale of the photographs constant. It was concluded that this technique had value in transposing the photographs to maps where changes in ground level were not abrupt. On another flight, the location of Island Lake, located near the Manitoba–Ontario boundary, proved to be 25 miles in error.

Still campaigning for replacement of the D.H.4, High River's 1925–26 report adds to its sins listed the previous year. "The rate of climb above 12,000 feet is between 50 and 100 feet per minute, which is not sufficient to counteract the bumps which often attain a vertical velocity of 100 to 500 feet per minute."[67] The cruising speed of 80 to 90 mph often limited ground speed to 30 or 40 mph, and sometimes winds were of such velocity that no progress could be made. The undercarriage rubber shock absorber was a source of trouble, as it deteriorated in cold, wet weather. A table of readings from two altimeters carried on the same flight showed differences as great as 400', a serious problem when maintaining a constant height was critical for good photographic results.

Concerned about the spread of wheat rust in Western Canada, the Department of Agriculture enlisted the assistance of both Victoria Beach and High River. On routine patrols, microscope slides coated in Vaseline were exposed, indicating the rate and spread of airborne spores.

Winnipeg operations continued to expand during 1925 and lowered its operating costs by utilizing new Avro single-seat aircraft for fire patrol. The seven Vikings on strength were used for photography and firefighting duties.

Norway House was opened again in 1925. Flying from the shallow and turbulent Saskatchewan River presented problems so the sub-base at The Pas was closed and a new one opened about 40 miles northeast at Cormorant Lake, near Mile 43 on the Hudson Bay Railway. The three bases were linked by radio with Winnipeg.

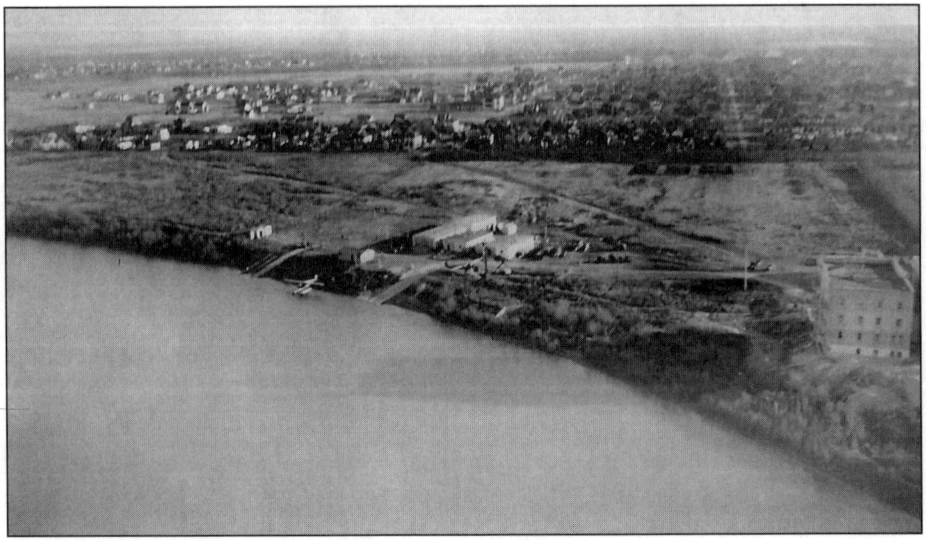

A wider ramp and workshops were constructed at the Brandon Avenue haulout facility on the Red River and used extensively by both RCAF and commercial operators to assemble and dismantle aircraft in the spring and fall, and for storage during the winter. 29 October 1925.
NAPL A949-37

67 Ibid. p. 58.

Low clouds or haze for most of the summer limited the planned 50,000 square miles of oblique photography over Manitoba and Ontario in 1925 to 38,000 square miles. Areas covered included parts of northwestern Ontario and eastern Manitoba, along the Saskatchewan River and other waterways westward into Saskatchewan, and the 400-mile route of the Hudson Bay Railway, then under construction, from "end of steel" at The Pas to Port Nelson.

Conditions encountered by the single Viking's crew on the railway survey were typical of the time, and the annual report indicated why Port Nelson was later abandoned in favour of the more northerly terminus at Churchill. The trip took nine days, seven of which were spent on the ground waiting for suitable weather. "On arrival over Port Nelson, the impression from the air was that the area provided poor facilities for the mooring and shelter of aircraft. The coast appears to consist of very wide mud flats with a comparatively narrow river channel approximately one mile from the northern shore." Risking a landing on the seaward side of a bridge to an island dredged for the port facilities, the Viking was anchored on the mud flats: "Fortunately, no strong winds or storms were experienced."

Refuelling at Port Nelson also presented problems. The fuel cache was at the RCMP barracks, nearly two miles from the anchored Viking. Two barrels of gasoline and one of oil were loaded onto a borrowed cart and pushed along a narrow-gauge railway as far as some flatcars, over which they had to roll the barrels. Another borrowed cart transported the fuel to the beach, but by this time it was 10 p.m. and dark, so refuelling the aircraft was postponed until morning. The "beach" consisted of large rocks, making it impossible to bring the barrels to the water's edge, so the fuel had to be carried in borrowed small buckets, and ferried to the Viking. Meanwhile, the flying boat had to be continually pushed farther out for fear of stranding as the tide ebbed. Evidently with relief, the account ends, "No trouble was experienced in getting away or off the water."

The Viking's 5,800 pounds all-up weight limitation was a constant concern. F/L Edwards, OIC at Victoria Beach, recommended that the four-man crew be reduced to three with the mechanic also taking on the duties of photographer. This would improve the aircraft's performance or permit more stores, particularly safety equipment, to be carried. Navigators, usually qualified surveyors, were considered essential, not only to guide the pilot over the maze of waterways and make notes to assist in the interpretation of photographs, but "to ensure the safety of the party in the event of their being stranded far beyond human habitation." He also recommended the provision of instruments that are today considered standard: rate of climb indicator, periodic compass, turn indicator, level, altimeter, air speed indicator, petrol gauge—"a reliable type is essential"— watch, and engine instrumentation.[68]

F/L Edwards then outlined his specifications for a photographic aircraft operating in his area of interest. He wanted a 1,000-pound payload, exclusive

68 Ibid., p. 64.

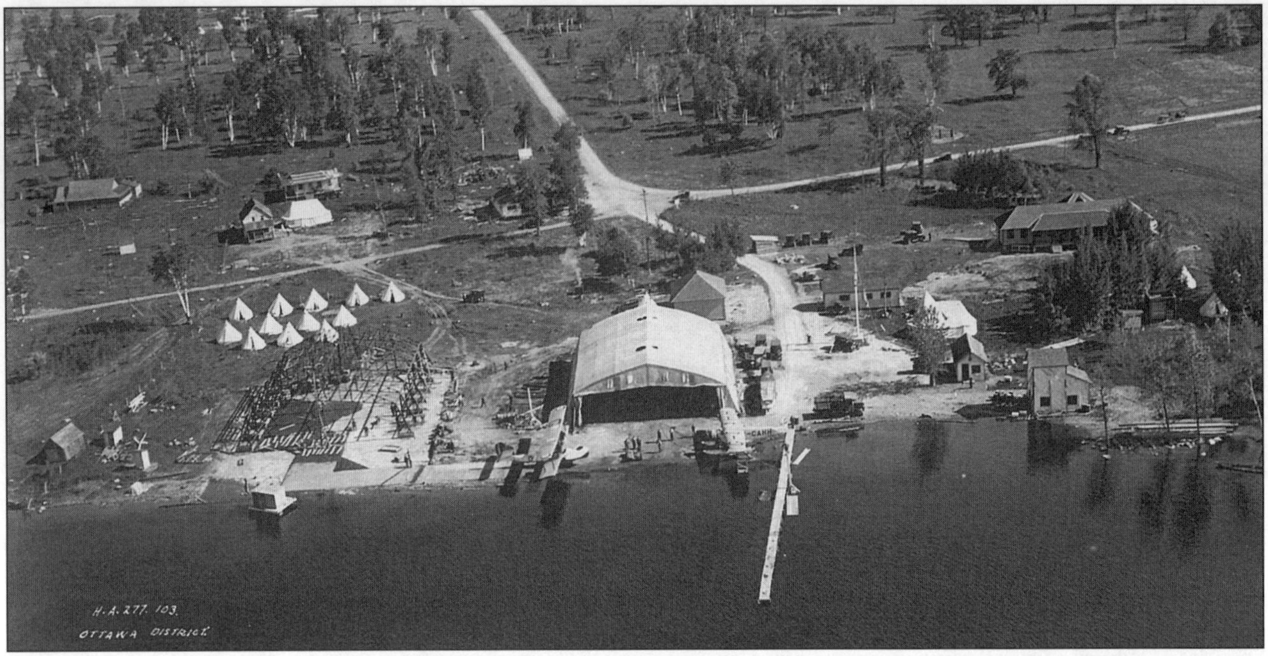

A base for waterborne aircraft was opened in May 1925 at Shirleys Bay on Lac Deschênes in South March, twelve miles west of Ottawa, because floating logs in the Ottawa River at Rockcliffe presented a serious hazard. The new facility, adjacent to the army's Connaught Rifle Range, was still on the Ottawa River, but with fewer floating hazards. A second hangar was being constructed on 3 June 1927. Photograph taken by surveyor/navigator John Carroll, DLS.
NAPL HA277-103

of full gasoline, oil and water tanks. He detailed the minimum load of three crewmen at 180 pounds each inclusive of personal and flying kit, photographic equipment of 135 pounds, emergency supplies of 101 pounds, and engine and airframe spares, all totalling 979 pounds, but did not include parachutes or spare film magazines, as many as twenty of which had to be carried on occasion at six pounds each. He wanted the aircraft to take off from rough or glassy water with a full load, climb to photographic altitude within thirty minutes, and have an endurance of five hours with a thirty-minute reserve.

Wheeled aircraft continued to use the airfield at Rockcliffe. Waterborne aircraft on strength at Shirleys Bay consisted of two Avro seaplanes and one each of HS-2L, Viking, Vedette and Varuna. The twin-engined Vickers Varuna flying boat, second of its type, was at Ottawa for acceptance trials. Although metal propellers were not available and wooden ones were temporarily installed, the Varuna, with an all-up weight of 5,300 pounds and a crew of five, showed promise. The Shirleys Bay base was used primarily for experimental work, trials of new aircraft and general duties in eastern Ontario and western Quebec, but photographic missions were also undertaken. The whole length of the Rideau Canal from Kingston to Ottawa was photographed in 7 hours and 20 minutes' flying time, and the Petawawa Forest Reserve was photographed for the Forestry Branch to test the feasibility of forest mapping from the air. Photographs were also taken of the Experimental Farm at Ottawa for the Topographical Survey Branch and, for other departments, obliques of harbours, hydroelectric power sites and historical sites in Niagara, eastern Ontario and western Quebec.

Four venerable HS-2Ls were dispatched in August 1925 for aerial sketching of the forest lying between Lake of the Woods and Port Arthur with unfortunate results. One made a forced landing and was stormbound before repairs were completed. The crew made themselves at home with a nearby survey party, but were unable to notify the other aircraft at Sioux Lookout. Anxiety increased during the next three days and search aircraft set out as soon as the storm abated. Perhaps prematurely, as conditions were still bad and strong gusty winds brought down one of them as it was landing to refuel at Savanne. The pilot was severely injured and the two crewmen were killed.

The two HS-2Ls at Dartmouth were handicapped by poor weather conditions. It was possible to complete only a small portion of the ambitious photography program planned for New Brunswick and Nova Scotia. A mosaic of forest lands near Fredericton belonging to the University of New Brunswick was completed, but a second operation in that area had to be abandoned after the aircraft sat on the water at Fredericton for almost a month waiting for the weather to change. One hundred-and-forty square miles around Yarmouth were eventually photographed for map revision, but other tasks in Nova Scotia remained unfinished due to cloudy and foggy weather. The inclement weather did give S/L Tudhope and his small staff of two officers and nine airmen some time to start assembling a new acquisition of ten ex-US Navy HS-2Ls—which turned out to be in poor condition. They had been built in 1917 and had remained in storage since hostilities ceased in 1918.

With encouragement from the RCAF during 1925, some of the provinces and the private sector were starting to play a much larger part in aerial survey and photography.

Ontario was essentially self-reliant and completed a base facility at Sault Ste. Marie for a fleet of 17 HS-2Ls and two Loening flying boats operated by the OPAS.

Dominion Aerial Exploration continued its work for the Quebec government and moved its headquarters from Roberval to the Laurentide Air Services facility at Three Rivers. Poor weather in Quebec hindered photography in 1925, although an HS-2L did take about 300 oblique photographs in support of a survey party travelling 150 miles up the Moisie River from Seven Islands. Forest mapping was also undertaken over the upper Ottawa River from temporary bases at Fort Coulonge and Lac Dumoine, and on the upper Péribonka River from nearby Roberval.

Following the loss of contract work for Ontario when the province formed its own air service, Laurentide Air Service ceased operations in 1925. Its pilots, however, continued their pioneering efforts, leading the way in remote northern operations and accumulating experience of considerable value to future expeditions. B.W. Broach formed Northern Air Services Limited in Haileybury, Ontario, taking over the Rouyn goldfield service. Another ex-Laurentide pilot, C.S. Caldwell, with Lt.-Col. Scott-Williams, supported a prospecting expedition of Northern Syndicate Limited in a Napier Lion-powered Viking for six weeks in the Yukon. The aircraft was shipped by rail to Prince Rupert and flown via Wrangel, Alaska, and Telegraph Creek to the main area of interest, Dease Lake, British Columbia, and as far north as the Liard River in the Yukon. The Viking was returned to Vancouver after a total of 96 hours' flying time, dismantled and placed in storage. Caldwell's taste for adventure unsatisfied, he piloted a tiny Avro 554 biplane, G-EBFE, on skis for the Newfoundland sealing fleet. *Antarctic Baby* was powered by a Le Rhône engine and had been built for Sir Ernest Shackleton's Antarctic expedition, but never used. Lowered to a handy ice sheet from the *S.S. Eagle*, Caldwell would scout the area, trusting to find both the ship and an adequate adjacent expanse of flat ice on his return. He flew reconnaissance for the fleet from the *Eagle* for the next two seasons and replaced G-EBFE with a land-based Avro Avian with longer range in 1928. Caldwell was not available in 1929. His replacement, Alex Harvey, was trapped in fog, endemic to the region, on his first patrol and crashed on the Newfoundland coast. Harvey was uninjured, but the Avian was wrecked and the looming Depression stalled plans for its replacement.

Brock & Weymouth of Canada Limited, a new aerial photography company, was established at Montreal in 1925. The parent Philadelphia company had developed a system of mapping from vertical photographs by the stereoscopic method and saw potential in the expanding Canadian market. The first contract for the Montreal-based company was a railway survey from O'Brien on the Canadian National Railway about fifty miles to Rouyn, Quebec. A Fokker C11 biplane with a 185 hp BMW engine was shipped by rail to O'Brien and operated on skis from frozen lakes in the early spring, without the benefit of any shelter. Canadian National's chief engineer, however, agreed with the view expressed by Henry Wicksteed in 1924 regarding the Southern Abatibi Railway to the effect that vertical photography was too expensive and too slow for railway reconnaissance. The great advantage he saw was that Brock & Weymouth's stereoscopic method did give reasonably accurate contours. His somewhat grudging support notes the expense of getting the aircraft and the film processing equipment to the site, and the necessity to accurately establish control points by ground survey. He discounted completely the need for aerial contour mapping on the prairies where many miles of ground survey could be completed in one day, but did recommend testing the method for the production of topographical maps. Brock & Weymouth also photographed an area on the Saguenay River near Chicoutimi where a new power station was being constructed. In all, Brock & Weymouth vertically photographed 167 square miles and completed accurate topographical maps of the two areas.

Fairchild continued photographic operations through 1925 with its Huff-Daland and Curtiss Seagull. Two hundred and sixty-eight hours were flown on

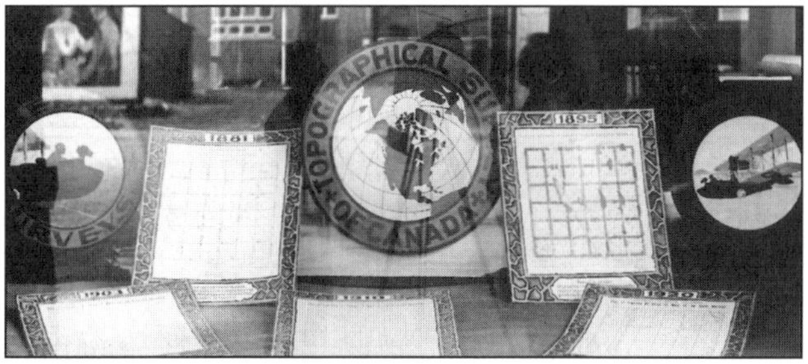

The Topographical Survey Branch took great pride in the progress made in aerial mapping. This display in the window of its office in the Labelle Building, 105 George Street, Ottawa, had its crest and five township plans showing progressive detail from 1881 to 1920 flanked by aircraft photographs. *NAC PA45655*

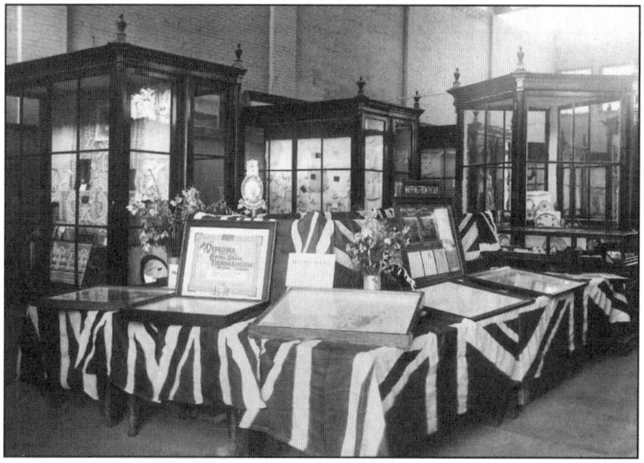

Among carpets, clothes and china, the Topographical Survey Branch displayed a relief map and examples of mapping from the air in the Arts Building at the Central Canada Exhibition, Ottawa, August 1925. *NAC PA45664*

a variety of tasks including forest type mapping in Quebec and the Maritime Provinces, topographic mapping, and pictorials for advertising, postcard and publicity purposes. The Seagull operated from Sheet Harbour, Nova Scotia, for several weeks of forest type mapping over the eastern districts of the province.

The division of duties—the Topographical Survey Branch having overall responsibility for aerial photography, map-making and distribution, and the RCAF in control of all flying operations— was working well. Farsighted individuals, notably John Armistead Wilson, secretary of the Air Board from 1922 to 1923 and then assistant director of the RCAF, ensured that aerial photographs were indexed and readily available. In 1925, this collection was formally established as the National Air Photography (now, Photo) Library and, by 1927, had 100,000 photos on file. Today, the collection has about six million photographs, completely covering Canada and subject to increasing demands from technical, commercial and private users.

J.A. Wilson, generally recognized as the leading figure in early Canadian aviation, never missed an opportunity to forward the interests of aerial photography. His 77-page article "Canada From The Air" in *The National Geographic Magazine*, October 1926, described 10,000 miles of flying between Sidney, Nova Scotia, and Prince Rupert, BC. A stirring travelogue profusely illustrated with fine aerial photographs, most credited to the RCAF, the article inspired many visits to Canada. Held in the archives of the National Aviation Museum is an impressive album, "Canada From The Air," containing 100 similar aerial photographs, taken in 1921–23 from Prince Edward Island to Prince Rupert. The accompanying "Lecture Notes" and attribution of the photographs to the RCAF indicate that the photographs were used for publicity purposes, perhaps by the RCAF: "Until very recently it was never possible to view the country from this aspect and many of the pictures will show a startling difference of aspect than those taken from the ground."

In the 1925–26 annual report, F/L H. Edwards, OIC at Victoria Beach, described the art at that time of map-making from aerial photographs:

> An aerial photographic traverse is made beginning at some place whose geographical position has been exactly located, following a suitable watercourse in a large loop of 200 to 400 miles, and ending at another point whose position has been exactly determined. An altitude of 5,000 feet is held and photographs are taken in groups of five every three miles, one straight ahead and two to either side, covering an arc of 90 degrees on each side of the line of sight. (In previous seasons, only three pictures were taken, one ahead and one on each side. The additional two pictures are found to give better control of the side pictures through the larger overlap.) These pictures and a rough plot from them of the waterways are then given to the surveyor, who, using them as a guide, makes a ground traverse over the course already flown fixing stations along the route by instrumental observations and marking his course and the stations occupied on the pictures. This gives complete control along the loop of the traverse. The area lying within the loop is then filled in by parallel photographic flights beginning, in this case, on the shore of Lake Winnipeg and ending on the controlled traverse. No groundwork is necessary in the interior of the blocks bounded by the traverse.[69]

Experience, F/L Edwards maintained, had demonstrated that the aerial photography method was equally accurate to ground-based surveys. He forecast the technique being used for a survey of the whole of Northern Canada by small ground parties transported and supplied by aircraft that would also take the photographs: "The infinite amount of time and labour a survey of this nature will save is only realized by those who have had experience of travel in Northern Canada."[70] Experiments were carried out by taking oblique photographs along an unsurveyed line 65 miles in length running west from a known point at Island Lake for about one hour. Subsequent checks by ground observation established that the maximum error in plotting from the oblique photographs alone was one-quarter of a mile, considered a reasonable compromise in view of the minimal time and cost involved.

Speaking at the annual meeting of the Association of Ontario Land Surveyors in February 1925, J.W. Pierce, DLS, emphasized the need for topographical maps, commenting that anyone attempting to travel through the country "following other than the main thoroughfares, the [existing] maps immediately prove to be not only incomplete, but so erroneous and misleading as to be worse than useless."[71] After a detailed review of the practical advantages of aerial photography, evidently from personal experience, he added a few advantages from the perspective of the ground surveyor:

> He is relieved from dependence on native guides with all the attendant delays and uncertainties so well known to you all. The location and extent of an obstruction in the nature of a falls or crossing from one lake to another, even the selection of a camp location for the following night (the

69 Ibid., pp. 64-65.
70 Ibid., p. 65.
71 *Association of Ontario Land Surveyors Annual Report # 40*, Proceedings of the 33rd Annual Meeting, Toronto, 17–19 Feb. 1925, p. 172.

party usually moves ahead each day), are all known and may be arranged for in advance, so that every member of the party is aware of what he is to encounter the following day. What is even more important, each man knows where he is to find camp in the evening.[72]

By 1925, the RCAF had a busy photographic section under the command of F/L J.R. Cairns, with one photographic officer, one warrant officer and thirteen airmen. Responsible for training photographers and oversight of all photographic work, the section provided cameras and film, processed the results and maintained a film negative file.

The PDs dispatched roll film magazines by express delivery to the section as soon as possible after exposure, together with information on location, aircraft altitude, weather, lens and exposure. Hypersensitized emulsions were used to develop 75-foot-long rolls of film, each yielding up to 115 7"x 9" negatives. Two contact prints of each negative were made, one for the customer department and one for the master file maintained by the Topographical Survey of Canada. In 1925, 255 rolls of film, each averaging over 100 pictures, were processed and some 50,000 prints made.

Of no direct impact on aerial photography, but of considerable interest to the PD crews, parachutes were officially placed on RCAF inventory in 1925. Largely through the influence of Col. William Barker, VC, who had seen many aviators die for lack of a parachute, F/O Albert Carter, MM, and Cpl. A. Anderson took the parachute riggers course at the US Army Air Technical School at Chanute Field, Rantol, Illinois. One hundred and seven parachutes were purchased and their use made mandatory. Carter and Anderson made the first RCAF parachute descents and trained another 16 airmen in their operation. They in turn trained other airmen and 35 practice "jumps" were made during the year.

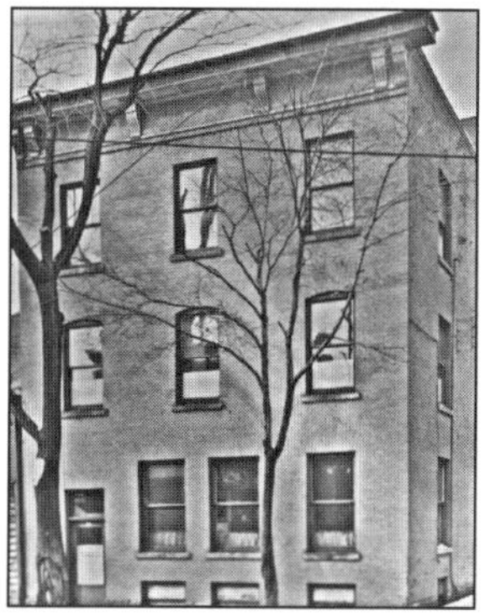

Ordinary film and plates were processed locally by the unit concerned, but all roll film exposed from photographic aircraft was returned to the photo section, familiarly known as Elgin Cottage, located at the corner of Albert and Elgin Streets in Ottawa. The RCAF boasted, "Nowhere in the world has there been concentrated a larger output of photography for aerial survey purposes."[73]
Elgin Cottage, Ottawa, 27 January 1927.
NAC PA62450

72 Ibid., p. 177.
73 *Report on Civil Aviation for the Year 1925*, op. cit., p. 78.

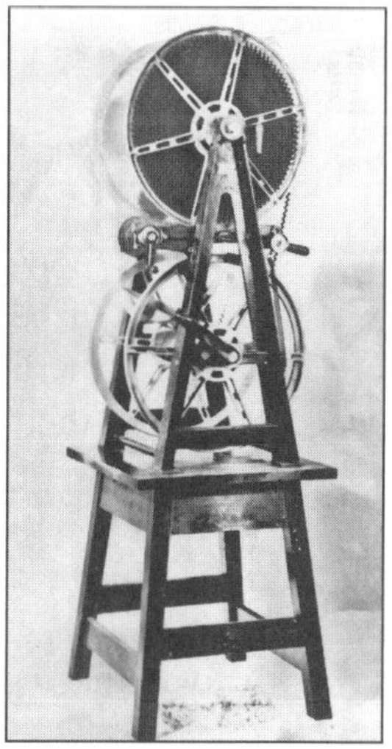

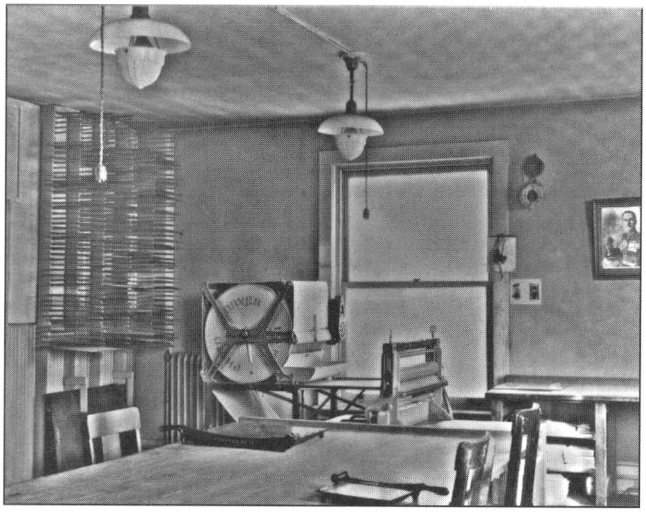

Film winding machine, Elgin Cottage, 15 January, 1925. (far left)
NAC PA62301

General workroom of the Photographic Section, Elgin Cottage, 27 January, 1927. The photograph on the wall is of G/C Stanley Scott, MC, AFC, Director, RCAF. (left)
NAC PA62447

1926-1927

THE 1926–27 *Report on Civil Aviation* played the same tune as in the previous year, noting that the progress of domestic aviation had been chiefly in the remoter parts of the country for transportation and survey purposes. Other nations, it claimed, due to generous support from the public purse, were leading the way in the establishment of airmail and passenger services: "Canada must, before long, take her share in this development, as she has in other methods of transportation."[74] The 1926–27 estimates, however, had not been approved when Mackenzie King's government was defeated in the October 1925 general election, requiring RCAF expenditures well into 1926 to be covered by Governor General's Warrants. The total amount appropriated for the year was $2.2 million, an increase of $317,150 from the previous year. By the end of the year the RCAF had almost reached its authorized strength of 95 officers and 375 airmen.

The government introduced organizational changes on 1 July 1926 that placed civil aviation under the Directorate of Civil Government Air Operations, becoming the Department of Transport in 1936.

The two HS-2Ls at Jericho Beach were feeling their age and it was possible to keep only one serviceable with the other held in reserve. Of 225 flying hours, 30 were devoted to vertical photography of approximately 4,000 square miles over southern Vancouver Island.

High River photographic operations were again handicapped by poor weather and smoke from forest fires, and the planned program could not be completed. The two D.H.4B aircraft covered 2,550 square miles of southern Alberta and central Saskatchewan in 132 flying hours, spending most of their time on the ground waiting for the weather to improve. From 12,000' they photographed the Turner Valley oilfields, the northern irrigation district, Lethbridge, Calgary, St. Mary's River, Waterton Lake, Battleford and Prince Albert for the Topographical Survey and Dominion Water Powers Branches. A motion picture was "shot" in Rocky Mountain Park (Banff) for publicity purposes.

75 *Report on Civil Aviation for the Year 1926* (Ottawa: Dominion of Canada, 1927): 5.

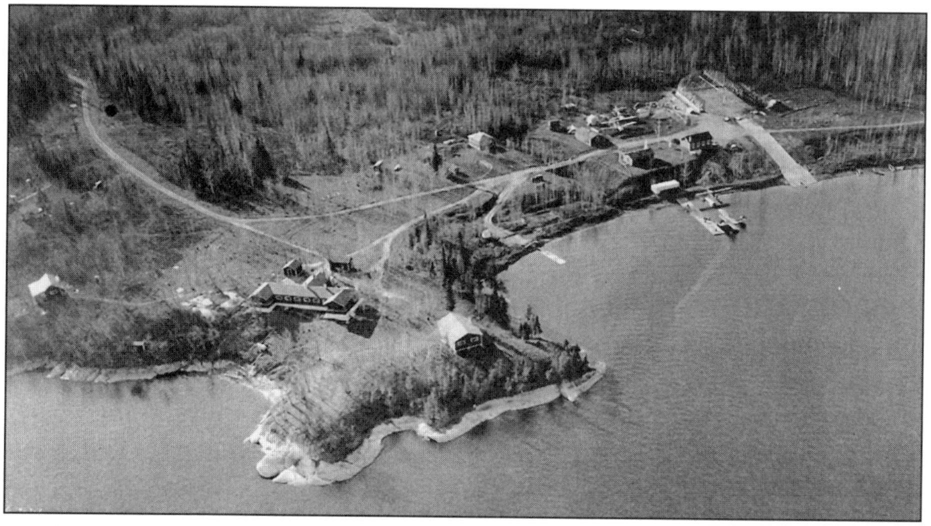

Lac du Bonnet Air Station looking south, 16 August 1928. The large building left of centre contained the messes for officers and airmen; to its right was the officers' quarters, and on far left was the CO's residence. The slipway at the far right led to a hangar under construction and a maintenance area alongside with a Viking on a trolley. The airmen's quarters were next left. At the left of the slipway were the orderly room, flagstaff and stores building. A Vedette was moored to the dock and a Fairchild FC2 sat on the beach.
NAPL A935-78

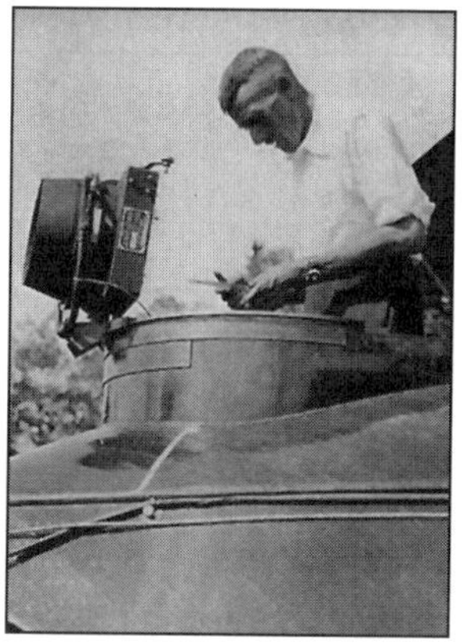

Eric Fry, DLS, installing a new mounting ring and Fairchild camera on a Viking, 1926. Photograph by R.D. Davidson, DLS, 1926.
ALEX MILNE

The majority of RCAF oblique photography in 1926 was conducted from the Manitoba stations. On 10 August, the main operational base was transferred from Victoria Beach to Lac du Bonnet on the Winnipeg River, about 70 miles northeast of Winnipeg. The existing sub-bases at Norway House and Cormorant Lake were maintained.

Over 3,000 square miles were vertically photographed in the more settled areas of eastern Manitoba and western Ontario, where the topography had little vertical relief. The preparation of maps was again greatly facilitated by the original cadastral surveys dividing the land into mile and half-mile squares marked by fence lines and travelled roads readily visible on the photographs.

For the next decade, the organization of RCAF photographic operations for topographical survey was based on the "Viking Flight" of two Viking flying boats from Winnipeg in 1926. The self-reliant establishment consisted of two pilots (L.R. Charron and G.F. Mason-Apps), two surveyor-navigators (R.D. Davidson and E.S. Fry), two photographer-mechanics (A.J. Milne and R. Marshall), one photographic officer and one cook. After completing 1,400 square miles of oblique photographs southwest of Lake Winnipeg from 3 to 8 June, the flight moved from Victoria Beach to Lake St. Joseph, about 90 miles north of Sioux Lookout, Ontario, where a fuel cache had been established during the winter. From this base camp, the main operation was oblique photography from 5,000' in completion of work started in 1925 to cover an area of about 25,000 square miles in the Red Lake District, east and north to the eastern end of Lake St. Joseph, and from Sioux Lookout east and north to the eastern shore of Lake Nipigon. On 14 July, the flight moved to Sioux Lookout, then to Collins on the Canadian National Railway near Lake Nipigon where it remained until the middle of September.

Weather was the constant enemy of detached operations and every second of "photographic days" had to be utilized. On 4 September 1926, the monotony in one Viking, G-CYET, flying "lines" at 5,000' was rudely disturbed when an aileron control rod broke near the pilot's feet. Fortunately, a small lake lay directly ahead and, by liberal use of rudder control, he was able to make a safe landing. Disdaining the delay, possibly taking weeks, to procure a new control rod, the crew found a suitable birch pole on the beach, and with four nails improvised a replacement. They quickly resumed photography and only made a more permanent repair when the inevitable rainy day kept them on the ground at Collins.

The Viking Flight took vertical photographs of the waterways in the area of Lac Seul then on to Cormorant Lake on 18 September for topographical and forest mapping photography north and west of Lake Winnipeg. This mission was scheduled for September to take advantage of the change in foliage of the birch and poplar, which would then show greater contrast with the coniferous types. Some 24,000 square miles were photographed by 5 October before the two Vikings moved to Norway House. Forest type mapping continued before heavy snow called a halt and the flight hastily headed via Lac du Bonnet to the Red River at Winnipeg on 22 October. One more flight was made on 27 October over the west shore of Lake Winnipeg from the Red to the Icelantic River, bringing the Vikings' total to 420 hours photographic flying, before the aircraft were brought ashore at Winnipeg for winter overhaul.

A second photographic flight was hurriedly formed late in the season at Winnipeg with two new Vedette flying boats. F/O C.R. Slemon, Surveyor Jenkins and Sgt. R.I. Barton in G-CYGW, and F/L B.N. Harrop, Surveyor Donnelly and LAC F.J. Ewart in G-CYGX left Ottawa on 10 September "in adverse conditions." They flew westward via North Bay, Sudbury, Como, Long Lac, Nipigon, Sioux Lookout and Minaki to Lac du Bonnet, where they had cameras installed. Despite bad weather and a total of 11 forced landings to fix problems, primarily caused by carburettor icing, they began operating from Norway House on 18 September. During the next six weeks, the Vedette Flight flew a total of 62 hours and 34 minutes, sketching forest types when the weather was not good enough for photography, before joining the Viking Flight at Winnipeg for winter overhaul. The Vedette crews reported: "The new types were found to be admirably suitable for the work."

A/V/M R.E. McBurney recalled that carburettor icing was still a problem in 1927 and forestry patrol Vedettes were ordered to stay below 4,000′ until a carburettor heater was designed and installed.

The work of the two photographic flights of the Manitoba Station in 1926 resulted in the preliminary issue of seven map sheets for forestry, mineral, power, construction and general development purposes. Taking their names from significant lakes or rivers, each sheet covered an area approximately 94 miles in length and 70 miles in depth. As a result of the previous year's oblique photography, topographical map sheets for Lac Seul, Point du Bois, Carroll Lake and Red Lake were also issued.

The Department of the Interior recognized the need by the growing aeronautical community for special maps. Experimental strip maps, and the first aeronautical chart, with a scale of eight miles to the inch (1:506,880), were published in 1926. Showing the Winnipeg District, the chart was of little real value to pilots as it did not show contours or elevations except for that of Lake Winnipeg. It did show the Winnipeg and Lac du Bonnet bases, marked by four-mm red circles, a symbol that is still used. During the next few years both military and civil organizations explored different versions of the strip map and chart, settling on the "eight-mile series" of aeronautical charts. Two hundred and twenty maps of these charts were needed to cover Canada at a scale of eight miles to the inch and all were published between 1929 and 1944.

Starting in 1958, the aeronautical charts were changed from a true scale of 1:506,880 to 1:500,000, but retained the same boundaries. From 1979, this series was replaced by a series of "Visual Navigation Charts," reducing the total number of charts to 51.

Floating logs from the Gatineau River still presented problems at Rockcliffe in 1926, so waterborne aircraft continued to be based at Shirleys Bay, west of Ottawa. The Manitoba pattern was followed in forming a photographic flight consisting of the prototype Varuna, the prototype Vedette, 'FS, and a new Vedette, 'GZ. From a forward base at Larder Lake, just west of the Ontario–Quebec border, they vertically photographed almost 6,000 square miles of the mineral-rich Rouyn area for the Geological Survey Branch in 331 flying hours. One of the Vedettes operated from Seneterre, northeast of Val-d'Or, to cover the eastern portion of the area. Photographs were taken of the cities of Hull and Ottawa, north shore of the Ottawa River, Grenville and Carillon Canals, Petawawa and the Gatineau Valley. In addition, Fairchild Canada was contracted to take 418 vertical photographs of the Batiscan area, east of Three Rivers, for the Topographical Survey Branch.

The Dartmouth base was not activated in 1926.

During the 1926 season, the Vancouver station contributed 1,546 photographs; High River, 5,289; Manitoba, 20,249; and Ottawa, 14,619, covering 59,000 square miles for mapping purposes. This brought the cumulative total of areas photographed to 152,560 square miles—a tremendous achievement, but only about four percent of Canada's total land area.

Up to the end of fiscal year 1926, the CAB and its successors had taken more than 100,000 aerial photographs. The increasing pace of aerial photography indicated that this number would grow to enormous proportions, especially if all Canada was to be photographed. Indexing, filing and retrieval was sure to be a huge task. The Topographical Survey Branch of the Interior Department had this responsibility. The index system introduced consisted of maps indicating the location of the photographs together with books containing additional information, such as camera type, lens, direction and tilt of the camera, elevation of the aircraft, and the location of the ground plumb point (i.e., immediately beneath the camera). The scales used were one mile to the inch for vertical photographs, and four miles to the inch for obliques. For the settled parts of Western Canada, township plans—scale one half-mile to the inch for verticals, three miles to the inch for obliques—were used. The photographs were stored in steel boxes, each with the number of an index map and containing about 200 photographs. Essentially the same system prevails today and the photographs can be viewed at the National Air Photo Library on Booth Street in Ottawa.

The BC government contracted with the RCAF in1926 for aerial photography of municipalities in the Vancouver area and for forest fire patrol duties, undertaken from the base at Jericho Beach. In contrast, the RCAF made use of bases constructed by Ontario throughout the northern reaches of the province for the Provincial Air Service. Quebec, also, was establishing its own air service, but through contracts with Canadian Airways and other companies.

Aerial sketching of forest resources was closely related to aerial photography. By the close of 1926, approximately 120,000 square miles had been covered in this manner, largely by commercial interests. The best available topographical maps were used, often of very limited, and sometimes non-existent, detail that influenced the type of sketch and the degree of accuracy possible. Apart from the commercial benefit derived, the most important result was probably the realization that the supply of wood was not limitless. Forest management and conservation policies improved significantly during the following years. F.T. Jenkins, Managing Director, James D. Lacey & Co. (Canada) Limited, in a 1927 address to the Canadian Society of Forest Engineers, considered aerial sketching a vital component of the forest industry. He compared the timber cruiser's traditional role of plowing through the woods, taking measurements at intervals and trusting to the good old standby law of averages for results, to an ant travelling through a forest of grass, "and sometimes has as much chance to see the surrounding countryside."

1926 was a busy year at Canadian Vickers. The Vedette was in production and satisfactory RCAF tests of the prototype Varuna I resulted in the order of seven Varuna IIs for 1927 delivery. Design was started on four new aircraft types: the amphibious Vanessa single-engine, four-passenger cabin biplane, aimed at the commercial market; the Vista single-seat flying boat powered by a 60 hp Armstrong-Siddeley Genet radial engine for economical forestry patrol; the Vigil for the same purpose, but built for operation in mountainous areas; and the Velos for photographic work. Vickers won the bid for the latter three new types in response to RCAF requirements.

A unique design, the Velos seated photographer, navigator and pilot in tandem cockpits. In order to get the necessary field of view for oblique photography, the lower wing of the biplane had a small span, providing minimal lift. The stub wing did provide a platform for access to the twin engines, and for mooring the amphibian when in a float configuration—although this was a hazardous occupation with the propellers revolving. In significant departures from traditional materials, steel tube replaced wire for wing bracing, the fuselage structure was welded steel, and most of the airframe was steel or duralumin, with wood used only for the upper wing.

The Royal Canadian Corps of Signals (RCCS), responsible for all military communications, provided equipment and personnel, who came under the RCAF station commander for general administration and duty assignment. The RCCS developed self-contained transmitting-receiving ground stations based on a three-kilowatt Delco generator with a 110-volt storage battery. Transmission at 500 to1,000 watts between ground stations required a 300-foot long antenna strung between two masts. Small airborne "wireless sets" were installed in aircraft engaged in forest fire patrols, but had only a one-way, air-to-ground capability, and required the deployment of a reel-mounted antenna. Radio communication was relatively unimportant on photographic operations, and wireless sets were not normally carried because of the additional weight involved.

The name of one of the pioneering aviation companies in Canada, Dominion Aerial Exploration Company, disappeared in 1926 when Canadian Airways Limited took over the facility at Three Rivers. Still with managing director Capt. H.S. Quigley at the helm, the new company also operated the base at Roberval for the provincial government. Four HS-2L flying boats were dispatched to Seven Islands to continue the previous year's successful support of survey parties in the interior.

Fairchild Aerial Surveys continued operations from Grand Mère with its Curtiss Seagull, Vickers Vedette, and Huff-Daland Petrel. Photographic surveys were completed in the Ottawa Valley, Gaspé, and for railway and hydro developments in Eastern Canada. The Vedette was employed for the season on forest type mapping in the Gatineau region, Anticosti Island, and an area north of Sudbury for the James D. Lacey Company, consulting engineers. The two companies had collaborated for several years. Fairchild photographed a total area of 8,536 square miles, processing the film in its own laboratories along with its ever-popular aerial view postcards and publicity pictures. Also, mosaics and line maps were prepared to assist with engineering projects, power developments and town planning. An associated company, Elliot-Fairchild Air Service Limited, in partnership with aviation entrepreneur Jack V. Elliot, planned to take over the passenger/freight service from Haileybury to the gold fields at Rouyn with a metal-hull, seven-passenger Aeromarine flying boat. Elliot Air Services was also formed in 1926 to start an airmail and passenger service, but neither partnership lasted past the summer. The name of the surviving company was shortened to Fairchild Air Transport and pilot Glen Burge was hired to replace Elliot.

C.S. Caldwell embarked on another adventure in 1926 with mechanic Irenée Vachon, supporting the Northern Syndicate Limited's mineral exploration in the Great Slave Lake district. Their Viking was shipped on the Alberta & Great Waterways Railway from Edmonton north to Lac la Biche on 6 June. With the help of local workmen, the aircraft was rigged and tested and then flown via Fort McMurray to Fitzgerald, their base camp on the Slave River in the Northwest Territories. The next base, some 125 miles northeast, was identified by only a sketch map, "almost useless as there are hundreds of lakes, all similar in appearance, dotted over the country."[75] The narrow lake was finally located, despite the fact that it lay north-south instead of the sketched east-west. Further bases were established, extending into unmapped "Barren Lands," before flying the Viking back to High River on 22 August. The biggest problem experienced in navigation over the immense number of unmapped lakes was the unreliability of the compass, "owing to local [magnetic] attractions."[76] Having to repair a cracked crankcase was not considered a serious problem and continuous daylight was an asset.

Several small companies entered the passenger and freight business. One of them, the Compagnie Aerienne Franco-Canadienne, under contract with Quebec, photographed 3,800 square miles from bases in the Gaspé. Two French-built, single 180 hp Hispano-Suiza engine, two-seat Schrek-FBA 17H.E.2 flying boats were used.

The provinces were gradually establishing their own independence, although British Columbia still relied on the RCAF base at Jericho Beach in 1926 for aerial photography of municipalities in the Vancouver area and for fire-protection duties. In Ontario, the OPAS was essentially independent and the RCAF was able to make use of its bases constructed throughout the northern reaches of the province. Quebec, also, was operating its own air service, but through contracts with Canadian Airways and other companies.

75 Ibid., p. 27.
76 Ibid., p. 29.

1927-1928

THE AIR ESTIMATES for 1927–28 were increased from the previous year by $1.7 million to $3.9 million. A large portion of this, $675,000, was specifically allocated for the purchase of new aircraft, and a commitment was made to erect a mooring mast at St. Hubert in support of the British experimental airship program. An additional item of $850,000 was included in the estimates for the Department of Railways and Canals to fund the Hudson Strait Expedition, in which the RCAF would have a principal role.

Popular belief in the future of aviation was given a huge boost in May 1927 when Charles Lindberg arrived in Paris, France, in his Ryan monoplane *Spirit of St. Louis* after his epic solo transatlantic flight. The achievement received additional publicity in Canada when W/C Lindsay Gordon flew a Varuna and cameraman from Shirleys Bay to photograph the arrival of Charles Lindberg at Lindberg Field (Ottawa Airport, now Macdonald-Cartier International) on 1 July 1927. The cameraman had just dismantled his camera after completing the task when two of the accompanying 12 Curtiss Hawks collided, killing one of the pilots.

The RCAF annual report again goes into great detail describing the generous support given by foreign governments for the worldwide development of passenger, mail and freight operations. A more optimistic view is taken of domestic developments than in previous years, but the opportunity is again taken to prod the federal government: "Development [in Canada] has not been rapid nor has it been forced into uneconomic channels by Government subsidies. What flying there is has been self-sustaining and has served a useful purpose. Otherwise, it would not have continued."[77] With the increasing segregation of civil and military flying, operations such as fishery and forestry patrol of the west coast were gradually taken over by private companies under contract to the federal and provincial governments.

Early in the year, a Jericho Beach HS-2L made a photographic survey of swampy ground around Harrison Lake on behalf of the Ministry of Agriculture's continuing study of mosquito control.

77 *Report on Civil Aviation for the Year 1927* (Ottawa: Dominion of Canada, 1928): 15.

Unusually wet weather in Alberta during 1927 reduced the forest fire incidence, but also inhibited photographic work by High River's two D.H.4B aircraft. Verticals and obliques were taken of approximately 2,230 square miles over Lacombe, Alberta, and Melford and Saskatoon, Saskatchewan. Photographs were taken of flood damage along the High River, part of an engineering study to mitigate future damage.

From the Winnipeg Station, two flights of two Vikings photographed in Manitoba, Saskatchewan and northern Ontario. An additional sub-base was activated at Ladder Lake, near the small community of Big River in central Saskatchewan. Only four of the five new 180 hp Armstrong Siddeley Lynx-engined Varunas assigned to Winnipeg for photographic and general duties were put to work. One, G-CYZU, sank in Cross Lake, Manitoba, after an emergency, downwind landing when F/O L.R. Charron had the frightening experience of losing aileron control. Over 600 hours of oblique and vertical photography covering 36,456 square miles were completed in areas around Rainy Lake, Kenora, Winnipegosis, Flin Flon and Wood Buffalo Park, but this was only about 45 percent of the planned program, mainly due to bad weather.

Viking G-CYET that had taken Basil Hobbs and his crew on their 1924 pioneering expedition broke up pulling out of a dive after exiting a cloud near Hillboro, Manitoba, on 17 July. Pilot P/O W.C. Weaver, navigator Frederick Hay Wrong, DLS, and AC1 J.T. Eardley fell to their deaths. Photo pilots, who lacked appropriate instruments, never voluntarily entered clouds. The resulting enquiry concluded that the aircraft had probably been flying above the cloud when it either experienced engine failure or a downdraft forced it to descend into cloud. Subsequent tests resulted in strengthening the cockpit area of all the remaining Vikings.

Not as serious, but an indication of the hazards faced by the photographic crews, was the experience of Vedette G-CYFS's crew flying out of Torrance, Ontario, to photograph the Muskoka area in July 1927. Pilot F/O R.H. Barber had just taken off when crewman LAC S.A. Greene climbed out of the front cockpit to check the wind-driven generator mounted on a wing strut. Reaching for the generator connections, the unseen whirling propeller blades lopped off two of his fingers. A horrified Barber looked up to see Greene swaying and in imminent danger of falling, so he made an emergency landing. Overshooting the lake, he put the aircraft down in a grassy clearing. Greene was repaired, but G-CYFS was written off. In June, G-CYGX was also written off after a hard landing on glassy water, but with no damage to the crew.

Ottawa continued the pattern of detached photographic flights instituted in 1926. Two Vedettes photographed about 3,000 square miles from a temporary base at Bala Bay, Lake Muskoka, and another 1,700 square miles from Ramsey Lake near Sudbury. The Varuna I and a Vedette photographed 154 square miles from La Tuque, Quebec before the Vedette diverted to Baie St. Paul to carry out exploratory photography across the unmapped region between the Batiscan River and Grand Lac Jacques Cartier. The two aircraft rejoined forces at Lac St-Joseph to photograph 142 square miles in the Batiscan and St-Anne districts. Moving to Roberval at the end of September, vertical and oblique photographs were taken around Chicoutimi, Jonquieres, Kenogami, Arvida, Bagotville, St-Alexis-de-la-Grande-Baie, Saguenay River, Ha Ha Bay, Peribonka River, the source of the Saguenay River at Grande Décharge, Lake St. John, Ashouapmouchan River, Mistassini River, and Lake Kenogami. The Varuna and Vedette called at Roberval before returning to Ottawa on 1 November.

Although some accusations were raised that the RCAF was undertaking work that should go to the private sector, the RCAF was keen to develop a Canadian commercial capability, and worked closely with industry. For example, Fairchild was paid in 1927 to test a British-built F-8 Eagle camera and Aldis aiming sight, loaned by the RAF, in its Huff-Daland Petrel while undertaking aerial photography for the Canadian International Paper Company. While generally satisfactory, mechanical problems were experienced with the equipment and the RCAF stayed with its established North American suppliers, Fairchild and Kodak.

One of the new Varunas was shipped by rail to Dartmouth and assembled there for photographic survey. Fog and low cloud inhibited vertical photography of the assigned 1,940 square miles from a sub-base at Shelburne, Nova Scotia, and only 1,728 square miles were completed. Oblique photography at Bridgewater was completed in better weather. On the ferry flight back to Ottawa in November, the Varuna sank in the Saint John River at Fredericton, New Brunswick, after a hard landing.

With improved methods of interpretation, greater use was made of vertical photography in 1927, particularly for the investigation of potential

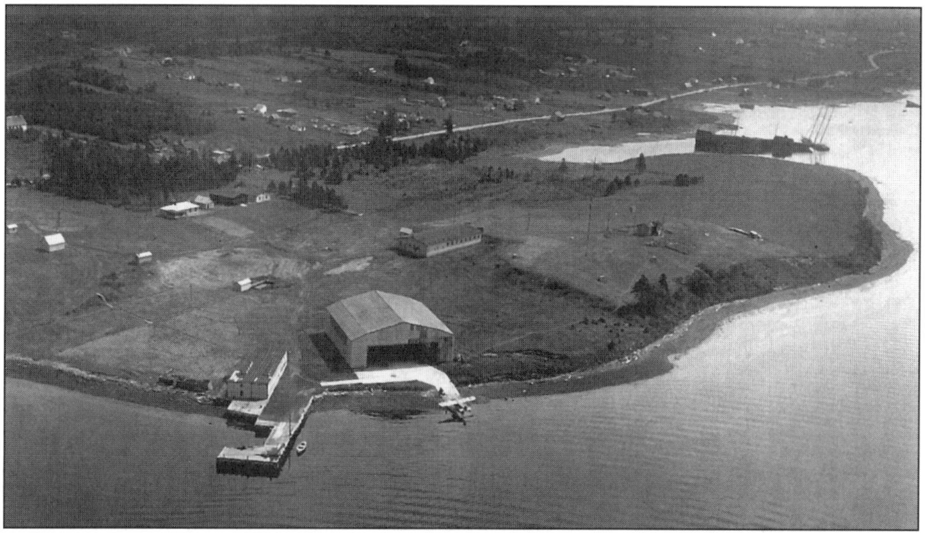

After being inactive in 1926, Dartmouth Air Station was reopened in 1927 and had taken on an air of permanence by 28 August 1929.
NAPL A1636-100

water power sites, water storage and other engineering projects such as power lines and railways. Stereoscopic study of vertical aerial photographs alone, it was determined, could usually ascertain the merits of a site, dispensing with the need for expensive, and sometimes fruitless, preliminary groundwork. Engineers found this new application of aerial photography of immense value in Canada where hydroelectric potential was so often located in forested and hard-to-access rugged terrain. Aerial photographs were also being used for similar work in reconnaissance and laying out power transmission lines and railway locations, "with results that could be obtained in no other way in the same time and with the same expenditure."[78]

Approximately 62,000 aerial photographs covering about 46,000 square miles were taken by the RCAF in 1927.

Prompted by a federal government decision to build the Hudson Bay Railway and terminals, the Department of Marine and Fisheries initiated an expedition in January 1927 to study ice conditions and the length of the navigation season for ocean-going freighters and grain ships passing through Hudson Strait. The expedition was also charged with establishing three locations for direction-finding stations to ensure safe navigation for vessels traversing the strait. The initial RCAF plan to use kite balloons for ice reconnaissance was rejected because the cost of balloons at $45,000 each was deemed too great. Fortuitously, as experience proved that it was necessary to fly below the almost-constant layer of low cloud in order to see the ice. Six Fokker Universals were purchased at a cost of $16,000 each, flight tested at Camp Borden, packed into crates, and shipped in a 15-car train to Halifax. Most of the aircraft spares, safety equipment and other supplies had been shipped direct to Halifax. All was loaded aboard a freighter, the SS *Larch*, which carried approximately 700 tons for each of the three bases to be

78 Ibid., p. 89.

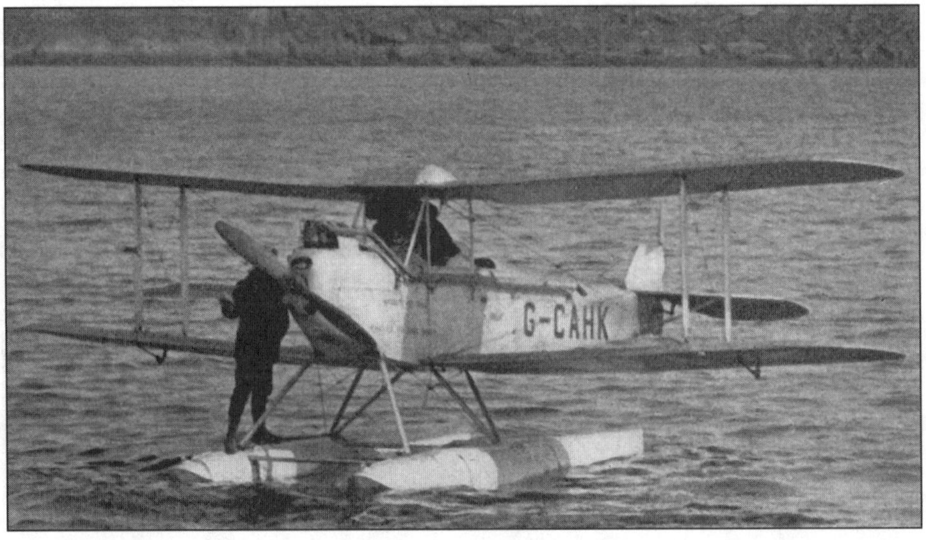

This diminutive de Havilland 60X Moth with a big name—*Spirit of the Valley of the Moon*—was invaluable for locating the three bases—Wakeham Bay, Nottingham Island and Port Burwell—but was wrecked at its Wakeham Bay moorings in August when a storm suddenly blew up. The seas were too rough to approach the little float plane that rode out the tempest for over 12 hours before it capsized. S/L T.A. Lawrence is standing on the float. *REPORT ON CIVIL AVIATION, 1927*

established at strategic locations in Hudson Strait together with crews to construct the bases. The icebreaker CGS *Stanley* transported 44 members of the expedition including six RCAF officers and 12 airmen, and a de Havilland Moth stowed on deck. The expedition departed from Halifax on 17 July.

F/L Thomas A. Lawrence was promoted temporary squadron leader for the duration of the expedition and made responsible for flying operations. Dr. Andrew McLean was in overall command, but was taken ill and returned to Ottawa. Leaving Lawrence in charge caused a little friction, as not all the civilians approved of having a military officer in command. As Lawrence intimated later in an interview for the RCAF Museum, the air force officers also were not entirely happy with their assignment. They had been told that, whatever the outcome of their dangerous task, the grain port at Churchill would be built to placate Western grain farmers.

Each base had a doctor, cooks and storekeepers provided by the Department of Marine and Fisheries. One officer and three other ranks from the RCCS had the responsibility for wireless communications. One RCMP officer was assigned to each of the three bases, and one civilian "cinematographer," George Valiquette, was located at Wakeham Bay. Lawrence recalled many years later that they only recruited the cook for Wakeham Bay just before leaving Halifax. F/O Carr-Harris brought along "a thin, wiry old gentleman about 70 years of age who said he could cook." They later realized that the expedition was his refuge because he was "just two jumps ahead of the Mounties for boot-legging." As it turned out, not only was he an excellent cook, but his talents as dynamiter were also called upon to break up the ice along the shore in the spring—although his habit of warming the sticks of dynamite behind the stove caused some concern.[79]

79 T. A. Lawrence, *The Aviators of Hudson Strait*, National Film Board Video (1973), O173 632.

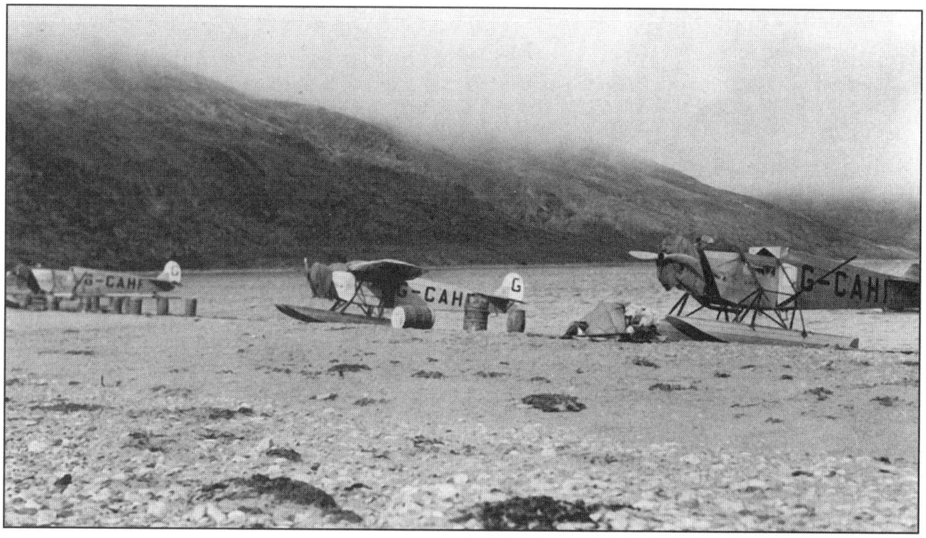

Six civil-registered Fokker Universals, purchased by the federal Department of Marine and Fisheries, operated from the three bases, and were fitted with wheels, floats or skis as required. Assembly of G-CAHF was not yet complete in the fall of 1927 at Wakeham Bay. *DNDPU RE13778*

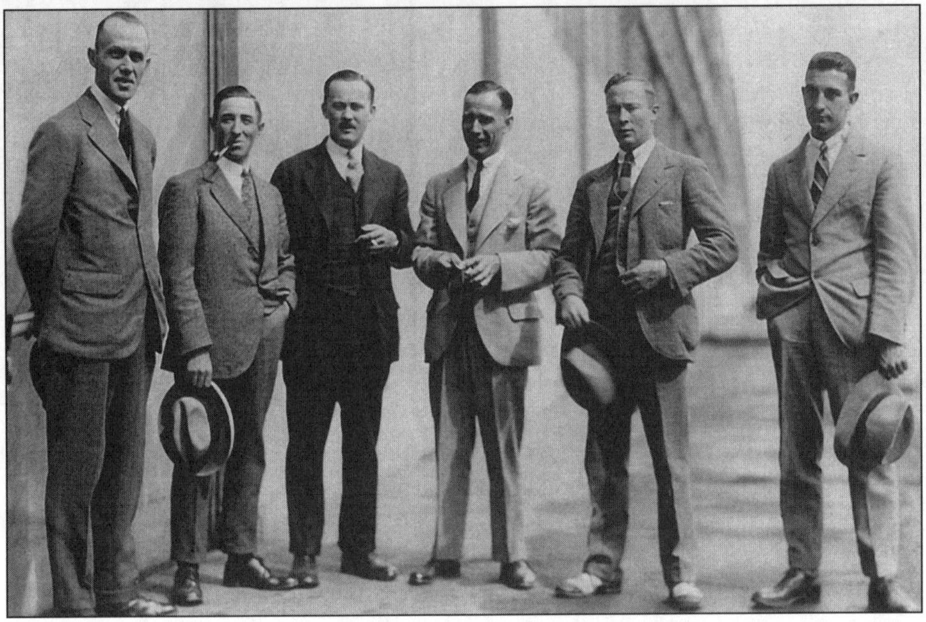

Pilots of the Hudson Strait Expedition, 1927. Left to right: F/L A.A. Leitch, F/O A.J. Ashton, F/L F.S. Coghill, S/L T.A. Lawrence, F/O B.G. Carr-Harris, F/O A. Lewis.
SUZANNE H. HIGGINS

Spare aircraft engines were issued to each base along with a 30-foot motor launch, skiff, Fordson tractor, radio station with a pair of 150-foot masts, marine railway, two derricks, aircraft fuel, coal, stoves, bedding, guns, rifles, ammunition and a miscellany of other stores, including provisions for 16 months. The main task was to establish the duration of the navigation season into Hudson Bay and the patterns of ice movement in Hudson Strait. This was achieved on daily patrols, weather permitting, by taking oblique photographs with hand-held cameras and plotting the results on hydrographic survey charts. The aircraft had to fly at a constant altitude, and the photographs taken with the same angle of deflection and with a sixty percent overlap to facilitate interpretation of the nature of the ice by stereoscope. Lawrence was also expected to report on the feasibility of aircraft as an aid to navigation in the strait and the suitability of bases and equipment for extended operations in the conditions experienced.

After unloading supplies and picking up the construction crews, the ships departed before the onset of winter, leaving the expedition to its own devices. The aircraft started patrols in September, but the expedition was primarily concerned with completing the bases before the onset of severe weather. By the end of December, they had constructed an "airfield" on the sea ice at each base and ski-borne operations commenced. Most of the flying was at low altitude, below the persistent low clouds, in order to photograph the ice. Navigation was primarily by dead reckoning, keeping in sight of land as much as possible. The only maps available were British Admiralty charts dated 1837. Each aircraft carried two compasses, but they could not be relied on and seldom read the same due to the iron deposits in the area.

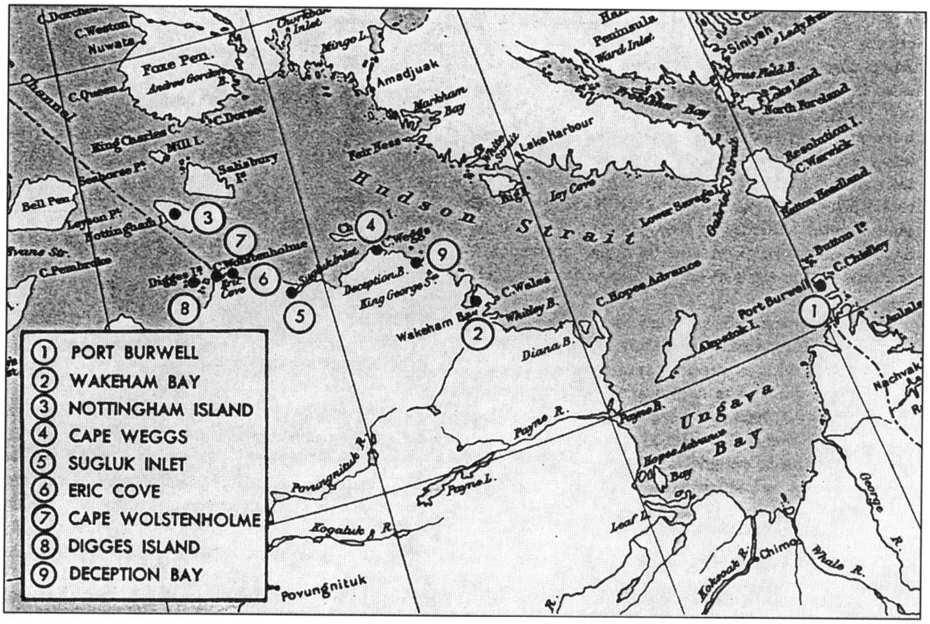

Key landmarks on the north shore of Quebec and the three bases established for investigation of ice movement in Hudson Strait during the winter of 1927–1928.
REPRINTED FROM THE ROUNDEL, VOL. 1, NO.8, JUNE 1949, P. 24

Emergency rations carried on each flight consisted of 12 pounds of bully beef, 15 pounds of hardtack (ship's biscuits), eight pounds of bacon, six pounds of pork and beans, three pounds of chocolate and one pound of tea—enough, it was estimated, to sustain three men for at least ten days. Each aircraft also carried a primus stove and coal oil, engine tool kit, lubricating oil, rubber raft, distress signals, rifle and ammunition, Arctic sleeping bags, rope, axe and knife. Other equipment was carried dependent on the season—anchor, bilge pump and silk tent for the short summer season; ice picks, snow knife, engine cover and blowtorch during the winter. Lawrence insisted that each aircraft carry an Eskimo passenger to assist if the aircraft was forced down, a procedure that probably save the lives of F/O Lewis and his crewman.

Flying in single-engine aircraft over the hostile and virtually uninhabited Hudson Strait, notorious for inclement weather, was extremely hazardous. The only link with the base, itself poorly equipped for rescue operations, was a temperamental one-way radio that required the streaming of an antenna. Messages could be sent by Morse key, but there was no way of knowing if they had been received until the aircraft returned to base. It is surprising that the accidents, which progressively increased in seriousness, did not end in any fatalities.

Returning to Nottingham Island from Erik Cove on Cape Wolstenholme, F/L A.A. Leitch encountered snowstorms and, low on fuel, decided to land on an ice floe and wait overnight for clearer weather. This involved draining the engine oil and heating both it and the engine with a primus stove carried for this purpose before replacing the oil the following day. With improved visibility, Leach determined that he had been way off course and made it back to Nottingham Island with only about one quart of fuel remaining and his crew suffering from frostbite. A few days later, S/L Lawrence put down at Suglet Inlet in a blinding snowstorm to wait for one day before getting as far as Deception Bay, where he had to wait for a stormy nine days before returning to Wakeham Bay.

The third incident came close to disaster. F/O A. Lewis flying from Port Burwell in March, with passengers F/S Terry and one-eyed Eskimo guide Bobby Anakatok, lost his bearings in a snowstorm, ran out of fuel, and made an emergency landing on an ice floe off the Labrador coast. G-CAHG was not damaged, but had to abandoned. Using their inflatable boat to cross many open leads, they reached the coast in eight difficult days. Emergency food exhausted, they survived on raw walrus meat for another five days until they came across an Eskimo and his wife who gave them food "of a kind" and returned them to Port Burwell by dog team after an absence of 13 days.

S/L Lawrence decided to leave his aircraft at Port Burwell for use by F/O Lewis. For the experience, he decided to return some 635 miles along the foot of Ungava Bay to Wakeham Bay by dog team accompanied by an Eskimo. Although he said he felt fitter than at any other time in his life, Lawrence admitted to relief after 36 days when F/O Carr-Harris landed and offered him a ride back to base. With typical articulate reserve, he qualifies his admiration of the Eskimo with advice to anyone travelling in his company: "He does, however, lack ideas of conservation of supplies and equipment and will unwittingly treat these items without thought of the morrow, travelling

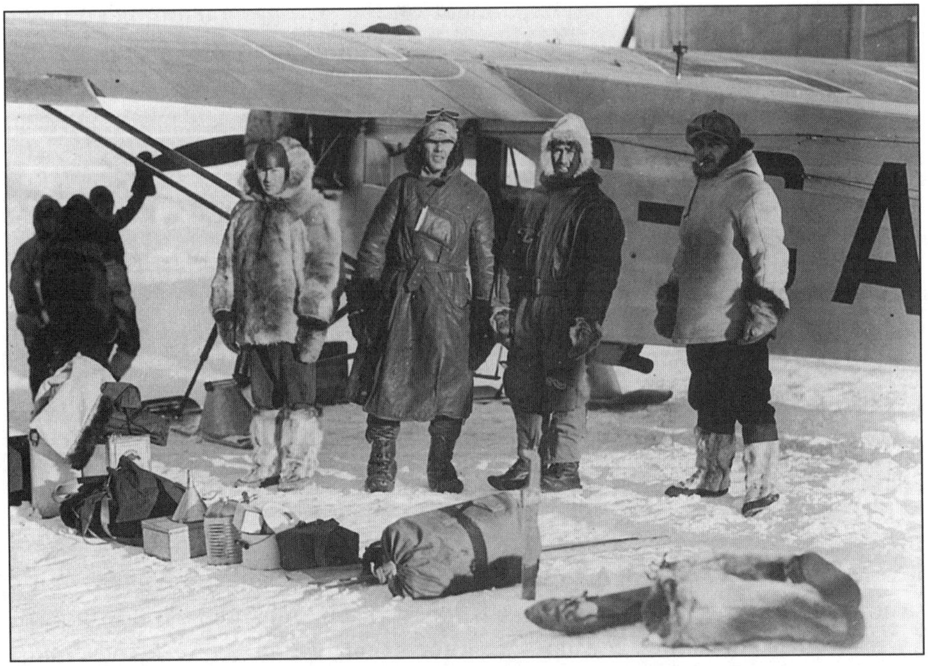

Checking safety equipment prior to setting out from Wakeham Bay in search of G-CAHG, 19 February 1928. Facing the camera, left to right: F/O Carr-Harris, F/L Leitch, S/L Lawrence and Fox News photographer George Valiquette.
DNDPU RE12407-7

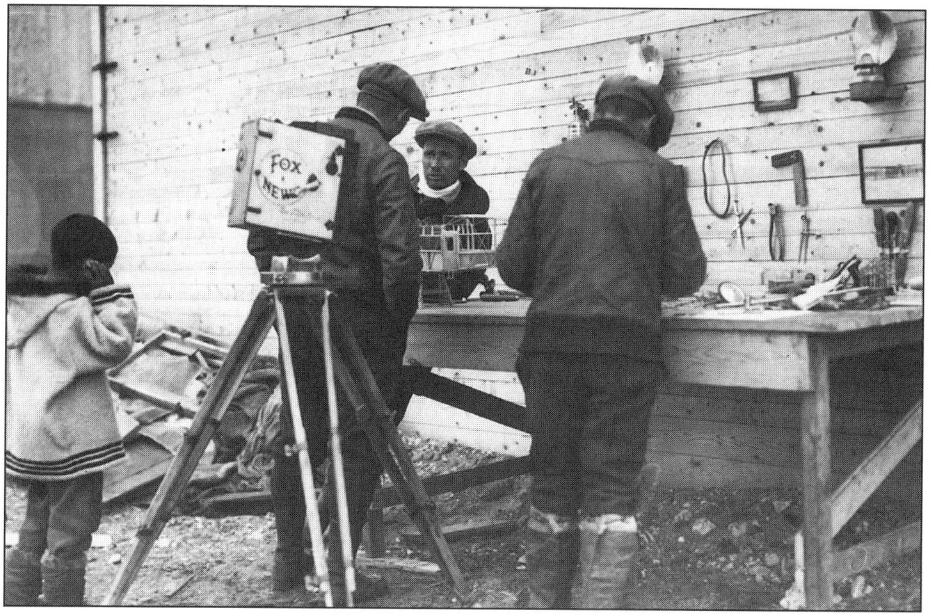

Movie camera in position to photograph a model Fokker Universal constructed during the winter of 1927–28 by Sgt. Andrew Caggie at Wakeham Bay. The six-foot wingspan model of G-CAHI was returned to Ottawa and is now a valued possession of the National Aviation Museum.
NAC PA55660

In 1973, the National Film Board compiled a 28-minute video, *The Aviators of Hudson Strait*, from Fox News photographer George Valiquette's 1927–28 16mm silent movie footage. A commentary was added by then-retired A/V/M Thomas A. Lawrence, CB, CD.
REFERENCE B0173632.

casually along while plenty of food is in evidence and with haste when running short, in an effort to reach a point of replenishment. The latter expedient is not always possible and might end disastrously."[80]

The last flight on skis was made on 17 June and the aircraft were fitted with floats. In place of the relatively smooth ice landing field, the crews now had to face the hazards of rough seas, ice floes and rock-infested beaches where every landing was fraught with hazard. The Universals were not equipped with water rudders and could only be steered with bursts of throttle. Once committed, the pilot had to rely on men immersed in the frigid water to capture and steer his aircraft to the relative safety of a 15-foot-wide cleared patch of beach.

Ice departed from Hudson Strait by July 1928, and the expedition made preparations to close the three bases and return home. The remaining five Fokker Universals successfully assembled at Erik Cove, near Cape Wolstenholme, the imposing headland marking the western limits of Hudson Strait. They planned to fly together down the eastern side of Hudson Bay, utilizing fuel caches established by the Hudson's Bay Company and Northern Aerial Minerals Exploration Limited. Three aircraft took off successfully, but a float support of F/L Coghill's aircraft, weakened by salt water corrosion, collapsed in the rough sea; the nose of the float turned inwards and was sliced by the propeller. The fifth aircraft was unable to take off due to engine trouble. F/O Carr-Harris landed and rescued the three occupants of Coghill's machine and S/L Lawrence radioed the *Larch* and *Montcalm* for assistance. Examination showed that the float fittings of the other aircraft were in poor condition and return by air was considered unduly risky. The *Larch* took aboard three

80 C.R. Dunlap, private records.

aircraft, and the two in best condition were flown with lightened loads to Wakeham Bay, dismantled and shipped out on the CGS *Canadian Voyageur*. The expedition members arrived in Halifax, with some relief, in October. Experience gained in 370 hours' flying time taking 2,285 ice photographs in very difficult conditions contributed greatly to the subsequent success of northern aerial operations and to navigation in Hudson Strait. Lawrence commented on the inaccuracy of existing charts of the area, "a large and dangerous factor in the development of the strait as a commercial water route." He recommended aerial vertical photography as being the most rapid and accurate method of preparing new charts.

The Hudson Strait Expedition used the Eastman A.1 Oblique, 4" x 5" negatives, and Fairchild Hand Oblique Camera, Navy Type, 5" x 7" negatives. F/L Lawrence recommended against use of the Fairchilds in strenuous condition, although, when functioning, he admitted that they gave the best results. He experienced "endless trouble, due to mechanical failures and their susceptibility to the cold, jamming and failing to function in even moderately cold temperatures." The Eastman cameras, fortunately, "never gave any trouble."[81]

In a speech to the 1981 CAHS Convention, Lawrence, then an air vice-marshal, paid sad tribute to the invaluable help given to the expedition by local Eskimos:

> All of us came out of the north with pleasant and thankful memories of our Eskimo employees and neighbours. They were a happy and co-operative people, friendly and unspoiled, always ready to take part in work and in play. They were completely courageous. What I saw of the Eskimo people across the high Arctic 30 years later was sad evidence of what the opening of the north and contact with civilization can do to our original natives.

The Compagnie Aerienne Franco-Canadienne and Fairchild Canada continued contract work for the Quebec government, joined by Canadian Airways Limited. Fairchild Canada lost its Vedette in a landing accident, but acquired three Fairchild FC-2 monoplanes from its US parent. Brock & Weymouth continued development of its stereoscopic mapping system, but had a quiet year. A group headed by James Playfair of Midland, Ontario, purchased a majority interest in Canadian Vickers. Ontario added four two-seat de Havilland Moths to its fleet and photographed 4,572 square miles.

The RCAF took a serious look at the design of the Varuna. Of the seven delivered in 1927, three had been lost, and a number of modifications were made to the surviving aircraft during the 1927–28 winter, primarily hull strengthening and weight reduction measures. The development of all the new aircraft types initiated by Canadian Vickers in 1926 was not proceeding as well

81 T.A. Lawrence, *Report to the Director Civil Government Air Operations, Department of National Defense, on the Hudson Strait Expedition 1927–28* (Ottawa: RCAF, 1928): 33.

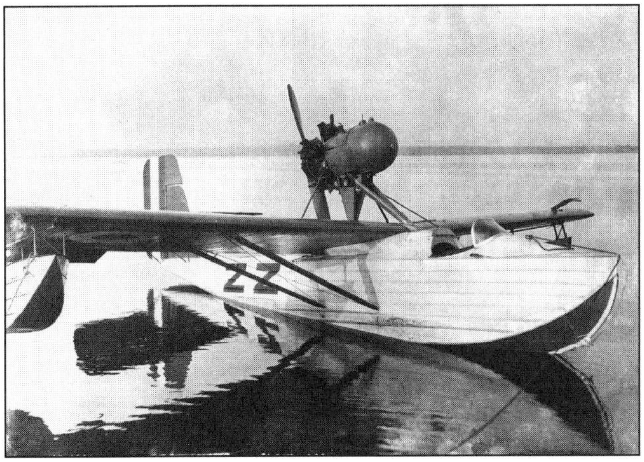

The tiny Vista could only be used in sheltered waters and the sole prototype was eventually shipped to Jericho Beach where salt water attacked the aluminum hull, so it was scrapped.
Canadian Vickers Vista G-CYZZ, Rockcliffe 31 October 1928
NAC PA62440

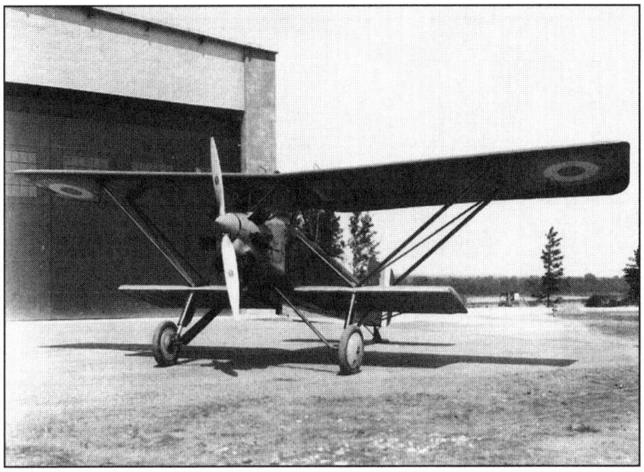

The prototype Vigil forestry patrol biplane failed to attain the altitude performance required for mountain work and no order was placed.
Canadian Vickers Vigil G-CYZW, Rockcliffe, 19 August 1929
NAC PA62573

or as quickly as expected. The prototype Vanessa passenger biplane was taken by S/L J.H. Tudhope to Rimouski in 1927 to pick up mail from the inbound *Empress of France*, but a float collapsed and, although salvaged, the aircraft was scrapped and none was ordered. Design of the Velos photographic amphibian was modified to enclose the crew compartment and to incorporate more powerful engines, but the aircraft did not go into production, largely because of its overweight condition and design shortcomings. Testing of the two smaller single-seaters, the Vista flying boat and the Vigil patrol aircraft, had commenced.

On a more encouraging note, six Vedette IIs were delivered to the RCAF, three with Wright J-5 engines and three with Rolls-Royce Lynx engines. The RCAF also purchased Vedette C-GYGA from the estate of Dalzell McKee who had been killed in the crash of another Vedette, 'GB.

The cumulative total of RCAF photographs taken by 31 March 1927 was 163,000, covering 199,220 square miles.

1928-1929

The 1928–29 *Report on Civil Aviation* again details aviation progress throughout the world, giving particular attention to foreign government subsidies for establishing passenger, freight and mail services, training pilots, building infrastructure and enacting regulations. Although there is implicit comparison with the relatively meagre support from the Canadian government, domestic airmail services were introduced, 16 flying clubs organized and an airship mooring facility built at St. Hubert, Quebec. Estimates increased from the previous year by $1.2 million to $5 million.

The growing number of aircraft accidents was identified in the annual report as a critical problem. Eighteen fatalities in 1928 versus four the previous year, most of which occurred in the relatively settled southern regions, not in the more hazardous northern and mountainous areas (one student fell out of his aircraft near Windsor, Ontario). "The underlying reason is probably the rapid expansion of the flying services, lowering the average efficiency of the flying and maintenance staff. Really experienced pilots are scarce today and operating companies have had to bring on a new generation. These, naturally, lack the experience of the wartime pilots, who have many years of training behind them."[82] It is significant that most of the Canadian fliers who pioneered aerial photography wore several World War I ribbons and were assigned to the most hazardous duties—but had the best safety record.

Veteran pilots formed the nucleus of the PDs formed in 1928 along the pattern established in previous years by the Manitoba and Ottawa stations. The PDs were expected to be self-reliant on photo missions for everything except fuel supplies, usually laid down the previous winter at strategic locations, and food supplies.

A flurry of correspondence between S/L R.S. Grandy, CO Ottawa, and S/L Leigh F. Stevenson, CO Winnipeg, early in 1928 concerned the PD equipment lists and personnel. Equipment lists generated "for criticism and recommendation" included axes, shotgun, saws, shovels, lanterns, cooking gear and tents to establish their own camps. Cameras for both vertical and oblique

82 *Report on Civil Aviation for the Year 1928* (Ottawa: Dominion of Canada, 1929): 6.

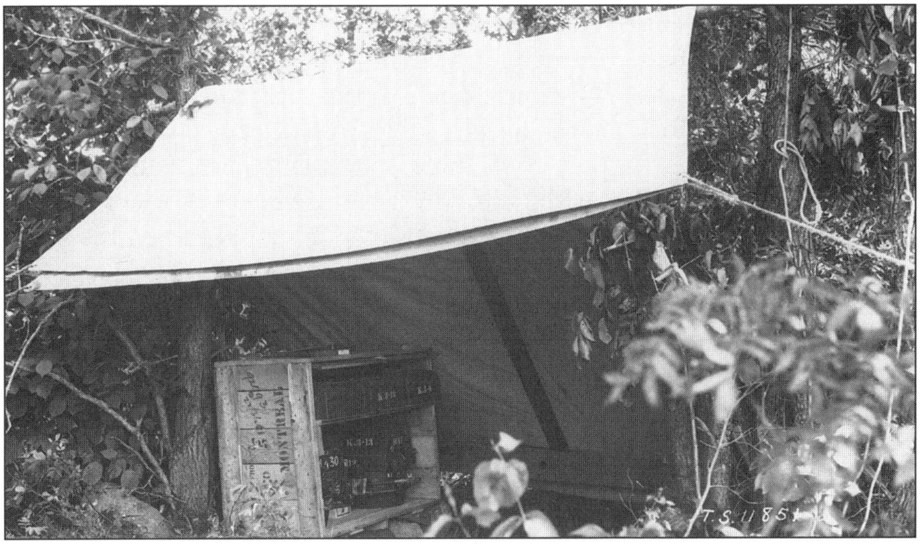

Fresh film had to be delivered to the PDs, and exposed film dispatched to Ottawa for processing. Storage was always a problem. This was one PD's way of keeping film and camera dry and cool during the summer of 1928.
NAC PA204443/ R.D. DAVIDSON

photography were included, together with a wide range of engine and aircraft spares, some of which would sound strange to modern aircraft mechanics—tire patches, solder, rope, cedar, mahogany, fabric, thread, leather, beeswax, glue, cleaning rags, sandpaper, shellac, soap, dope, varnish, wire, nails, screws, pins and thimbles. Chewing gum was also an indispensable aid in repairing leaks. The exhaustive quality control procedures now the norm in aircraft construction did not exist and absolute reliance could not be placed on factory procedures. In November 1928, a suspicious No.3 PD at Winnipeg disassembled and examined elevator control rods on Vedette 'YF to find, "the same appalling workmanship as on Vedette 'YE."[83]

Organizing the PDs for a timely start each season was a vital task. Each PD had a nominal establishment of one flight lieutenant pilot as officer in charge (OIC), a second pilot, two cameramen and two mechanics. In fact, very few OICs were flight lieutenants, although some were "acting" in that rank. In the lean days of the 'thirties, several PDs were under the command of flight sergeants or sergeants. The establishment also varied considerably with the availability of personnel and the task to be undertaken.

W/C J.L. Gordon, director of Civil Government Air Operations, issued detailed instructions covering the provision of fuel caches, maps, photo film, hospitalization, personnel replacement, equipment lists, reporting and recording procedures, and pay and allowances. Provision was made for OICs of Photographic Detachments to draw an advance of $200 for "unforeseen expenditures, such as employment of civilian casual labour, rental of motor boats, hire of boats, rental of buoys, etc."[84]

The PD OICs were well aware that they needed "authority" to requisition any item from stores, or to make a purchase, if they hoped to be reimbursed. They all tried to anticipate requirements and lists abounded. F/L

83 NAC RG24 Vol. 4889.
84 Ibid.

L.R. Charron, CO at St. Hubert and destined to lead a PD, drew up a list of food requirements for seven men for a week. The basics included 45 pounds flour, 20 pounds bacon, 20 pounds sugar, 30 tins Carnation milk, rice, jam, dried fruit, canned fruit and canned vegetables. Also specified were Palmolive soap, Lux soap flakes, laundry soap, three bottles lime juice, four rolls toilet paper and four cans of Flit with guns. Much of the winter was spent ensuring that approvals were in place to expeditiously draw supplies when the spring flying season started. Nevertheless, the unusual nature of the requirements resulted in many frustrations with the bureaucracy in general and from "Stores" in particular.

Many items, of necessity, had to be drawn from army stores giving rise to its own set of problems. In May 1928, S/L S.G. Tackaberry placated Maj. D.E. Dewar, Office of the Quartermaster General, over a PD demand at High River for shovels not authorized in "Scales of Issue for Barrack Equipment." He apologized for overlooking the need to advise district ordnance offices of the special scales of equipment for PDs and, in addition to shovels, took the opportunity to add egg beaters, scrubbing brushes, dish cloths, knives, forks, pails, pans, egg slicers, soup strainers and baking tins.

Photographic aircraft could carry little more than required for operational flying, but a considerable range of additional supplies was necessary for detached operations. These had to travel by ground transportation and were a continuing headache for the busy PD OIC. For several weeks before the flying season started in 1929, S/L Tackaberry had a complete set of equipment laid out on the floor of the Labelle Building in Ottawa to help with the design of suitable packing cases for the ground transport of supplies.

It took several years to arrive at an acceptable inventory of packing cases for ground transportation. On 27 February 1930, Tackaberry formally requested recommendations from the PDs. F/L F.C. Higgins at Rockcliffe must have been prepared because the following day he specified six boxes for general stores, one for engine stores, one for engine spares and one for stationery. Some idea of the volume of stores needed is given from the fact that he expected the first two categories to each carry 400 pounds, and the engine spares box 300 pounds. On 14 January 1931, competitive bids were sought for 60 boxes 4'x2'x1'6" deep with 400-pound capacity, 10 boxes 2'8½"x1'8"x1'4" deep with 400-pound capacity, and 10 boxes 3'x1'5"x10" deep with 300-pound capacity. C.R. Hutchinson, Superintendent of Hugh Carson Company Limited, 72 Albert Street, Ottawa, responded:

> We would suggest that the box be made of ½" three-ply veneer lumber, covered on both sides by hard vulcanized fibre, and securely cemented to the veneer. The edges would be bound with a heavy vulcanized fibre angle, and the hardware—that is, the corners, valances, shoulders,

> hinges and bolts—would be of a heavy steel, as the hardware we use on commercial trunks. This hardware would be put on with solid rivets, burred on the inside, The trunks would be fitted with two solid leather handles, of sufficient strength to allow the trunks to be moved about and the bottom of each trunk would be re-enforced with four hardwood slats rivetted to the trunk, and 4 rollers which would enable the trunk to be moved without being lifted, The outside of the trunk would be finished with two coats of water-proof varnish, and the inside would be fibre with two coats of Shellac.[85]

Hutchinson quoted \$32.50 for the largest trunk, \$30.50 for the smaller, and \$29.50 for the smallest. He offered to make up a sample set before going ahead with the main order. On 11 February 1931, Tackaberry recommended the purchase of one set of eight trunks for evaluation during the coming season. Hutchinson reluctantly agreed to maintain his price but, "our costs will be considerably higher than were we favoured with the whole order." Carson gave good measure, nevertheless: all the dimensions exceeded specifications by at least 1 1/2".

On 15 October 1931, F/L William M. Emery, OIC No.7 PD, reported on the use of a set of eight trunks during his 1931 season. He included an account of the journey commencing by express rail from Ottawa to Huntsville, then by truck from Huntsville to Fairy Lake and a temporary base. The return trip was by express from Huntsville to Coral Rapids, over a construction railway to Moose River Crossing, and by hand about 500 yards to a second temporary base, where they were stored for the greater part of the summer. At the close of the season they were moved back to the railway on a wheelbarrow. The engine spares box and one large trunk were forwarded by construction truck to Coral Rapids, by express to Moonbeam, by truck to Remi Lake and thence by freight to Ottawa. The remaining six trunks were shipped by rail to Moose Factory and on to Ottawa.

Emery concluded that the trunks stood up well in transit. He thought five large trunks would suffice, instead of six, and recommended two partitions be included in one of them for more secure storage of liquids. He commented that on arrival at Huntsville, he discovered that the cork had blown out of the shellac tin and the contents had run over everything in the trunk. The only structural damage suffered was to hasps, several being bent and one broken off on arrival at Huntsville. Emery recommended that flush "trunk type locking hasps" be fitted. He concluded by noting that each of the eight trunks required a different key, and each of the two camera systems required eight keys, bringing the total to 24 plus the aircraft keys. He recommended that one key fit all trunks for each detachment. Tackaberry agreed and noted on the headquarters file copy, "Something might be done to reduce number of keys for camera cases also."[86]

85 Ibid.
86 NAC RG24 Vol. 4890.

The shipment of unaccompanied supplies and equipment posed a perpetual problem. On 8 May 1928, F/L G.V. Walsh, CO High River, noted that stores often could not be shipped from one destination to another "collect," as instructed, because there was usually no one at the destination to pay. His recommendation that freight warrants be issued to PDs was grudgingly accepted, with the proviso that they could only be used to prepay shipments to other PDs, to their own temporary bases, and to such points where shipments could not be forwarded collect.

Arrangements also had to be made for gasoline and oil supplies to be on hand at the advanced bases. Sometimes excess from the previous year was available, but had to be checked and "topped up" for the coming season. Supplies remaining from the previous year for PDs 3 and 4 operating in southeast Manitoba and southwest Ontario, for example, were at Fort Frances—482 gallons gasoline, 392 gallons ethyl and 38 gallons Castrol; Wabigoon—800 gallons gasoline, 30 gallons Castrol; and Kashabowie—no fuel. The Winnipeg Air Base had the responsibility to top up these quantities, where necessary, to the authorized amounts: Fort Francis—1,000 gallons gasoline, 92 gallons oil; Wabigoon—500 gallons gasoline, 35 gallons oil; and Kashabowie—840 gallons gasoline, 60 gallons oil. The same PDs also had fuel caches laid down at Lac-la-Rouge, Stanley Mission and Pelican Narrows in Saskatchewan.

The quality of photographs taken remained a mystery to the PDs until a report was received, usually by telegram, after the photographic section in Ottawa had completed the processing. A major problem, such as photographs ruined by sun glare, would entail flying the line again. Relatively minor problems would invariably show up and require resolution, followed by, in the best civil service tradition, a report to Ottawa. The response of F/O Slemon, OIC No.4 PD, on 30 June 1928 was typical of many made to Ottawa. He explained that obstructions showing on all film exposed in a particular camera was found upon investigation to be due to excessively long motor drive wires protruding in front of the magazine plate. The camera had been received from Vancouver, where the problem evidently originated, but No.4 PD shortened the wires and continued using it. Another problem, film fogging, was found to be caused by a warped magazine preventing the lid from closing completely. This problem was diagnosed by placing an electric torch inside the magazine in a dark tent. Therefore, as a precautionary measure, Slemon covered the slots in all his magazines with electrical tape.

In response to another complaint from Ottawa, Slemon examined shutters, pressure plates and glass in his camera to explain why some film he had sent in was completely blank. He found no cause, so he closed the report by politely suggesting that the fogging, and the blank film, could have resulted from the age of the film, issued the previous year, or previous handling, a possibility indicated by one paper leader missing and another torn.

The "paperwork" involved in launching the PDs each season must have been a source of great frustration, detectable even in the formal communications. For example, on 18 May 1928, S/L L.F. Stevenson, Winnipeg CO, sent a telegram to the CGAO on the subject of PD stores:

1. It is noted that although ammunition has been provided on the above establishment, no provision has been made for rifles or shotguns.
2. There are not sufficient rifles or shotguns at this Unit to equip Photographic Detachments, and it is requested that suitable arms be provided by Headquarters.

The same day, S/L Tackaberry shipped five 12-gauge shotguns from Ottawa. Headquarters, as always, liked to have the last word and he also sent a message: ". . . your stock sheets show nine (9) shotguns being on hand at 31 March 1928." Stevenson then had an inventory problem.[87]

Perhaps reflecting misuse, all PDs were instructed on 6 June 1928, "When morphine tartrate tablets included in the medical kit are used, a certificate stating the circumstances in which used signed by the Senior Officer of the Detachment or Aircraft must be forwarded to the Officer Commanding the Unit to which the Detachment is attached for accounting purposes."[88]

On 14 June, S/L Grandy tried to simplify the stores situation and recommended that each station should use its own judgement on issuing supplies to PDs. Despite best efforts, bureaucratic problems persisted. For example, "Stores" responded with an "Inability Demand" regarding the 25-pound bench vise included in the engine repair kit and requested a substitution. A PD officer suggested that a quick trip to the nearest hardware store would solve the problem. The request for "mosquito bars," which must have been essential equipment for the northern campers, was summarily rejected with the advice, "get from Ordnance."

Getting down to the important items by 14 June 1928, Headquarters authorized the issue to each PD of multitudinous copies of 22 official forms, ranging from "Fragile" labels to demand vouchers and patrol reports—of course accompanied by a phalanx of pencils, rubber stamps, binders, wrapping paper, rulers and "Jewel" clips.

The large amount of equipment necessary for independent "detached" flights gave rise to a plan for "mobile" operations from established bases. This was soon rejected, largely because it was impossible to forecast the weather with adequate accuracy to enable the aircraft to reach the area of operations in time to do any useful work. There was no real alternative to living at the site, ready to take advantage of every photographic minute.

In response to a proposal by S/L Stevenson, photographic pilots and headquarters staff convened in Ottawa on 1 March 1928 to plan an ambitious program. S/L Grandy, CO Shirleys Bay; F/Ls Charron, Mason-Apps, Higgins

87 NAC RG24 Vol. 4889.
88 Ibid.

S/L Stevenson was sympathetic to the difficulties experienced by the PDs and had early knowledge of surveying, as his father was crown land surveyor for Kent County, New Brunswick, during the closing years of the 19th century. He held a commission in the militia in 1914, but believed his cavalry regiment would never see action so he enlisted as a private in the infantry. Surviving two years in the trenches, he obtained a commission in the RFC and joined the CAB on his return from France. From command of RCAF Station Borden he was sent to Britain in 1939 as air officer commanding the RCAF Overseas. Back in Canada in 1940, A/V/M Stevenson organized and commanded No.4 Training Command.

and Harrop; and F/Os Foley, Gilbert, Ivey, Phinney, Emery, and Graham reached general agreement on the allocation of resources and the division of work. W/C J.L. Gordon, director, Civil Government Air Operations, authorized the formation of eight PDs on 1 April 1928. All PDs came under the control of the Ottawa Station, but No.1 was attached to High River and Nos. 2, 3 and 4 to Winnipeg, nearer their area of operations, for stores and accounting purposes.

With the exception of No.8 PD, which had an RCAF navigator, all PDs included a single civilian navigator-surveyor from the Topographical Survey Branch. For safety's sake, the normal practice on long flights was to fly in aircraft pairs, so only one navigator was considered necessary.

No. 1 PD received two new Fairchild FC-2Ws with 410 hp Pratt & Whitney Wasp engines giving better high altitude performance than the FC-2 with its 220 hp Wright Whirlwind engines. Two pilots, one surveyor as navigator, and two airmen mechanics flew the FC-2Ws from High River over the Rockies to the west coast. They took 299 oblique photographs in the Vicinity of Victoria and Esquimalt, and 11,988 verticals at Squamish, Mud Bay, Sicamous, along the Fraser and Thomson Rivers, and in the Fraser Lake district. For good measure, a four-hour customs patrol was made along the coast on behalf of the Department of National Revenue.

No. 2 PD from Winnipeg, with two Fairchild FC-2s, two pilots, a surveyor and two mechanics, took vertical photographs over central and northern Manitoba and western Ontario, a total of 6,135 photo miles.

No. 3 PD, also from Winnipeg, with two Vickers Vedettes, two pilots, a surveyor and four mechanics, took vertical and oblique photographs along the Churchill River and obliques of western Ontario, northern Saskatchewan and Manitoba, a total of 5,900 photo miles.

No. 4 PD with two Vickers Vikings from Winnipeg, two pilots, a surveyor and five airmen took oblique photographs along the Manitoba–Ontario border and western Ontario, northern Saskatchewan, and northern and central Manitoba.

In a noteworthy deviation from the military tradition of, "do it first, then complain," F/O C.R. Slemon, OIC No.4 PD, queried his orders of June 1928. Only the extreme seriousness of the issue would prompt a junior officer to take this action that, even if justified, would certainly prejudice his promotional prospects. Slemon, however, had already demonstrated the painstaking attention to detail that put him in a senior position with Bomber Command in World War II, and he may have saved the world from nuclear conflict during the Cold War. He was directed to proceed with two single-engine Viking flying boats to Wood Buffalo National Park in northern Alberta and to the Northwest Territories. He was also to transport a Hudson's Bay Company representative, "at such times as it will not interfere with photographic operations." His assessment illuminates an early stage in the career of one of

Canada's most respected airmen and describes the conditions, both geographic and bureaucratic, in which the PDs operated.

In three single-spaced foolscap pages, Slemon detailed to W/C J.L. Gordon, director of Civil Government Air Operations in Ottawa, his objections, which fall just short of calling the proposed mission foolhardy, even suicidal. He drew on his experiences the previous year to describe the area and its hazards. Aircraft would have to be based part of the time at the McMurray settlement on the junction of the Athabaska and Clearwater Rivers. He described the Athabaska as very dangerous with a very swift current, numerous rock reefs and ever-changing sand shoals that could not been seen through the muddy water. In the spring, the water was full of floating timber and debris and, as the water level receded, even experienced boatmen found it difficult to navigate small boats safely. Takeoff could only be achieved in the downstream direction, between high and precipitous banks: "The place should be avoided by aircraft." His directions to photograph approximately 70 miles of the Clearwater River valley at an altitude of 2,000' inspired the comment: "140 miles at low altitude . . . along a stream where a safe landing in case of an emergency is almost impossible."[89]

Similar references were made regarding suggested bases on the Slave River. He thought Hay Island, four miles upstream from Fitzgerald, was the best operating base of a very poor selection. Yet, even here, he forecast a dangerous situation for the fully laden Vikings that would have to take off upstream against a very swift current or downstream over 18 miles of rapids.

Chipewyan on Lake Athabaska offered a somewhat different set of hazards. This very shallow lake was notorious for strong winds that could vary the water level by as much as six feet, leaving aircraft stranded at their moorings in what had shortly before been the deepest water in the vicinity. Fortunately, Slemon wrote, only a short stay there would be necessary before moving to Moose Lake. Transportation flights would have to serve Moose Lake from Fitzgerald or Fort Smith, crossing sixty miles "devoid of water and landing places," as were large areas to be photographed.

Slemon described the territory to the west of Slave River as very dry with swamps that did not afford safe emergency landing places. "In the case of a forced landing and damage to an aircraft the nature of the country is such that travel would be extremely difficult and the chances would be against the crew reaching safety." He devoted a paragraph to the perils of operating single-engine aircraft, which should, he wrote, have a reserve of power in case of an emergency and be capable of a quick takeoff and rapid climb. "In my opinion, the Vickers Viking is entirely unsuited for the carrying out of these operations in the country as I have seen it, and I consider it fortunate that the small amount of work carried out there last season was attended by no serious injury."

Having explained the tactical dangers, Slemon then queried the strategic value of proceeding at all. He believed that smoke from unchecked forest fires

89 NAC RG24 Vol. 4915.

blanketed the area from the middle of July until the rainy season commenced in the fall, making aerial photography impossible most of the time. "The early part of the season during May, June and perhaps early July are known to be the most favourable as regards weather conditions." He recommended that "the urgency with which these operations are required be compared with the undue risk involved in their being carried out this season and that the advisability of the attempt be carefully reconsidered."

Slemon also pointed out that one Viking was due for a major engine overhaul in 55 hours' flying time and that 40 hours would be consumed simply getting to the area and returning. He also requested the assignment of a second navigator as insurance that should one aircraft go down, both crews could act independently.

A copy of Slemon's letter went to S/L L.F. Stevenson, CO Winnipeg Air Station, who conferred with F/L Charron, who had been in charge of operations in the Lake Athabaska area during 1927. Charron is quoted in Stevenson's 30 July memo to DCGAO as concurring with Slemon whom he believed to have portrayed the situation there "very conservatively and fairly." Stevenson suggested that the whole operation entailed unnecessary risk. He noted that Slemon was proceeding reluctantly, and went out on a limb to note that he assumed headquarters accepted all responsibility in case of accidents to aircraft or personnel.

Meanwhile, S/L Godfrey had referred Slemon's memo to F.H. Peters, the surveyor general. On 16 July, Peters first covered himself by denying that his office would insist on work being undertaken that the RCAF considered too dangerous. He then went on to say that he saw no good reason not to undertake the work. His selective answer ignored the smoke problem, endemic in the area after July, and minimized the low water hazards. He noted that all the information in Slemon's report was known when agreement was reached, "therefore, there does not seem to be any good reason for changing the plans at this late date after commitments have been made." He contended that the area contained numerous large lakes suitable for the landing of planes and made a point of wanting the Clearwater River photographs completed. He concluded, from the comfort of his Ottawa office, "From our knowledge of the district we would not have considered it an extra hazardous undertaking, provided of course it is completed before the low water of the fall months."[90]

Faced with this letter from the Interior Department, W/C Gordon had little choice. On 19 July, Slemon was directed by telegram to proceed immediately, but his instructions did give him an "out" in stating that his primary objective was to obtain information on flying conditions and to locate suitable bases. The Topographical Survey Branch could not supply a second navigator and Slemon was advised to put the sole navigator with the most inexperienced pilot and let the other pilot "do the best flying possible in the circumstances." The almost time-expired engine was changed in Winnipeg,

90 Ibid.

but both Vikings made bad starts to the expedition. En route to Winnipeg, one had to make a forced landing near Minaki to repair a broken water connection and the tailplane of the second had to be replaced after damage on a test flight. The two Vikings departed from Winnipeg for the Wood Buffalo National Park area on 25 July. Fuel caches had been laid down at Fort Chipewyan (464 gallons gasoline, 48 gallons Castrol "R") and Fitzgerald (4,608 gallons gasoline, 48 gallons Castrol "R," 148 gallons Mobile "B").

Rain, sleet and snow were the common fare for Slemon's reluctant detachment en route to Fort Chipewyan on Lake Athabaska, and after they arrived on 5 August. There was little opportunity for photography, and the detachment moved south to Cooking Lake at the end of the month where both aircraft were unserviceable for several days and had to wait for parts to be flown in from Edmonton, 20 miles to the west. The bad weather continued and the detachment returned safely to Winnipeg, but without accomplishing any significant work.

By 18 August, W/C Gordon was ready to respond to Stevenson's memo regarding Slemon's trip. Quoting regulations which defined hazardous flying he emphasized: "A decision, therefore, regarding hazardous flying on the actual operation rests entirely with the pilot doing the work." On the subject of Slemon's apprehension regarding conditions at the sub-bases, he stated that driftwood, logs, etc were only a problem in the spring so would not be a problem in August. Sandbars, he felt, might cause "a certain amount of anxiety . . . but good landing places should be easily found." Stevenson had queried the reason for assigning Topographical Survey's Mr. Russell as navigator rather than R.D. Davidson, who had been part of the 1927 expedition. He was told that this had no bearing on whether the operation should proceed. Stevenson was also chastised for his temerity: "It is pointed out that Headquarters always accepts the responsibility of aircraft and personnel detailed on any operation."[91]

Slemon lost the skirmish, but won the battle. As OIC of the expedition, he had the last word, and the RCAF's annual report tersely states, "It was considered that to photograph this area involved undue risk to personnel, and the operation was cancelled by the Topographical Survey. No photography was done. The flying time was 4 hours, 20 minutes."[92] Slemon also won the bureaucratic war, reaching the rank of air marshal and becoming chief of the Air Staff.

The Slemon saga of 1928 continued as a typical example of the problems faced by the PDs. Headquarters ordered Slemon's No.4 PD to Cormorant Lake, Manitoba, on 5 September for photography over the eastern provincial border. For a change, there was a flat calm at Cooking Lake and the aircraft could not take off. Continued serviceability problems with Viking 'VE further delayed the transfer and Ottawa got a little testy on 10 September: "Advise rush telegram whether both aircraft serviceable and if not definite date to be

91 Ibid.

92 *Report on Civil Aviation for the Year 1928*, op. cit., p. 56.

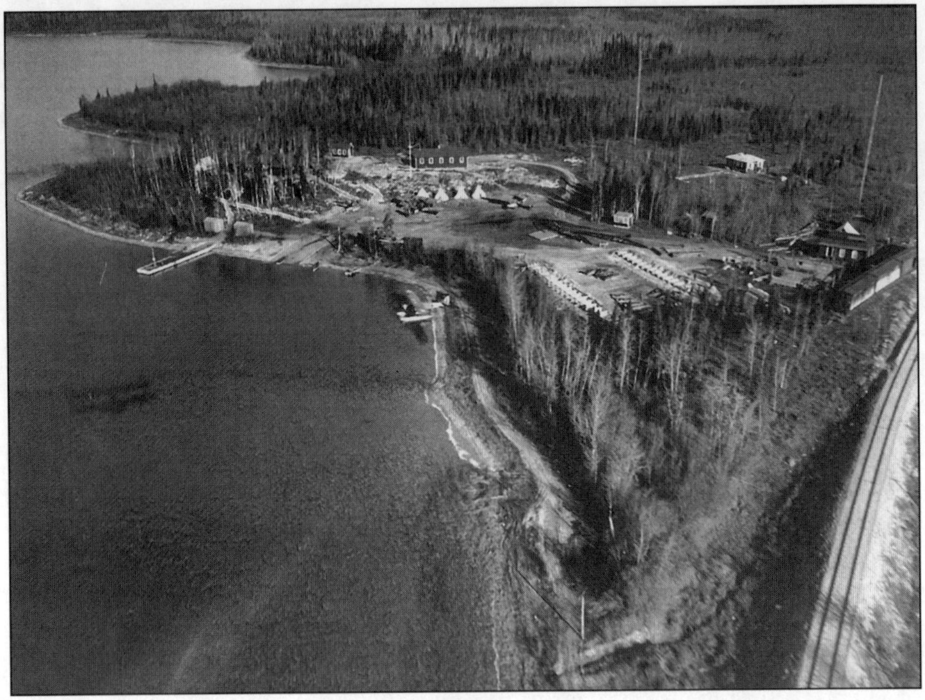

Cormorant Lake sub-base looking northwest, 15 October 1928. The foundations of a hangar can be seen between the tents and the Hudson Bay Railway siding.
NAPL A925-13

serviceable as you are to proceed to Cormorant Lake without delay." The best they received was an indication that the move would be possible "within the week." Surveyor Russell, who had hardly been welcomed by Slemon, was taken to hospital in Edmonton, and R.D. Davidson, as he originally requested, was promised as a replacement.

Slemon's two Vikings left for Cormorant Lake on 12 September and started photography over the Ontario–Manitoba boundary on 13 September. The following day, however, the main scavenger oil pipe of 'EU fractured, necessitating a delay at Big River, Saskatchewan, for six days until a replacement part was flown in. Snow immobilized both aircraft for most of the remainder of September. October saw better weather and photography continued. Slemon, then at Lac du Bonnet, requested instructions from Ottawa on 17 October as most of the lakes in his area had frozen. He was told to use his own discretion, but Winnipeg told him to continue. Slemon sat it out in overcast weather until 1 November when he decided that it was "unsafe to wait longer for suitable weather." By this time, a Vedette had replaced the troublesome Viking 'EU. The operation closed down on 3 November and the aircraft were flown to base for dismantling and storage, the Vedette at Lac du Bonnet and the Viking at Winnipeg.

No. 5 PD went to St. Donat with two Vedettes, two pilots, one surveyor and six airmen to vertically photograph over 4,000 square miles north of the Ottawa River and over south-central Quebec for the Topographical Survey Branch. No.5 PD also undertook an assignment for the Department of Public

Works, taking obliques and verticals of all the Gatineau Company's property in Ottawa.

No. 6 PD, based at Ottawa with two new Fairchild FC-2Ws, two pilots, one surveyor and four airmen, took 49 obliques of the Dominion Observatory in Ottawa and 1,901 verticals in eastern Ontario and western Quebec.

No. 7 PD, with two new FC-2Ws, two pilots and one surveyor, but only one airman, was delayed departing from Ottawa until logs and debris could be cleared from the lake at L'Orignal, Ontario, on the Ottawa River. F/O S. Graham and cameraman F.H. Lindop suffered extreme fatigue during a photographic mission without an oxygen supply at 13,500' over the Laurentians on 14 May. Subsequently, 10,000' was usually the maximum height required for three-hour missions; however, even this imposed severe discomfort. In June, No.7 PD moved to Ste. Agathe des Montes, Quebec, and took a total of 15,530 verticals in eastern Ontario and western Quebec. Special flights were also undertaken to take obliques north of St. Michel des Saints along the Flamand and St. Maurice Rivers, and a reconnaissance flight in preparation for the following year was made over the Lac Clare region.

One of No.7's FC-2Ws had an ignominious and unusual return to Ottawa at the end of the 1928 season. In an effort to complete the assignment, G-CYXP was stranded at Ste. Agathe des Montes when the lake started to freeze

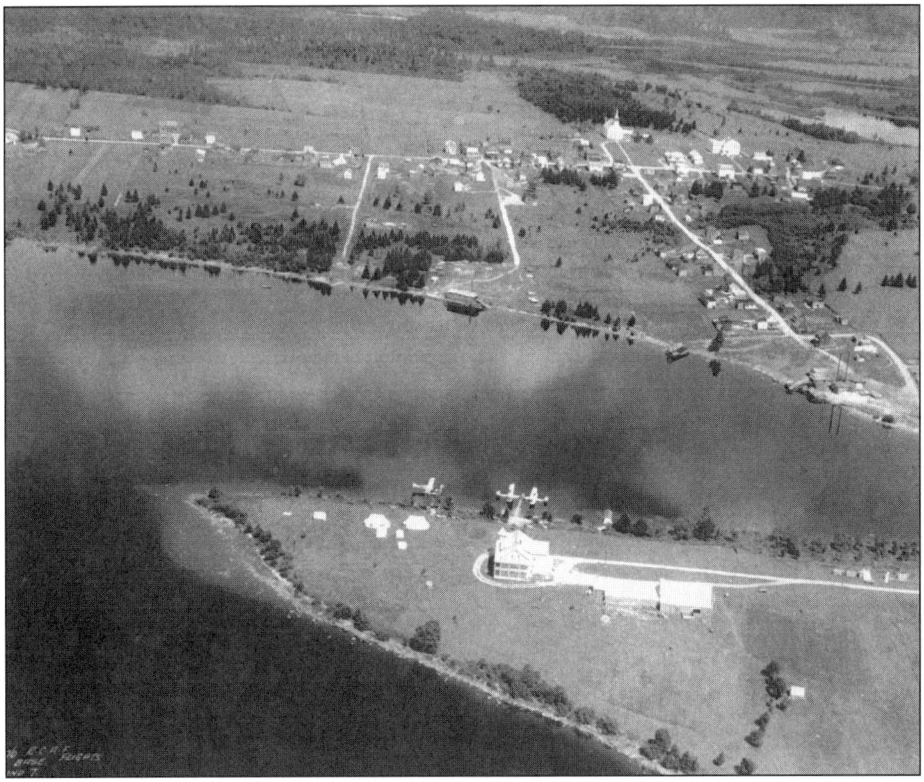

Most of the photo operations in western Quebec were conducted from a temporary base in the grounds of Manoir Bilodeau at St. Donat on Lake Archambault in the heart of the Laurentian Mountains. Two Vedettes of No.5 PD and an FC-2W of No.7 PD are moored in the sheltered bay, 13 August 1928.
NAPL A471-96

F/O Stuart Graham, second pilot of No.7 PD, had his wife Madge and three children living with him at the tent camp. His son Bob recalls, "Both Stuart and Madge spoke French, which gave their proximity to the village an added advantage, because no one there spoke English. Madge made friends with an affable and portly Madame Thibault, whose hearty, affectionate, exuberant embraces were unforgettable." It was their last summer camping together as Stuart resigned from the RCAF in August to be civil aviation inspector for Eastern Canada.

F/O S. Graham's Fairchild FC-2W G-CYYU, 1928, probably at Lac-à-la-Tortue.
R. GRAHAM

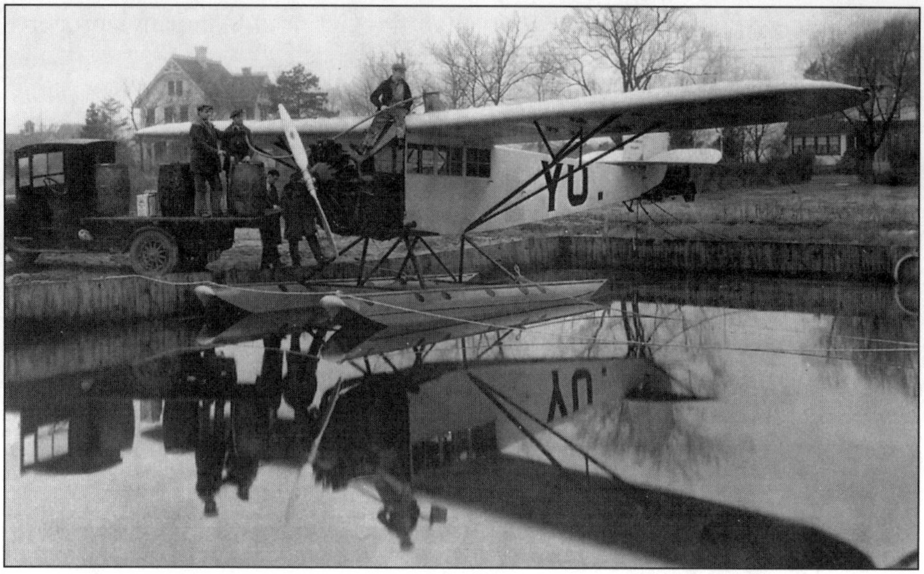

No.7 PD had a habit of communicating with its aircraft using an imaginative phonetic code—G-CYYU was "Your Uncle" and G-CYXP was "Xanatic Paladin." When an exchange of signals indicated the approach of "Young Woman," expectations of intrigue were dashed to discover nothing more romantic than the arrival of G-CYYW.

over. A set of wheels sent by truck from Ottawa did not help much as there was no clear ground of adequate extent and the ice was insufficiently thick to support the aircraft. The wings were folded back and a self-conscious crew tied 'XP behind the truck and towed it back to Ottawa.

No. 8 PD with two Fairchild FC2s, three pilots and four airmen from Ottawa took over 5,300 photographs around Shelbourne and Cape Breton in Nova Scotia, and Saint John and the Peditcodiac River in New Brunswick.

The Shirleys Bay facility, with an establishment of 12 officers and 45 airmen, kept busy during the summer months of 1928. Three Vedettes, one Varuna, three Fairchild FC-2s, three de Havilland 60X Moths, one Keystone Puffer, and one Douglas seaplane were used for training, transport, experimental work, spruce budworm dusting, and mail and passenger route investigations. About 1,000 square miles were photographed during training operations.

The PD files were relatively quiet, except for routine reports, during the summer of 1928 after the detachments dispersed across Canada. The battle with headquarters, however, never ceased. On 28 September 1928, F/L G.V. Walsh, High River CO, took it upon himself to comment to CGAO regarding "considerable delay in receiving various supplies and equipment required by No. 1 PD." S/L Tackaberry replied on 9 October, "From a scrutiny of our records it is found that the only delay which occurred was in the delivery of Emergency Equipment, and that this delay was due entirely to the fact that the Emergency Equipment originally intended for this flight was used by your Unit to equip the Peace River Detachment. If any other delay in deliveries were experienced, it should be drawn to our attention where the matter will be investigated." Driving home the lesson, Tackaberry reminded Walsh on 17

October that he was waiting for an answer. Walsh explained on 18 October that HQ letter of 18 April said PD Emergency Equipment would be supplied by Ottawa, therefore he gave his stock equipment to the Peace River Detachment, which required it much earlier than No.1 PD. Tackaberry would have none of this and responded on 26 October, "May a reply now be forwarded to our letter dated the 14 (actually 9) ultimo and hastener of the 17th instant." There is no reply on file from Walsh.[93]

PDs, as "Johny on the spot," were expected to show the flag. No.8 PD was directed to fly over Lake Rossignol in August, landing if feasible, during a Boy Scout tournament, "consistent with the work on hand." The request from W.G. Ernst, member of Parliament for Bridgewater, for the loan of rifles to the Boy Scouts, however, had to be refused on the grounds that it was against the rules, "and the Department would not wish to take the risk of placing fire-arms in the hands of untrained lads without military supervision."[94]

On 19 October, S/L Grandy suggested that the PDs should retain the majority of stores, other than aircraft and engines, for use in the 1929 program. "This would eliminate a great deal of the stores work and accounting and should also prove more convenient for the Flights concerned."[95] On 15 January 1929, W/C J.L. Gordon went one step further, noting the desirability of placing as much equipment as possible in northern caches to eliminate the necessity of transporting it two ways by air.

The Ottawa Photographic Section moved in May 1928 from Elgin Cottage to the eighth floor of the Jackson Building, on the corner of Bank and Slater Streets, where it was housed until a special building was completed at Rockcliffe in 1936. The staff in 1929 consisted of two officers and 18 airmen,

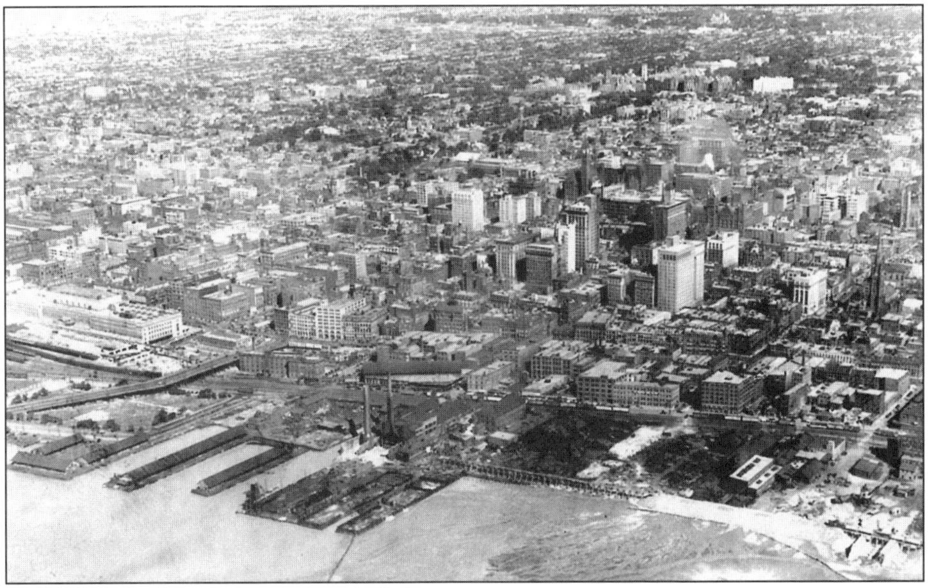

Aerial view of the Toronto waterfront, 9 April 1928.
NAC PA62454

93 NAC RG24 Vol. 4889.
94 NAC RG24 Vol. 4917.
95 NAC RG24 Vol. 4889.

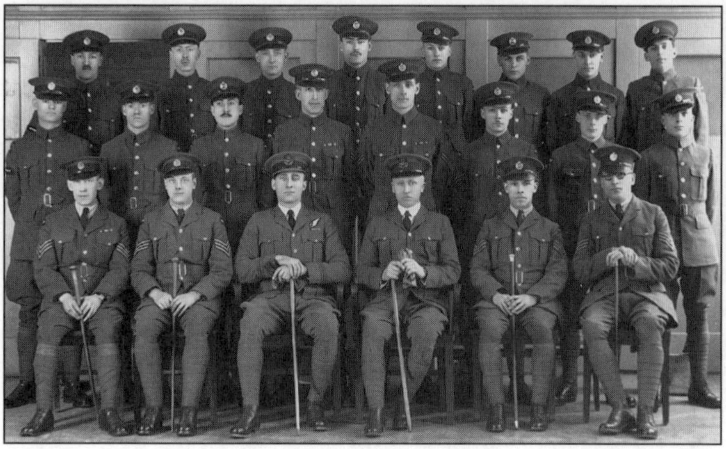

Staff of the RCAF Photo Section, Jackson Building, Ottawa, 1929. F/L E.R. Owen front centre, left. *NAC PA204444*

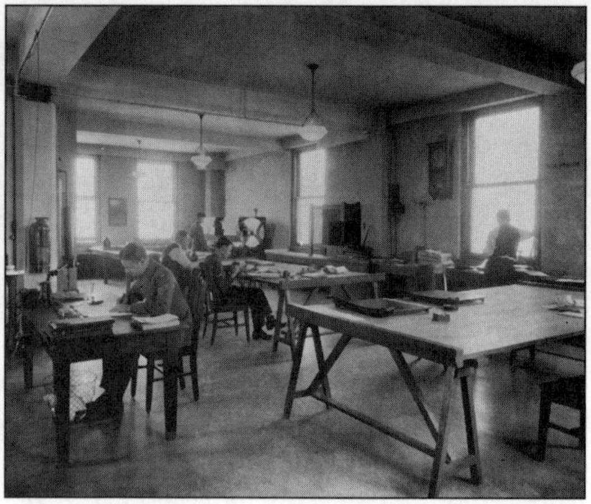

General workroom, Jackson Building, 7 February 1929. *NAC PA62663*

supplemented by seven civilian employees during the busy season. They processed 1,002 rolls of film sent in by the PDs during 1928, almost double the previous year's total of 645 rolls. From the 88,000 negatives, 180,000 prints were made for the Topographical Survey Branch and 18,000 for commercial concerns and private individuals. Fairchild Canada added another 4,015 photographs to the national inventory, which totalled 254,000 by 31 March.

Commercial operations grew apace in Canada during 1928 with many new entrants in the passenger, freight, mail and prospecting sectors. Canadian Airways, absorbed that year by International Airways Limited, and Compagnie Aerienne Franco-Canadienne continued to satisfy Quebec's needs, photographing more than 6,000 square miles. Ontario had its own air service, photographing 4,800 square miles and sketching 4,300 square miles. Fairchild,

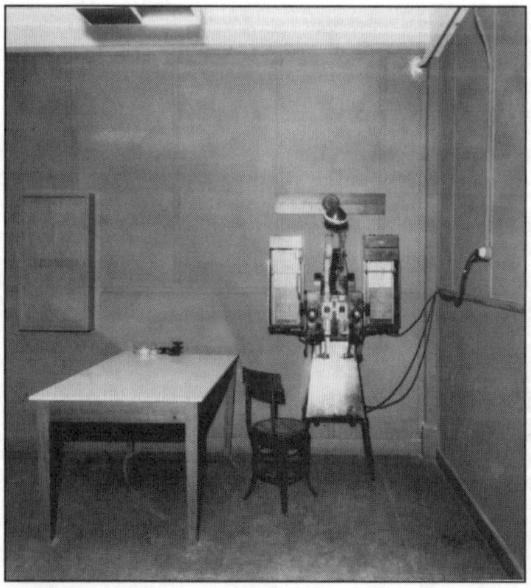

Film printing room, Jackson Building, 7 February 1929. *NAC PA62665*

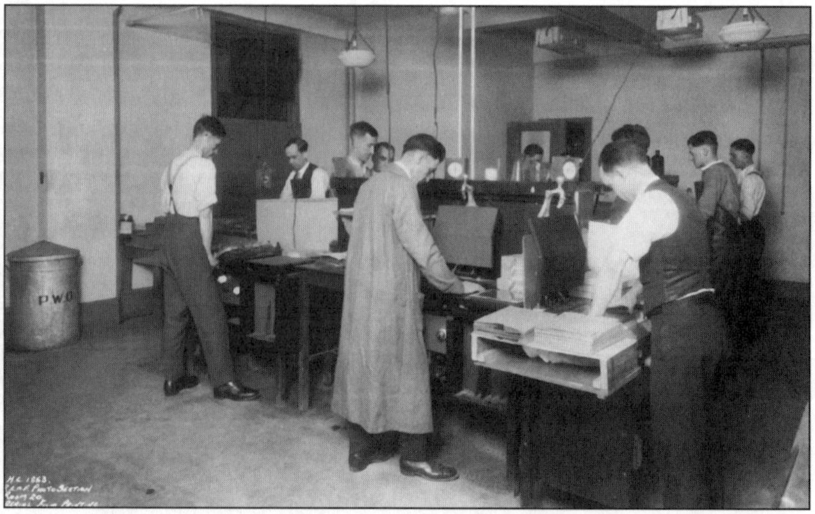

Aerial film printing room, Jackson Building, 7 February 1929. *NAC PA62638*

now with six Fairchild FC-2 monoplanes, one Huff-Daland float plane, one Curtiss Seagull flying boat and one Aeromarine AMC No.1 all-metal flying boat, prospered with its aerial photography and associated activities. Fairchild continued its profitable postcard and advertising photography, and completed 2,130 square miles of vertical photography, 3,000 square miles of obliques and 5,000 square miles of forest sketching.

There were also significant advances in the Canadian aircraft manufacturing sector during 1928. Canadian Vickers moved into metal airframes and started production under licence of Fokker Super-Universal transports carrying two pilots and six passengers. Vickers delivered another seven Vedette IIs to the RCAF, bringing the active total to 16 based at three locations across Canada. The British de Havilland Aircraft Company established a Canadian subsidiary at Mount Dennis, near Toronto, and delivered 60 Moths and two larger D.H.61s. Among the other entrants was Pratt & Whitney Aircraft of Hartford, Connecticut, with an engine assembly and servicing facility at Longueuil, Quebec, that now employs 8,000 Canadians and is the world's leading supplier of aircraft small gas turbine engines. Encouraged by an RCAF order for improved FC-2s (425 hp Pratt & Whitney Wasp engine and an enlarged fuselage, designated FC-2W), Fairchild Aircraft Ltd. commenced construction of its new 38,800-square-foot plant near Longeuil on the St. Lawrence River.

The remaining faithful D.H.4s were written off the books of the RCAF on 30 November, the surplus World War I aircraft having performed exceptionally well long beyond their expected life, a tribute to their maintenance crews.

1929 - 1930

FINANCIAL ESTIMATES for 1929–30 increased by $878,269 from the previous year to almost $6 million. Economies were projected from combining both seaplane and landplane operations at one location, instead of splitting the work between Vancouver and Camp Borden, so a supplementary vote of $250,000 was included to commence preliminary work on a new training station at Trenton, Ontario. The RCAF, however, was still short of its approved establishment although its strength had grown to 175 officers and 669 airmen.

Of great long-term benefit to photo operations, fruitful co-operation between the RCAF and the National Research Council resulted in the establishment of a wide range of aeronautical research facilities. A testing tank for aircraft hull and float models was followed within a few years by wind tunnels, engine and fuel testing laboratories, and materials and instrument laboratories.

As the 1929 season commenced, S/L Stevenson in Winnipeg documented his concern regarding available navigation skills. He reminded Ottawa of No.4 PD's annual report stating that a trained navigator is essential, "if good results are to be obtained." On 12 March, S/L Godfrey advised him that the Topographical Survey Branch of the Department of the Interior intended, "as far as possible, to attach a Navigator to each oblique photographic aircraft."[96] By this time, however, RCAF pilots had developed the necessary expertise, and there may have been an undercurrent of discomfort at having a civilian included in what was otherwise an all-RCAF group.

In 1929, the delivery of 13 Vedettes from Canadian Vickers brought the total available for the eight PDs to a total of 23, six having been written off. One of the write-offs, G-CYWR, had a very short career, crashing nine days after acceptance and killing F/Os P.G. Stanley and J. McLaughlan of the RCAF's Aeronautical Inspection Directorate. G-CYZC lasted only one month before being wrecked by pounding waves in Manitoba while trying to ride out a storm, but the crew, P/O A.D. Ross and Cpl. A.V. Greene, escaped unscathed. The series of accidents culminated on 10 September with F/O

96 Ibid.

G.A.A. Le Moine's fatal crash of Viking G-CYZO while returning to Lac du Bonnet from a forestry patrol; crewman A. Malyk was seriously injured although M. Novrot was thrown clear and uninjured. The increasing need to train flying boat pilots for photography and fire detection duties resulted in four new Vedettes being shipped by rail to Vancouver where the seaplane school could enjoy a longer season.

High River was closed during the winters of 1928 and 1929 except for a "care and maintenance" staff, and the base's aircraft establishment was reduced to six de Havilland Moths. No.1 PD operated again from there into British Columbia during the 1929 season with two Fairchild FC-2W aircraft, two pilots and three airmen taking about 18,000 oblique and vertical photographs. A preliminary survey was flown for the Geodetic Survey Branch from a sub-base at Summit Lake, north of Prince George, of the little-known region up the Parsnip, Finlay and Liard Rivers as far as Dawson in the Yukon Territory.

No.2 PD with two FC-2Ws and Nos. 3, 4 and 5, each with two Vedettes, all under Winnipeg control, accumulated 38,420 photographs of Saskatchewan, Manitoba and northwestern Ontario. Major areas they covered were around Miminiska Lake on the Albany River, Quetico and Rainy Lakes, the Winnipeg and English Rivers, Flin Flon, Kississing and Granville Lakes, and the Mudjatic River.

One of the Vedettes, 'XZ, was ferried from Rockcliffe to No.3 PD at Sioux Lookout in May via refuelling stops at North Bay, Sudbury, Oba Lake and Orient Bay, a total of some 850 miles. The pilot was required to "take no unnecessary risk," and to request weather reports from railway dispatchers at each landing place. F/O C.R. Dunlap made a similar ferry flight with Vedette 'ZB to Cormorant Lake in northern Manitoba, a distance of 1,600 miles. This was a proving flight for the pilot navigator whose longest cross-country flight to that date had been from Camp Borden to Rockcliffe.

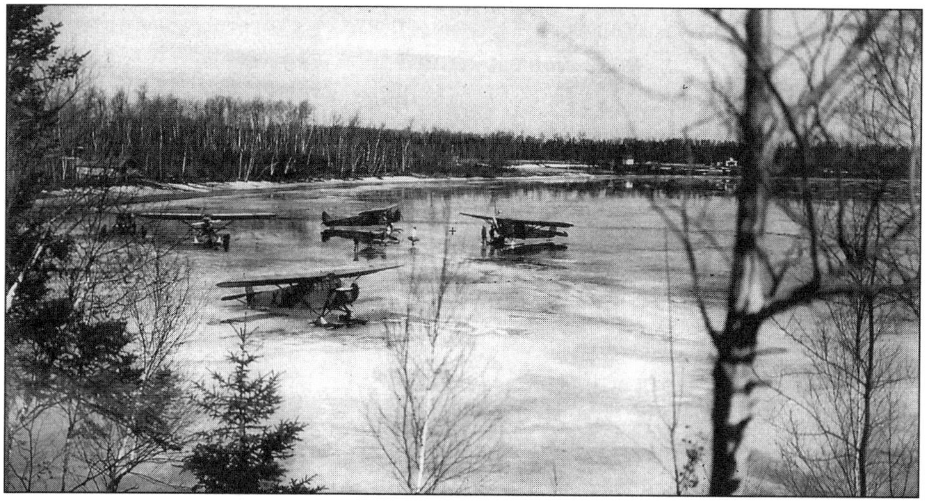

A flock of Fairchilds on ice at Lac du Bonnet, 29 March 1929.
NAPL A1008-5

Sickness in the PDs was almost unheard of, perhaps as well because everyone had to pull his weight and there was no room for "passengers." This is surprising considering their poor accommodation, irregular meals and long, uncomfortable, stressful hours at high altitude without additional oxygen, deafened by the roaring engine. The bare bones of official files give an indication of the disruption resulting from the few instances were men could not perform their duties. The OIC of No.3 PD advised the secretary, Ministry of National Defence, by telegram that LAC H.E. Barry, a fitter, had been placed in the isolation hospital at Prince Albert suffering from diphtheria, a doctor was coming to inoculate the rest of the detachment, and he urgently needed a replacement fitter. The replacement must have arrived and Barry did return to duty, but the only acknowledgement of the telegram on file is another telegram from Ottawa chastising the OIC for directing his telegram to the Ministry, not the Department, of National Defence.[97]

F/O E.A. Copp, OIC No.8 PD, had a similar problem. On 4 November, he reminded Ottawa that Cpl. Winship had been operated on for appendicitis on 1 November and that no replacement had yet arrived. He suggested that Cpl. Bradford be sent as replacement as there was an urgent requirement to repair torn fabric on one aircraft. He added, with a note of desperation, "Bradford has the keys for boxes; will I break boxes or wait for keys?"[98]

F/O Copp ran into more trouble in September. The paymaster rejected an amount of $16.20 from his travel claim for ferry trips between Dartmouth and Halifax: "This action has been taken as it has been usual in the past for officers on command at Dartmouth, and operating from that place, to secure accommodation there. Consequently, if an officer on duty at Dartmouth prefers to secure accommodation at Halifax, ferry charges in respect of the journeys to and from that city cannot be admitted as a charge against the public."[99] There is no evidence of a response from Copp.

The two Vedettes of No.5 PD left Rockcliffe at 1020 hours on 6 June 1929 for Kashabowie, about 60 miles west of Port Arthur (now Thunder Bay). F/O B.G. Carr-Harris, Sgt. Keighley and Cpl. Green crewed Vedette 'GA; F/O B.F. Johnson and LAC DeSeguin Vedette 'ZL. Carr-Harris's report on the 850-mile trip illustrated the typical difficulties that had to be overcome before photography even started. One hour after takeoff, they encountered heavy rain and both aircraft landed at Petawawa until the weather cleared in mid-afternoon. Each aircraft took on 20 gallons of gasoline in the hope of reaching Sudbury, but more heavy rain forced them down near North Bay where they overnighted, arriving at the Ontario Provincial Air Services base in Sudbury the following morning. The flexible fuel pump drive on 'GA had failed before reaching North Bay, necessitating the use of a hand pump to feed the engine as far as Sudbury. It took three days for a replacement drive to arrive, but the time was put to good use hauling fuel six miles from the local Imperial Oil station. An early start on 10 June at 0830 only took them as far as

97 NAC RG24 Vol. 4915.
98 NAC RG24 Vol. 4917.
99 NAC RG24 Vol. 4915.

Missanibi, where they put down in one-mile visibility. They were able to continue later in the day, following the Algoma Central Railway to Oba. With indications of better weather on 11 June, they left for Orient Bay, but clouds descended to 500' and the two aircraft lost sight of each other. 'GA landed at Loop Portage to wait out the weather, but 'ZL proceeded to Long Lac, and on to Orient Bay the following day where it was caught up by 'GA. Refuelled, both aircraft proceeded on 3 June to Port Arthur and on to Kashabowie. Each Vedette accumulated approximately 14 hours' flying time during the 8-day journey.

PDs 6 and 7 from Ottawa still had their pairs of FC-2Ws and took a total of 21,450 photographs of Ontario and Quebec in the vicinity of Lake Nipissing, Blue Sea Lake, Senneterre, Rouyn, Doucet, Lake Olga and Quebec City. The season ended badly with F/O S.R. Sunnocks being injured and his aircraft destroyed following an engine failure at Lake Dibo, Quebec.

F/O N.G. Forester, OIC No.6 PD, had his share of problems with "The System." While the Royal Bank readily accepted the deposit of his cash advance for the season in the name of No.6 PD, the local branches would not allow him to withdraw any cash until the paymaster in Ottawa provided authorization for him to:

> . . . sign and make all or any cheques and orders for the payment of money, and to pay and receive all moneys, to negotiate with, deposit with, or transfer to the Bank (but for credit of said detachment's account only) all cheques or orders for the payment of money, and for the said purpose to endorse the same, also from time to time to arrange, settle, balance and certify all books and accounts between the said detachment and the bank.[100]

Forester's reaction can only be imagined.

In constant conflict with the paperwork, in July Forester had to correct some flight times in his report, following a reproof from the director of civil air operations: "Care should be taken in checking of flying time before patrol Reports are sent in to these Headquarters." In August, S/L Godfrey returned a flight report with a demand that the missing headquarters order number be added. Forester was also chastised, when at Senneterre in October, for not compiling his reports "in accordance with Headquarters Order and Instruction Order No. 29,062, dated 17-6-29, para. 2, in that the main Operational Order No. is not classified under one of the following divisions mentioned thereon."[101]

In common with the other PD OICs, Forester was requested by local people or other federal government representatives to take aerial photographs "while in the area." He asked for guidance in this respect from Ottawa when approached by J. Carroll of the Topographical Survey Branch to detach one aircraft from Sonneterre to take photographs near Clova in northern Quebec.

100 NAC RG24 Vol. 4916.
101 Ibid.

He was no doubt relieved when S/L Godfrey directed him not to undertake the mission because "the primary object to a two plane detachment is safety and if these aircraft are separated, the object is defeated. Therefore, as far as possible, the Detachment is to work from one base always."[102]

No.8 PD with two FC-2s took obliques and verticals at Prescott, Ontario, before heading, again, for the maritime provinces with orders to fly over Lockport and West Pubnico on Labour Day, "subject to non-interference with photographic operations." Photographs were taken in the Moncton area and several small areas in Nova Scotia.

While the RCAF Limited and Peace Establishments for PDs called for a flight lieutenant pilot to be in charge, with a flying officer as second pilot, very few had ranks higher than a flying officer in command. On 27 May 1929, G/C Gordon listed the eight OICs in PD numerical order— F/L A.L. Morfee, F/O L.H. Phinney, F/O T.M. Shields, F/O C.R. Slemon, F/O B.G. Carr-Harris, F/O N.G. Forester, F/O W.M. Emery, and F/O J.E. Jellison. He recommended the promotion of the seven F/Os to temporary F/L effective 1 April 1929, but had to issue a sharp reminder in December to have them put forward in the *Gazette* immediately.

Most of the inherent problems of independent operations by a military organization were solved over the years, but no really satisfactory arrangement was ever reached to pay the men in the field or to cover their travel and living expenses. Before a standing advance was reluctantly instituted, the officers in charge often had to finance local expenditures for food and transport out of their own pockets—and had endless problems in obtaining reimbursement. Following protracted correspondence and discussion, an instruction was issued on 17 April 1929 allowing each officer in charge of a PD a $500 advance to "meet all travel and subsistence expenses of all the personnel under his command." Accounts were to be submitted "at least once a month where practicable."[103]

Tightening up on discipline on 29 May 1929, W/C G.O. Johnson, CO Ottawa, directed that PD OICs "be instructed that uniforms are to be worn by all officers whilst on duty in the vicinity of any base or settlement."[104] However, they preferred to wear civilian clothes that "didn't matter" and preserve their expensive uniforms for other occasions—for good reason. Viewing any photograph of the period gives a clear indication that the aircrews suffered oil-, gasoline- and water-stained clothing, to say nothing of the grime accumulated in cooking, eating and sleeping in tented encampments.

On 18 July 1929, DCGAO gave notice to Winnipeg that a Mr. Wilson of the Motion Picture Bureau would be leaving Ottawa about 19 July to take moving pictures of PDs 3, 4 or 5 operating in the Reindeer Lake District. The intent was to "cover all activities of photographic detachments operating under canvas away from permanent bases."[105] Presumably the film was at least commenced because, on 26 April 1930, Winnipeg queried its status, to be told it was not yet finished. The subsequent response to a query of 22 September

102 Ibid.
103 NAC RG24 Vol. 4889.
104 Ibid.
105 Ibid.

1930 states that Wilkinson had been delayed by the R-100 airship visit. Another "Not yet completed" was the answer to Winnipeg's last query of 22 September 1931. Ernest Ralph Wilkinson was a photographer for the Motion Picture Bureau with several movies to his credit but, unfortunately, no record could be found of his PD work.

The separation of land and water operations between Rockcliffe and Shirleys Bay was uneconomical and inconvenient and it had been decided in 1927 to combine them at Rockcliffe. An additional 22 acres was obtained at Rockcliffe, where buildings were constructed, and Shirley Bay was vacated early in 1929. In addition to a wide range of experimental and test flying undertaken at the new facility, a Mobile Photographic Training Flight was formed of experienced personnel who undertook on-the-site training of PD personnel.

DND correspondence at the conclusion of the 1929 season outlined the economic benefits of supporting triangulation surveys by aircraft, thereby enabling more than double the amount of work to be completed: "This increased progress would justify the use of planes on this work at a charge of $130 per flying hour." The standard rate at that time was $100 per hour. A "wish list" was also indulged in:

> Low-powered aircraft should not be used, a 'plane as good, or better than, the Wasp-powered Fairchild is essential, and a 'plane with better visibility, such as the Bellanca, would be a great improvement, or for

Spring high water levels were a problem at most stations. Shirleys Bay, 21 May 1928.
NAPL A30-32

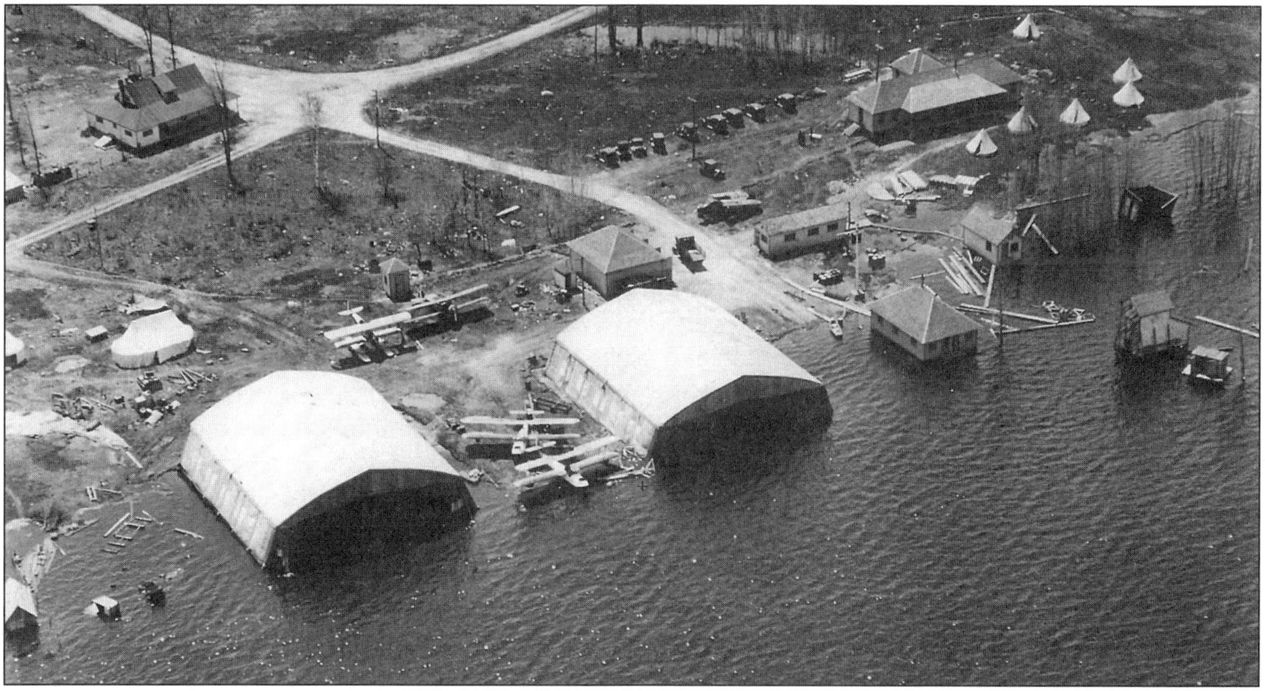

> drier areas a powerful tri-motor. Power is essential for landing and take off from small lakes. Specially designed canoes which could be carried under the fuselage between the pontoon struts should be provided for the next season for the transport of triangulation parties, as well as radio communication between 'plane, base and parties.[106]

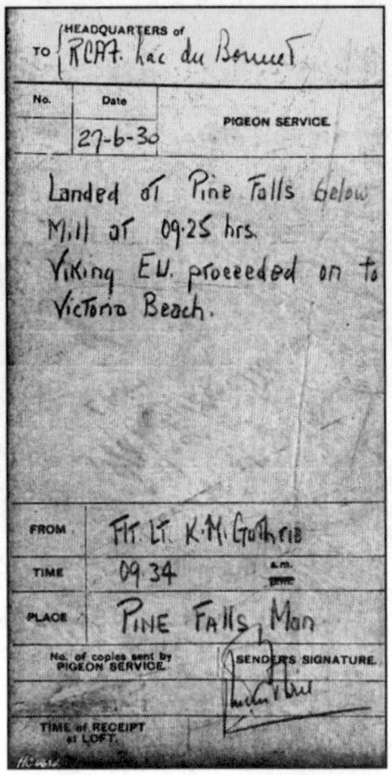

TO HEADQUARTERS of RCAF. Lac du Bonnet
No. | Date 27-6-30 | PIGEON SERVICE
Landed at Pine Falls below Mill at 09.25 hrs. Viking EU. proceeded on to Victoria Beach.
FROM Flt. Lt. K.M. Guthrie
TIME 09.34 a.m.
PLACE Pine Falls, Man
No. of copies sent by PIGEON SERVICE | SENDER'S SIGNATURE
TIME of RECEIPT at LOFT.

The between-wars photo detachments never did realize the benefits of three-engine reliability and continued to place reliance on pigeons for communications. Pigeon message from F/L K.M. Guthrie to his base at Lac du Bonnet, Manitoba, 27 June 1930. *NAC PA204445*

In an expanding commercial photography sector, Fairchild Canada continued photographic operations during 1929 from Grand Mère. The US parent of Fairchild purchased a controlling interest in Kreider-Reisner Aircraft Company of Hagerstown, Maryland because its Challenger series of aircraft were considered the best available for aerial photography. Re-designated as the K-R series these models were subsequently manufactured by Fairchild and played an important part in the future of the Canadian company. Brock and Weymouth faded from the Canadian scene after completing small photographic missions in Quebec, Manitoba, Saskatchewan and Alberta totalling 485 square miles. Interprovincial Airways photographed sections of Ontario and Quebec totalling 7,772 square miles. Compagnie Aerienne Franco-Canadienne, headquartered in Montreal, photographed 7,100 square miles along the St. Lawrence River, southeast of Quebec City and northwest of Montreal, for the Quebec government, and established a photographic laboratory in Quebec City. Canadian Airways, based at Point au Tremble, photographed a total of 1,335 square miles north of Georgian Bay, in southern Ontario, and in the vicinity of Weymouth, Quebec. Western Canada Airways, with headquarters at Vancouver and with many other bases throughout the western provinces, undertook airmail, fisheries patrol and other missions for the British Columbia government. Western Canada Airways took about 4,000 photographs on missions related to mineral exploration and forest resources. Continental Aero Corporation photographed 640 square miles from its base at St. Hubert, Quebec. The Ontario Provincial Air Services, now with 25 aircraft, photographed 1,900 square miles. Aerial photographs were also taken and sketches made by other companies in connection with mineral exploration, forest resources, railways, town planning and power development.

In concert with worldwide growth, the Canadian aircraft manufacturing industry continued to expand. Early in 1929 the Boeing Airplane Company of Seattle, Washington, entered into a joint agreement with Hoffar-Beeching Shipyards of Vancouver to build Boeing flying boats and mail-passenger aircraft. Armstrong-Siddeley in Ottawa and Aero Engines of Canada in Montreal established assembly, test, repair and spares facilities for, respectively, the Armstrong-Siddeley and Wright engines then in common use in Canada, supplementing Pratt and Whitney's similar plant operating at Longueuil since January 1928.

De Havilland moved to its current home at Downsview, North York, delivered 130 new Moths, and demonstrated the four-to-six-seat Hawk Moth.

106 Ibid.

Curtiss-Reid, established in 1928 as Reid Aircraft Company by W.T. Reid and Martin J. Berlyn, both previously designers with Canadian Vickers, enlarged its facility at St. Laurent, Quebec, and started production of its two-seat Rambler. The Ottawa Car Manufacturing Company, already an agent for the US Consolidated Aircraft Company, built 25 Avro Avian two-seaters under licence from A.V. Roe in the United Kingdom, and entered into a joint venture with Armstrong-Siddeley, UK, to build a plant for manufacturing all-steel aircraft of the Siskin and Atlas type, five being completed in 1929.

In a complicated series of transactions during 1929 under the guidance of Hubert Pasmore, Sherman Fairchild sold his controlling interest in Fairchild Aviation to "the Vickers Syndicate," a consolidation of Canadian aircraft industry inspired by James A. Richardson who had been an early investor in the company. Pasmore reorganized the airfield and seaplane base at Longueuil on the St. Lawrence River to produce the Fairchild 71 with a 425 hp Pratt and Whitney Wasp, the two-place KR23, and the three-place KR34. A footnote to this reorganization was the hiring of Col. William Barker, VC, DSO, MC, with the nominal title of president, probably because of his RCAF contacts. The appointment, however, lasted only a few weeks before his death at Rockcliffe on 12 March 1930 in the crash of a KR-21 he was demonstrating.

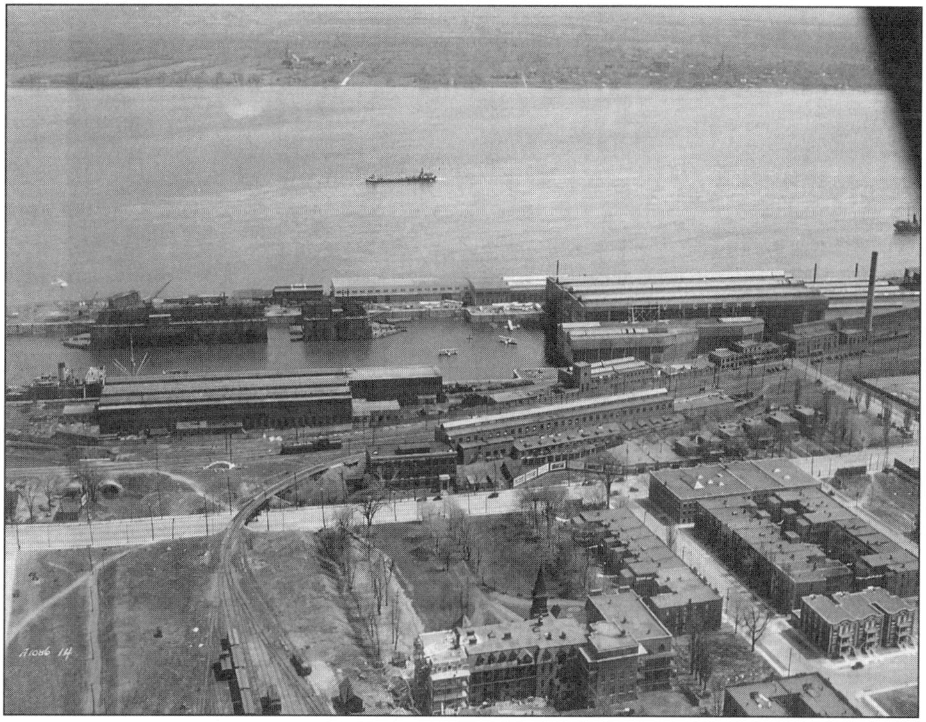

Canadian Vickers delivered 22 workhorse Vedettes during 1929 to various customers in variants that included an amphibian, and a metal-hulled version for the RCAF. Fifteen Fokker Super-Universals were built under licence, and the RCAF received the first two-engine, seven-passenger Vancouver, a replacement for the disappointing Varuna. The busy Canadian Vickers plant on 14 May 1929. Dry docks are to the left in the "Air Harbour." Stranraers, Cansos and Hampden aircraft were built here during World War II, but the whole area had fallen inactive by 1995. *NAPL A10086-14*

1930-1931

THE LOOMING Depression applied brakes to the growth of civil aviation in Canada in 1930. Funding for the RCAF, however, increased by $1.5 million from the previous year to $7.5 million. This was the last year of growth in RCAF appropriations until the buildup in anticipation of World War II.

As had become the established procedure, a meeting of concerned federal government departments was held during the 1930–31 winter to plan the photographic program for the coming year. Crews and aircraft had to be assigned and fuel caches laid down in order to get the earliest possible start in the spring. S/L Tackaberry instructed the Ottawa CO on 30 Jan 1930 to organize three additional PDs (Nos. 9, 10 and 11) before the commencement of spring operations. Orders on 23 May 1930 established the stores and accounting units for the coming season: Vancouver for No.1 PD; Winnipeg for 2, 3, 4, 5 and 9; and Ottawa for 6, 7, 8, 10 and 11.

Specifications for supplies to be carried were still being ironed out. On 3 March, F/L F.C. Higgins, OIC Photo Training Flight at Ottawa, recommended the supply of adhesive tape be increased from two to six rolls so film canisters, prone to light leakage, could be adequately sealed. He also recommended an issue of lampblack, "to facilitate filling in as necessary the engraved cross at the optical centre of the focal plane glass of aero cameras."

On 5 May 1930, S/L N.R. Anderson, CO Winnipeg, who had the interests of the PDs very much at heart, suggested that small portable typewriters be supplied to ease the reporting burden. S/L Tackaberry responded from Ottawa on May 12, "I am directed to advise that it is not proposed to supply typewriters, but consideration is being given to covering the air risk of privately owned typewriters used by officers commanding detachments. When a decision has been reached, you will be advised." This, effectively, told the detachment commanders, setting out on the season's work, the familiar story, "You're on your own, Jack." There is no evidence of further official action regarding typewriters, but F/O H.C. Foley, OIC No.3 PD, typed a report while at Buffalo Park sub-base, Alberta, in August 1931. Most subsequent correspondence from the detachments is typed. However, there is a memo from S/L R.S. Grandy, CO Winnipeg, dated 27 January 1936 rejecting the issue of typewriters: "Detachments usually have ample time to make up returns."[107] Much the same attitude applied in other areas: A/V/M R.E. McBurney, for example, recalled that he always carried his own .22 rifle and 14 gauge shotgun on his forest patrol and photographic operations.

No.1 PD, with two Fairchild FC-2W aircraft, vertically photographed over 5,500 square miles of British Columbia including the south end of Vancouver Island, the city of Vancouver and along the Fraser Valley. A busy program included 15 hours' patrol at the height of the fire season and 104 hours searching the northwest coast for a missing civilian pilot.

No.2 PD, commanded by F/O H.M. Kennedy, started the season with two Fairchild FC-2s, replaced in May with a pair of Fairchild 71s. Nos. 3, 4 and 5 PDs each had two new Vedettes, and No.9 a pair of Fairchild FC-2Ws. The five detachments, all administered from the Winnipeg station, took about 50,000 photographs during 37 operations stretching from Lake Nipigon to Great Slave Lake, concentrating on Lake Nipigon and English River in Ontario; Porcupine Hills and Granville Lake in Manitoba; and Dillon, Green Lake, and Ile-à-la-Crosse in Saskatchewan. In a salutary reminder of the dangers of the profession, early in the season F/O H.W.P. Carew and Sgt. A. Richards were killed in the crash of a Fairchild at Regina en route to a photographic assignment at Dundern Forest Reserve.

A need existed for a northern base to support transport and photographic operations in the Northwest Territories, so No.9 PD was dispatched to locate

107 NAC RG24 Vol. 4890.

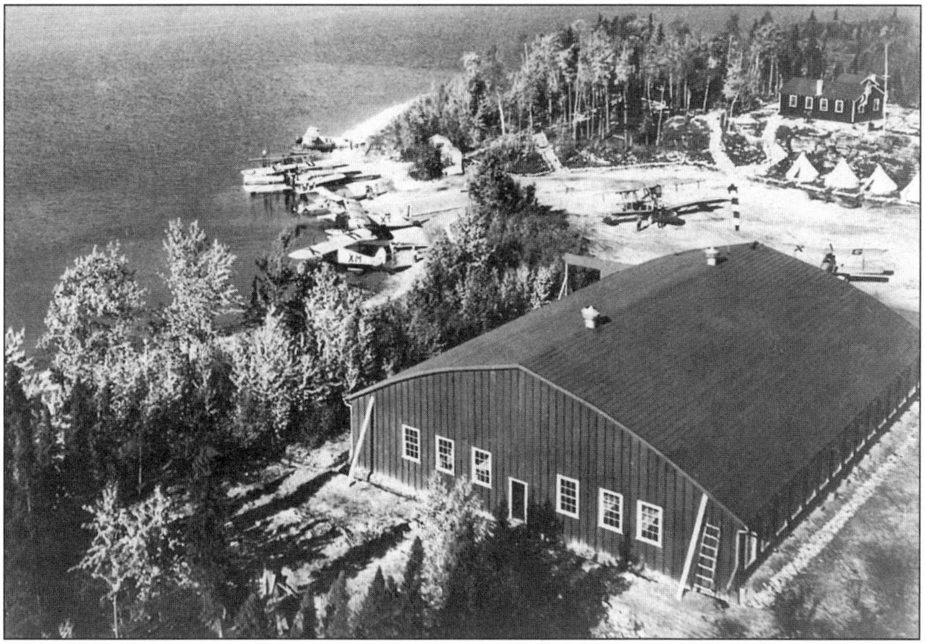

A busy day at Cormorant Lake in 1930.
DNDPU RE68-881

one. OIC F/O C.R. Slemon selected a site on Ryan Island in the Slave River, near Fitzgerald on the Alberta–Northwest Territories border, 510 miles northwest of the Ladder Lake sub-base. A small house at Fitzgerald was rented for an officers' mess and the airmen camped alongside in tents. No.9 PD crews deferred their departure until 6 October to assist with the erection of permanent buildings and were further delayed by bad weather, only getting as far as Cormorant Lake before winter set in and the lakes froze over. The aircraft had to be left in the hangar there while Slemon and his crew completed their journey to Winnipeg by train, forcibly demonstrating that aircraft should not remain in the north later than 1 October..

PD Nos. 6, 7, 8, 10 and 11 were each delighted to receive two new Fairchild 71s. The five Ottawa detachments photographed over 23,000 square miles of eastern Ontario, Quebec and the Maritime provinces. They photographed along the shore of Lake Huron, Guelph, Rice Lake and Lachine for the army's Geographical Section, and around Coral Rapids, Lake Nipissing, Algonquin Park and Orillia in Ontario and Calumet Island, Notasgan Lake, Blue Sea Lake and Mont Laurier in Quebec for the Interior Department. No.8 PD proceeded to the Maritime provinces where it photographed around Newcastle, New Brunswick, and undertook other smaller tasks, including "scenic obliques" of all the Dalhousie University buildings. No.8's daily report on 1 August 1930 from Natasquan Point on the north shore of the Gulf of St. Lawrence included mention of what must have been a thrilling sight: "Dirigible presumably R100 observed at 0500 hours proceeding westerly direction at approximately 500 feet two miles off shore."[108]

No. 10 PD OIC, F/O B.F. Johnson, with F/O F.A. Samson, Carpenter/Rigger Sgt. Bradford, and Fitter/Camera Operators Cpl. MacLaughlin and LAC Wiseman found economical room and board in comparative luxury at Blue Sea Lodge, Quebec, but once the tourist season started prices went beyond their allowance and they moved to an isolated rented cottage and hired a civilian cook.

More than 3,400 photographic and 800 reconnaissance hours were flown by the 11 PDs in 1930, taking 98,275 photographs covering 54,100 square miles. By the end of 1930, the Geographical Section of DND was advertising 207 topographical maps of 1"= 1 mile scale at 25 cents each; fourteen of 1"=40 miles at 40 cents each; and one experimental aeronautical strip map from Montreal to Morrisburg of 1"=40 miles for 25 cents.

The Geodetic Survey Branch had made strong representation in 1929 in favour of providing wireless three-way communication between the aircraft, base and triangulation parties on the ground. It was emphasized that, without communication, additional costs were incurred if an aircraft had to make additional flights when events delayed the ground survey past the planned completion date, or if they completed the work early and had to wait for the aircraft to pick them up. Potential savings of $100/hour for aircraft and

108 Ibid.

$50/day or more for the ground party, it was argued, would go a long way to providing wireless sets. Dr. J.D. Craig, director general of surveys, eventually received advice in February 1930 from the assistant director of Signals to the effect that the required communication could be achieved using some equipment on hand plus new equipment at a cost of $600 per set. Also required would be a gasoline-powered generator for which delivery was three months, so immediate authority to proceed would be needed. The RCCS, Geodetic Survey and the air force spent the summer of 1930 discussing the type of equipment needed and, of paramount importance, who would pay for it. They reached no conclusion, probably to the relief of the air force participants who were concerned about the additional loads to be carried.

Canadian Airways Limited, which absorbed Western Canada Airways, took aerial photographs of 12,625 square miles of British Columbia. Compagnie Aerienne Franco-Canadienne took obliques of 5,600 square miles and verticals of 200 square miles in the St. Maurice basin for the Quebec Department of Land and Forests. The Ontario Provincial Air Service, now with a fleet of 33 aircraft, primarily engaged in forestry work and transportation, flew 659 hours on photography.

Most of the northern lakes were infested with floating logs and, worse, deadheads skulking just beneath the surface. Although the water was clear, the dark debris-filled lake bottoms made these hazards virtually invisible from the air. Vedette G-CYWL flown by P/O H.W.P. Stewart hit a submerged deadhead while landing at Golden Lake, Ontario, in May 1930. His laconic report tells the story: "The hull was temporarily patched the next morning, and the Vedette re-floated. It was then flown to Lac du Bonnet, where it was taken out of the water for further examination. Further repairs were carried out at Lac du Bonnet and the hull is now serviceable." This type of accident occurred often enough to become routine, but it was a little more alarming for the crew of G-CYWL when the exhaust stub from No.3 cylinder fell off, damaged the propeller and holed the hull. Again, local repairs were made and the Vedette completed its assignment before permanent repairs were made at Lac du Bonnet.

As recounted by Air Marshal C.R. Dunlap in his address to the Canadian Aviation Historical Society in June 1972, he was instructed in aerial photographic operations at Rockcliffe by F/L Earl Jellison, "tops in the field." He considered himself fortunate to be then assigned to Jellison's No.11 PD with an extensive survey program planned for Ontario and Quebec. Other detachment members were Carpenter/Rigger AC1 E.F. Bullis, with camera operating skills, and Fitter Aero Engine/Camera Operators Cpl. S.C. Dearaway and LAC J.C. Mirabelli. They had two new Fairchild 71s, G-CYWF and 'WG, both on wheels. Dunlap recalled:

> Photo survey called for precision flying of the highest order. It was necessary to fly straight and level for sixty miles and then turn and do a

> parallel course in the reverse direction a mile or so removed from the first, and so on throughout the day. Overlapping vertical photographs were taken as one progressed along the line . . . also if all went according to plan there was lateral photographic overlap inasmuch as distance between the lines took care of this, provided one's navigation was precise. Experienced photo pilots seemed to acquire a special faculty in that they were able, without the benefit of instrumentation, to determine the degree of drift on any given heading . . . and a very necessary thing this was, for one was expected to remain on track within a tolerance of a couple of hundred yards. On top of this requirement was the need to keep constant altitude within a tolerance of 100 feet or so for hours at a time, and to be absolutely level each time the pilot's signal light flashed its five second warning prior to the taking of each picture. It was a tough league, for standards were set by the pilot of the other aircraft, known as 'No Gap' Jellison."

Friendly rivalry sprang up between the two pilots—master and pupil— to see who could go the longest without a gap in the coverage. The spring of 1930 remained cloudless for many consecutive days over the area between Lakes Ontario, Erie and Huron. From Mohawk Aerodrome near Desoronto, the competitors ploughed their furrows, day after day, at over 10,000', often for seven hours at a stretch and taking more than 1,000 photographs in one day. They completed a task expected to take all summer in eleven days. Although exhausted, the crews moved on to St. Hubert, Quebec, on 11 May to photograph the vast area south of the St. Lawrence River from the Ontario border to the State of Maine. Again "blessed" with fine weather, they had the job finished by 22 June. Jellison was generous enough to say they were about even in their competition with one gap apiece, but Dunlap acknowledged that his gap was the biggest.

Dunlap then transferred to British Columbia. Photo weather left no time for relaxation and he started work immediately after completing a four-day train journey. He was assigned 'XQ, a Fairchild FC2W, with No.1 PD and photographed large slices of coastal and interior British Columbia before moving to the Queen Charlotte Islands.

The Japanese were flexing their muscles in Manchuria and it was believed that Japanese naval officers masquerading as fishermen had been active in the Charlottes for several years. It was time to map the little-known islands. No.1 PD was used to living in tents, but looked for alternative accommodation in a region notorious for having two seasons—wet, and wetter. They found quarters at the single hotel in Queen Charlotte City, the first guests for seven years. "Believe me, those mattresses were damp, hard and lumpy and the rooms were far from fancy . . . but no matter how primitive the accommodation and the plumbing, it certainly beat living under canvas in that climate."[109] The weather was so overcast in the Charlottes that it took three years to complete the assignment.

Two flights blazed the trail to the north in 1930. In July, a pair of Fairchilds from Winnipeg, under the command of F/L J.C. Uhlman, left McMurray, Alberta, with Lt.-Col. E. Forde of the RCCS and two officials of the Indian Affairs and Agriculture departments. They inspected posts along the Mackenzie River Valley all the way to Aklavik and on to Herschel Island, Yukon Territory, in the Beaufort Sea. In 140 flying hours over 20 days, the detachment covered 11,365 miles, most of which was new territory for the RCAF.

An even more ambitious venture started out from Ottawa on 6 July 1930. F/L F.J. Mawdesley flying Fairchild 71 G-CYVX, and F/Sgt. H.J. Winny with Vedette G-CYWS departed from Rockcliffe on a trail-blazing flight to the Northwest Territories. The party included C.S. Macdonald, DLS, and mechanic/photographer A/Sgt S.C. Dearaway. Their objective was to explore and photograph the main water and air routes in the Northwest Territories and to inspect 32 fuel caches laid down throughout that vast territory. Gasoline, oil and refuelling equipment consisting of a pump, funnel and felt strainer had been deposited during the previous year at strategic, but often poorly identified, locations. Containers varied from 45-gallon drums weighing over 400 pounds to 10-gallon drums, the latter favoured because of their relative

109 C.R. Dunlap, Lecture to the Canadian Aviation Historical Society 3 June 1972 (Willowdale, Ontario: CAHS, 1972).

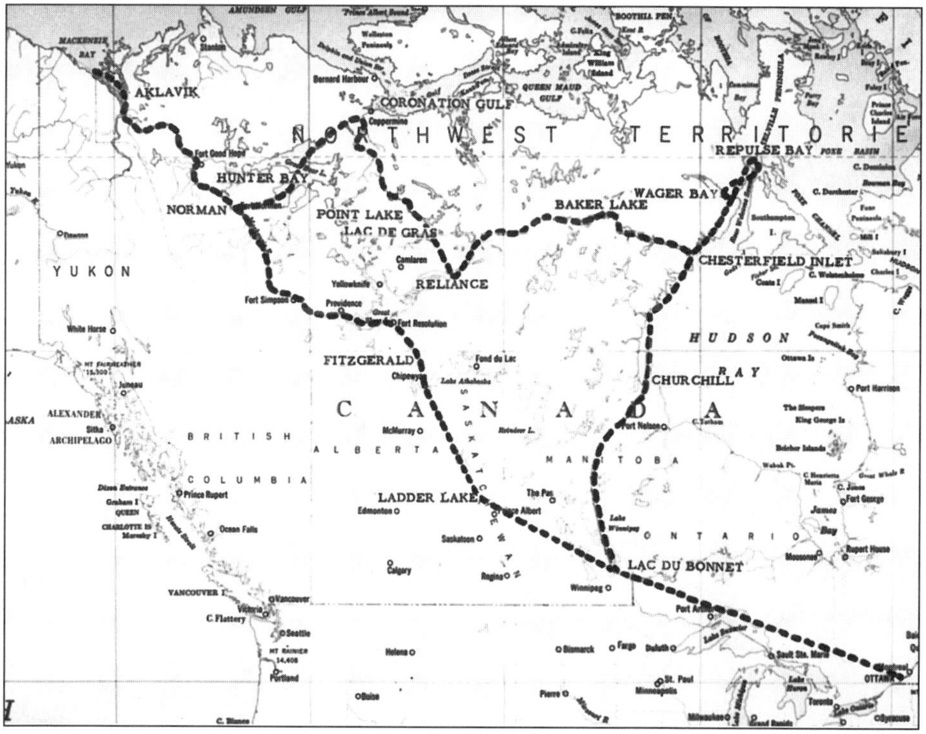

Pioneering flight to the Yukon Territory, Northwest Territories and Hudson Bay by the RCAF, July–October 1930.
DNDPU RE-16267

ease of handling. Where possible, a portable, lockable metal shed was brought in; otherwise, temporary shelters were built from local materials. Finding the cache was the first problem to be solved, as nature soon camouflaged the site. The top layer of drums then had to be lifted down to gain access to the lower level and the contents of each container measured and tested for any water content. Finding water meant a filtering operation and topping up the container.

Via Lac du Bonnet, Manitoba, and Ladder Lake, Saskatchewan, they reached Fitzgerald on the Slave River on 11 July, then followed the Mackenzie River from Fort Providence to Aklavik, arriving on 19 July. Persistent fog foiled their attempt to reach Herschel Island, and they retreated to Aklavik on 21 July. Several times it was necessary to set down on lakes to wait for an improvement in the weather. Returning up the Mackenzie to Fort Norman, they turned eastward to Reliance on Great Slave Lake. They took photographs of the Lac de Gras area, northwest of Reliance and beyond the tree limit, before the weather closed in. A leg of 600 miles northeast through the Barrens brought them to Baker Lake, although the Vedette ran out of fuel en route and the Fairchild had to fetch an emergency supply from Beverley Lake. Then to Chesterfield Inlet on Hudson Bay and north through Roes Welcome Sound to Wager Inlet and Repulse Bay, the limit of the inspection tour. Back to Chesterfield Inlet on 12 September and then southward along the western shore of Hudson Bay bringing them to Churchill on 21 September. Continuing southward through Manitoba to Lac du Bonnet and then east through Ontario, they touched down at Sudbury on 30 September. There, they were able to shave and make themselves presentable for the welcome at Rockcliffe

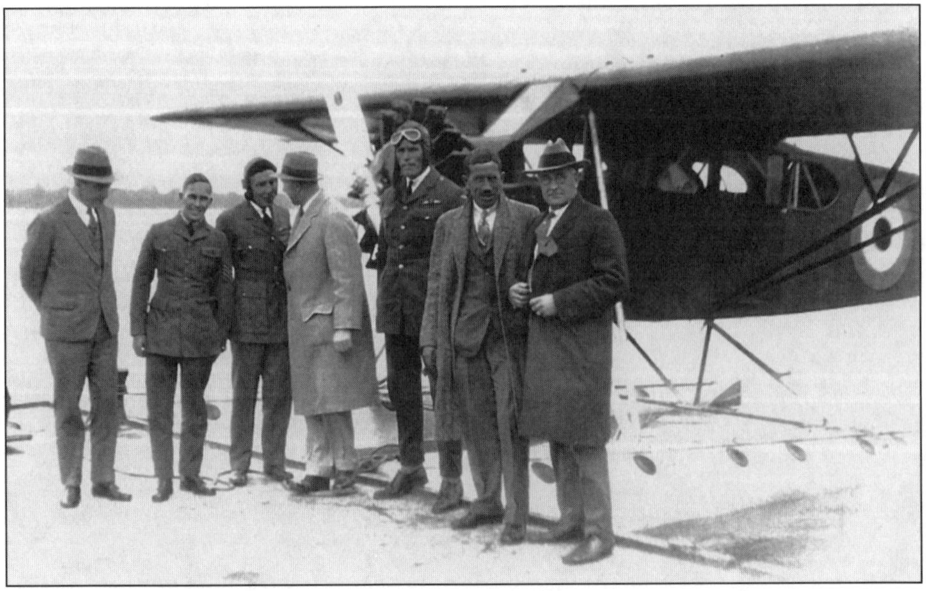

G/C G.O Johnson (left), W/C A.E. Godfrey (centre) and A.M. Narraway (right), controller of surveys, welcome (left to right) A/Sgt. S.C. Dearaway, F/L F.J. Mawdesley, F/Sgt. H.J. Winny and C.S. Macdonald, DLS, on their return to Rockcliffe from their epic northern expedition, 1 October 1930. Mawdesley's Fairchild 71B G-CYVX is in the background. The lives of the four aircrew are profiled in Part II. *RODERICK W. MACDONALD*

on 1 October. The detachment flew approximately 13,000 miles and tested the reliability of equipment and the viability of detached, northern operations. They took 3,100 photographs, most of which were used to produce strip maps of the travelled waterways, and to identify potential bases for future detached operations throughout the territory. Strip maps usually followed a waterway and required the surveyor to be put ashore at the beginning and end of each strip in order to set up a triangulation system by star or sun observations to establish the location of the photographs.

In the earlier days of the PDs, airmen were usually expected to cook for the detachments. As it happened, more civilian cooks were hired in 1930; perhaps as a comment on their culinary abilities—or lack of same. The employment of civilians, however, invariably created problems and this was no exception. R.P. Brown, financial superintendent, wrote a convoluted letter to G/C J.L. Gordon, DCGAO, on 12 July 1930 noting the employment of cooks by No.6 and No.10 PDs at a monthly salary of $100 plus $30 subsistence allowance. He suggested that, as the cook was living as part of the detachment, which had a subsistence allowance per man of approximately $60 per month, the cook should reimburse the unit at a rate of $30/month. In the circumstances, it is tempting to call the accountant a "bean counter," but he did recognize the merits of the detachment living as a team and suggested that, in future, cooks be hired at a stated salary per month plus "subsistence at the public expense." Gordon tersely complied by directing that cooks received subsistence allowance in accordance with conditions that applied to airmen under RCAF P. & A. regulations.[110]

The accounts department was always nervous about young officers succumbing to temptation or inexperience and misusing their financial advances while on detachment out of reach of more mature authority. On 30 November 1930, before many PDs had wound up their summer program, an accounting was sought of the following outstanding amounts:

No. 1 PD	F/L L.H. Harris	$200
No. 2 PD	F/O H.M. Kennedy	200
No. 3 PD	F/O M.K. McGregor	200
No. 4 PD	F/O E.A. Copp	123.35
No. 5 PD	F/L B.G. Carr-Harris	200
No. 6 PD	F/O R. Sunnucks	200
No. 7 PD	F/L W.M. Emery	200
No. 8 PD	F/O F.M. Carter	200
No. 9 PD	F/L C.R. Slemon	200
No. 9 PD	F/O L.H. Weedon	200
No.10 PD	F/O B.F. Johnson	200
No.11 PD	F/L J.E. Jellison	200

110 NAC RG24 Vol. 4889.

"Accounts" kept a tight rein, evidently not relaxing even during the Christmas holiday. A reminder went out on 29 December to Johnson ($134.00), Sunnucks ($141.95), McGregor ($200.00), Kennedy ($200.00) and Weedon ($36.87). All paid promptly, except for Johnson who was reminded again on 12 January 1931 that he still owed $229.81, evidently after a further advance. He was still delinquent on 13 February, but was excused because he was "in hospital."[111]

Despite the onset of the Depression, the aircraft manufacturing sector had a good year. The RCAF accepted the first of a new type, the twin-engine Canadian Vickers Vancouver, and five more were built in 1930, plus 11 Vedettes. The RCAF was so impressed with a new Vedette variant, the Mk 6, with a new-design metal "Alclad" hull and an enclosed cockpit for the pilot and navigator, that plans were made to convert all the existing wooden-hulled "boats." However, the modification added weight and the pilots were not impressed, so only one conversion was, in fact, competed. De Havilland delivered 53 light aircraft. Curtiss-Reid built 22 Ramblers. Ottawa Car manufactured 22 Avro Avians. Fairchild produced four 71Bs (the "B" designating Canadian manufacture of the US design) and redesigned the FC2 with a larger cabin capacity and more powerful engine, designated the Model 51. Boeing built four C-204 Thunderbird flying boats, and five 40H-4 mail planes at the Hoffar-Beeching Shipyards of Vancouver. Saynor and Bell in Montreal designed and built the single-seat Canadian Cub. US aviation entrepreneur Reuben H. Fleet incorporated Fleet Aircraft of Canada and built a new airfield and manufacturing plant at Fort Erie, Ontario. Floats, skis and landing gears were also manufactured in Canada and the engine sector moved from assembly into manufacture. Despite all these achievements, by far the most visible aviation event in 1930 was the arrival of the British airship R100 at St. Hubert, near Montreal, on 1 August.

111 Ibid.

1931-1932

AVIATION WORLDWIDE progressed in 1931, despite the Depression. Canadian public interest was stirred with many transatlantic and cross-Canada flights, including Wiley Post and Harold Gatty in their Lockheed Vega, *Winnie May*, en route to a nine-day round-the-world record. The Trans-Canada Air Pageant toured Eastern Canada during the summer. Nevertheless, the aviation climate in Canada was generally gloomy, although a few rays of sunshine broke through the overcast. Canadian government support of airmail development virtually disappeared but, of immense long-term benefit to aviation, the National Research Council wind tunnel was completed at Ottawa. The only significant domestic increase in flying occurred in the north, where mineral discoveries provided the incentive. The sale of eight surviving RCAF Vedette IIs and one Vedette V to provincial governments for one dollar each, on condition that released RCAF personnel be hired to fly them, proved a sign of the times.

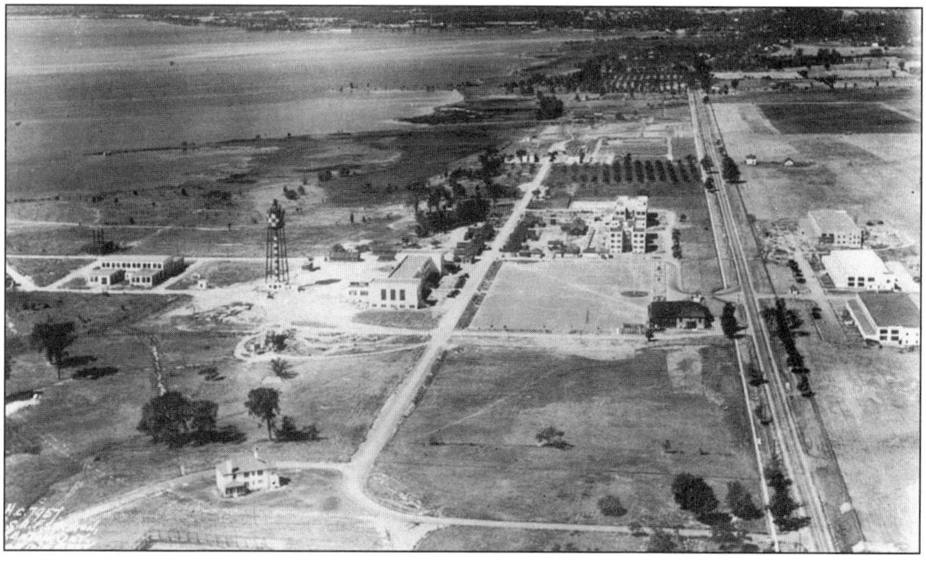

The new base at Trenton was activated in June 1931 as High River entered a "care and maintenance" status.

RCAF Trenton on the shore of Lake Ontario, looking west, June 1936.
DNDPU HC-7957

Signalling a turn in the tide of regular increases since 1922, funding for the air service was reduced by $2.1 million to $5.3 million for 1931–32. The RCAF evidently took the need for economy seriously because $1.2 million of this reduced amount remained unspent at the end of the year. A reduction in the total permanent force to 878 from 906 during the year was not very significant compared with placing the Reserves on inactive status, "except for such special courses as may be authorized."[112]

The provisional list of posting to PDs, dated 10 April 1931, shows that junior officers, many with non-permanent commissions, and NCO pilots were expected to undertake the duties of more senior personnel. A similar downgrading for other ranks was indicated, and the PDs were expected to prepare their own meals:

Establishment/Classification			**Appointee**
No.1 PD – Two Fairchild FC-2W			
F/L	Officer Commanding		F/O (T/F/L) L.H. Phinney (NP)
	F/O	Pilot	F/O C.R. Dunlap
	Sgt.	Fitter A.E.	Sgt. A.H. Warner
	Cpl.	Fitter A.E.	LAC G.R. Lortie
	Cpl.	Carpenter A.R.	Cpl. M. Squire
	Cpl. or Civ.	Cook	—
No.2 PD – Two Fairchild 71			
F/L	Officer Commanding		F/O H.M Kennedy (NP)
	F/O	Pilot	F/Sgt. R.F. Gibb
	Sgt.	Fitter A.E.	Cpl. W.C. Attwood
	Cpl.	Fitter A.E.	AC1 J.A. Odell
	Cpl.	Carpenter A.R.	LAC M.A. Hardy
	Cpl. or Civ.	Cook	—
No.3 PD – Two Vickers Vedette			
F/L	Officer Commanding		F/O H.C. Foley (NP)
	F/O	Pilot	Sgt. G.E. Cherrington
	Sgt.	Fitter A.E.	LAC J.G. Pytlik
	Cpl.	Fitter A.E.	LAC H.E. Barry
	Cpl.	Carpenter A.R.	A/Cpl. L.G. Saunders
	Cpl. or Civ.	Cook	—
No.4 PD – Two Vickers Vedette			
F/L	Officer Commanding		F/O E.A. Copp (NP)
	F/O	Pilot	F/O E. Hickson (NP)
	Sgt.	Fitter A.E.	Sgt. E.B. Turland
	Cpl.	Fitter A.E.	A/Cpl. J.C. Mirabelli

112 F.H. Hitchins, *Air Board, Canadian Air Force, Royal Canadian Air Force*, p. 248.

Establishment/Classification			Appointee
	Cpl.	Carpenter A.R.	AC1 J.R. Fraser
	Cpl. or Civ.	Cook	—
No.5 PD – Two Vickers Vedette			
F/L	Officer Commanding		F/O M. Costello
	F/O	Pilot	Sgt. G.V. Miscampbell
	Sgt.	Fitter A.E.	Cpl. G.F Walker
	Cpl.	Fitter A.E.	LAC H.B. Baxter
	Cpl.	Carpenter A.R.	LAC R.F.E. Kempster
	Cpl. or Civ.	Cook	—
No.6 PD – Two Fairchild 71			
F/L	Officer Commanding		F/O S.R. Sunnucks (NP)
	F/O	Pilot	A/Sgt. F.J. Ewart
	Sgt.	Fitter A.E.	A/Sgt. L. Poulin
	Cpl.	Fitter A.E.	AC1 L.A. Rhodes
	Cpl.	Carpenter A.E.	Cpl. A. Malcom
	Cpl. or Civ.	Cook	—
No.7 PD – Two Bellanca Pacemaker			
F/L	Officer Commanding		F/O (T/F/L) W.M. Emery (NP)
	F/O	Pilot	F/O R.M. Wynd (NP)
	Sgt.	Fitter A.E.	Cpl. J.E. McLaughlin
	Cpl.	Fitter A.E.	AC1 J.J. McNamara
	Cpl.	Carpenter A.R.	Cpl. T.N. Lindop
	Cpl. or Civ.	Cook	—
No.8 PD – Two Fairchild 71			
F/L	Officer Commanding		F/O F.M. Carter DFC (NP)
	F/O	Pilot	F/O G.C. Upson (NP)
	Sgt.	Fitter A.E.	Sgt. W.T. Winship
	Cpl.	Fitter A.E.	AC1 R. Gauthier
	Cpl.	Carpenter A.R.	LAC A.G Morison
	Cpl. or Civ.	Cook	—
No.9 PD – Two Fairchild 71			
F/L	Officer Commanding		F/O L.M. Weedon (NP)
	F/O	Pilot	F/Sgt. E.C. Tennant
	Sgt.	Fitter A.E.	A/Sgt. B.F. Pritchard
	Cpl.	Fitter A.E.	A/Cpl. L.A. Parmenter
	Cpl.	Carpenter A.R.	Cpl. A.R. White
	Cpl. or Civ.	Cook	—

Establishment/Classification			Appointee
No.10 PD – Two Fairchild 71			
F/L	Officer Commanding		F/L A. Carter MM
	F/O	Pilot	F/O D.F. MacDonald
	Sgt.	Carpenter A.R.	Sgt. A.J. Bradford
	Cpl.	Fitter A.E.	LAC J.A. Wiseman
	Cpl.	Fitter A.E.	AC1 K.N. Nelson
	Cpl. or Civ.	Cook	—
No.11 PD – Two Fairchild 71			
F/L	Officer Commanding		F/O (T/F/L) J.E. Jellison (NP)
	F/O	Pilot	F/O R.J. White (NP)
	Sgt.	Fitter A.E.	A/Sgt. S.C. Dearaway
	Cpl.	Fitter A.E.	LAC G.W. McCrea
	Cpl.	Carpenter A.R.	AC1 J.H.R. Beliveau
	Cpl. or Civ.	Cook	—

Although efforts were made to maintain the same crews for the season, circumstances dictated several changes prior to departure. Among them were F/O E. Hickson who replaced F/O A.H. Grace, pilot of No.4 PD, because "his abilities are still to be tested." Also in No.4, Cpl. H.D. Burley, carpenter, replaced J.R. Fraser because "Burley has now settled his family worries." Fraser replaced Cpl. White in No.9 PD, and F/O L.M. Weedon commanded No.2 PD.[113]

During 1931, some detachment designations were changed to General Purpose Detachment (GPD), reflecting a revised operational concept. Whereas the Photographic Detachments (PDs) of the 1920s consisted of two photographic aircraft, usually flying boats, the GPDs usually had one photo aircraft and one dedicated to supply functions. The GPDs were more flexible and self-sufficient and could be assigned tasks other than photography.

The responsibility for natural resources on the prairies was transferred to the provinces in 1931, bringing to an end the extensive forest patrols conducted by the federal air service since 1920. As a result, High River was placed on a "care and maintenance" status, and the substation at Norway House and four forestry bases were closed. In addition to the main task of photography for the Topographical Survey Branch, and in accordance with the ongoing policy of making aircraft useful to as many government agencies as possible, work was undertaken for the Dominion Water Power and Hydrometric Service, Northwest Territories, and National Parks and Forest Services Branches of the Interior Department. Other organizations that made use of the service were the Departments of Mines, Public Works, Marine and National Defence, and the Government of Nova Scotia.

Stores and accounting bases in 1931 remained unchanged: Vancouver for No.1 PD; Winnipeg for PDs 2, 3, 4, 5 and 9; Ottawa for 6, 7, 8, 10 and 11.

113 NAC RG24 Vol. 4890.

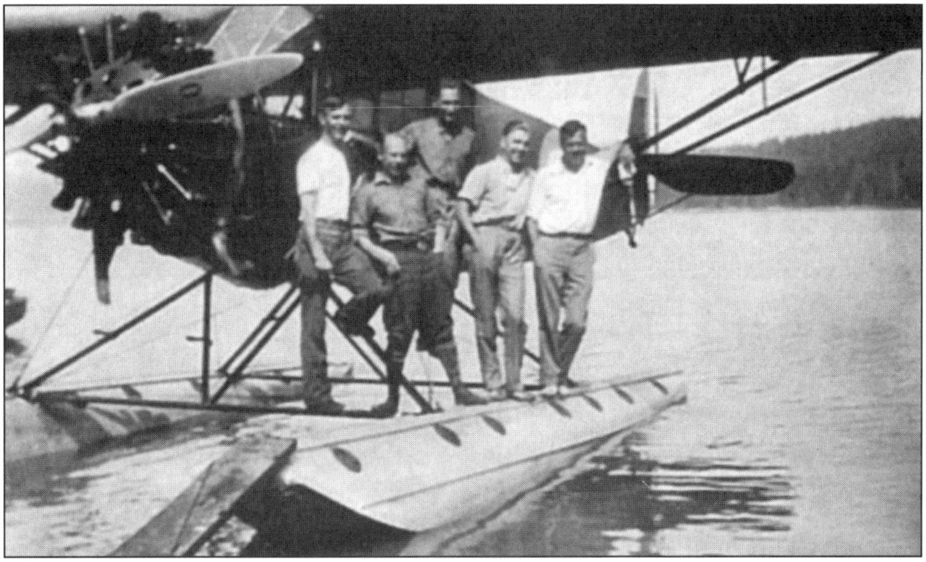

No.1 PD returned to British Columbia, photographing in the Quesnel area, Garibaldi Park and northern Vancouver Island for the Geological Survey Branch. Here, "the gang" pose on the pontoon of Fairchild 'XQ. (L to R) Cpl. M. Squire, LAC R.G. Lortie, Sgt. A.H. Warner, F/O C.R. Dunlap and F/L L.H. Phinney. The latter flew in World War I, earned a law degree and then joined the RCAF until the early 1930s, when he started a law practice in the Yukon and Northwest Territories. Lortie retired as a squadron leader and Dunlap as an air marshal. *C.R. DUNLAP*

Areas covered, from east to west coasts, were more extensive than the previous year, with photographs taken covering more than 76,000 square miles in 4,821 flying hours, an increase of more than 1,300 hours.

In Manitoba, 4,000 square miles were photographed around Nelson House, and in Saskatchewan, 9,200 square miles at Stony Rapids and Reindeer Lake. The largest 1931 effort, however, was in the northern hinterland where seven detachments were engaged at various times in photographing the vast territory from James Bay northwest to Great Bear Lake, for which existing maps showed only blank spaces. Photographs were taken along the east coast of Hudson Bay and James Bay, and the Belcher Islands, and support was provided to ground parties in the area. Photographs were taken of 14,000 square miles in the Northwest Territories from Great Slave Lake to Great Bear Lake and along the Yellowknife and Thelon Rivers. These were essential for

F/L R.E. McBurney, in Bellanca Pacemaker 'VG, and F/S H.J. Winny, in Vedette 'WS with navigator E.S. Fry, DLS, and photographer/fitter LAC Joseph Fortey comprised a typical GPD operating in 1931. A civilian cook was part of the initial establishment, but he was withdrawn after three weeks. Photographs were taken from the Vedette while the Pacemaker was kept busy transporting film and supplies.

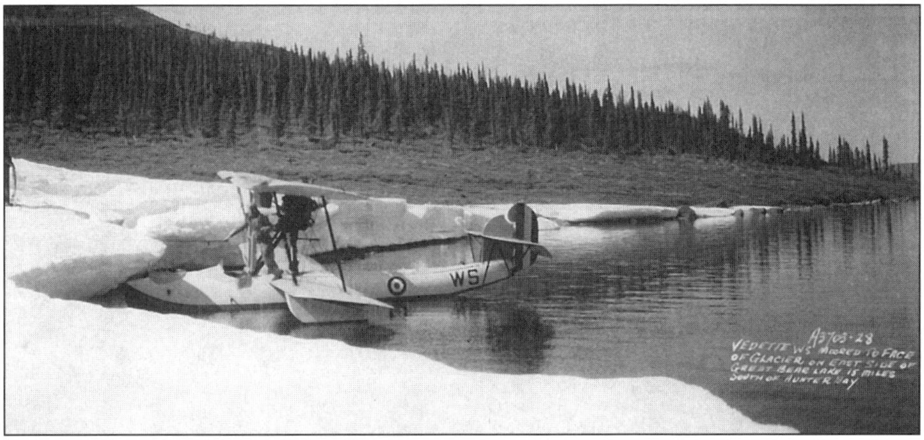

No.2 GPD lived on canned goods supplemented by fish they caught and a caribou that Winny shot. Here, they take a break moored to a glacier 20 miles south of Port Radium on the east side of Great Bear Lake, 14 July 1931. *NAPL A3703-28/J. FORTEY*

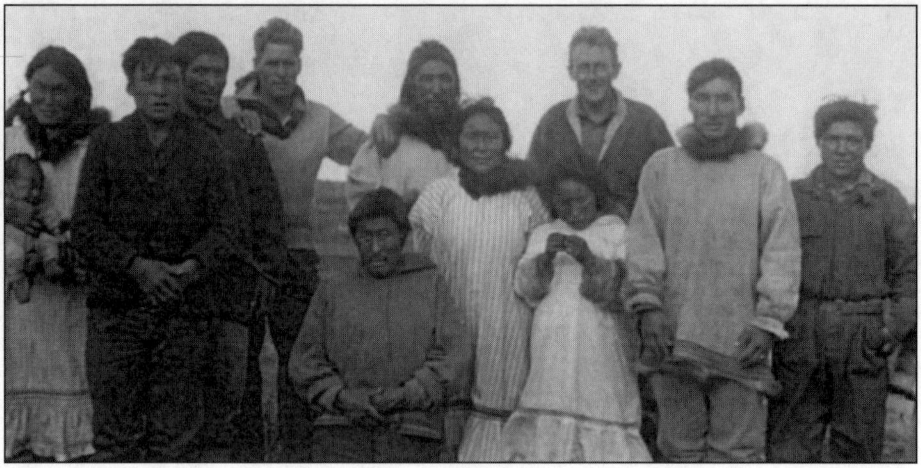

Northern fliers developed a great liking for the native people who were invariably friendly, cheerful and co-operative. Written on the back of Harry Winny's photograph is, "Some of our Eskimo friends on the Arctic coast." Winny is fourth from the left, Eric Fry, DLS, is third from the right. Probably at Burnside Harbour, NWT, 1931. *DIANE MACDONALD*

the plotting of air route maps for which there was heavy demand due to the discovery of radium and silver near Great Bear Lake.

From a base at Buffalo Air Park near Fitzgerald on the NWT–Alberta boundary, No.2 GPD completed six "strip maps" of the major air transportation routes in the Northwest Territories. The area covered stretched from west of Great Bear Lake to Baker Lake and north to Bathurst Inlet on the Arctic Ocean. The route generally followed Mawdesley's route the previous year, utilizing the same fuel caches, although one at the junction of Thelon and Hanbury Rivers was never located. Ralph McBurney recalled that another, at Wholdaia Lake, had been mysteriously moved from the beach about 100 yards inland and covered with moss.

Nearly 5,000 square miles were photographed along the Peace and Athabaska Rivers in northern Alberta. Large areas totalling nearly 6,000 square miles were covered in northwestern Ontario around Nakina, Lake Nipigon and Port Arthur. Photography of other extensive areas of the

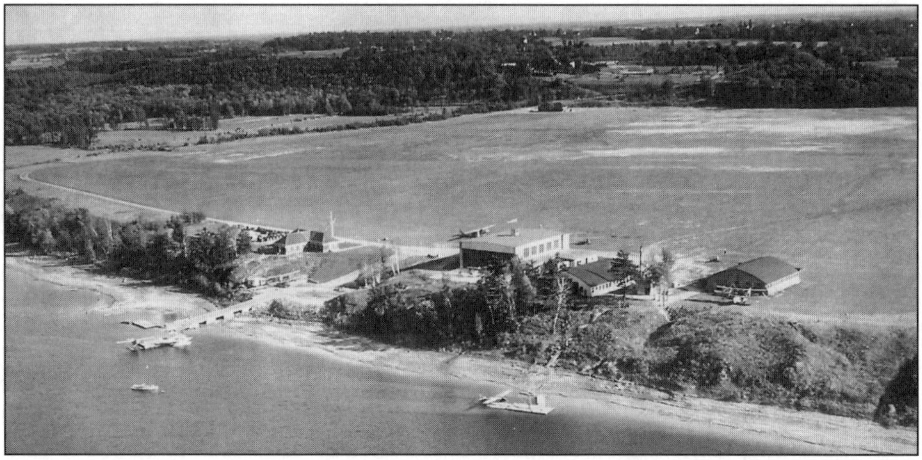

Ottawa Air Station, Rockcliffe, 2 October 1931. Administration building on the far side of the slipway, hangars and machine shop to the right. *NAPL A4369-4*

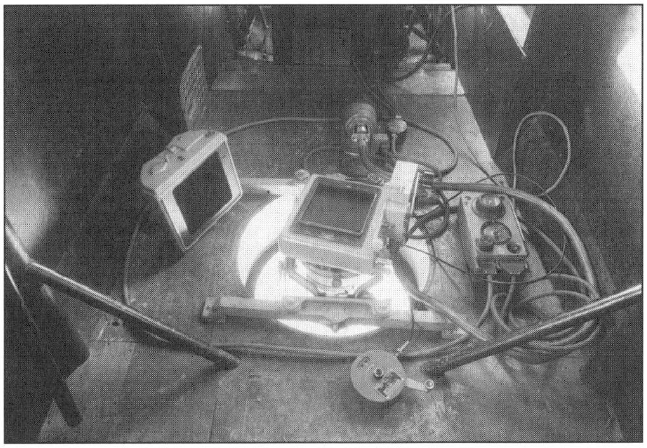

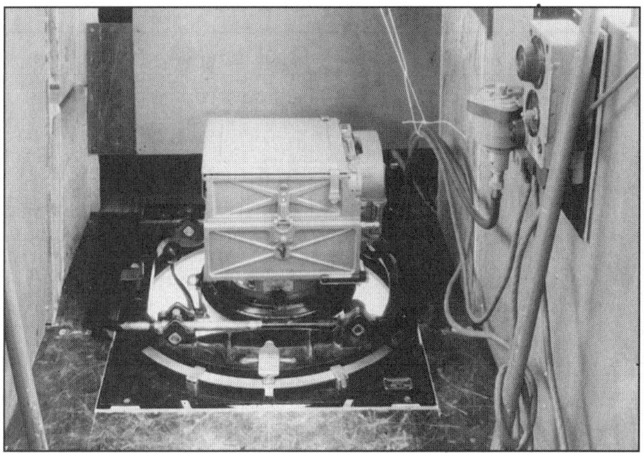

Experiments continued at Rockcliffe. Different aerial cameras were evaluated, but the Fairchild models remained in common use.

Left: RAF Service Model F.24, 16 November 1931.
NAC PA62943

Right: British Williamson, Rockcliffe 7 July 1932.
NAC PA63033

Manitoba–Ontario boundary, Riding Mountain Park and Banff in the south to Rankin Inlet, Great Slave Lake, Liard River, Great Bear Lake and Coppermine River in the Northwest Territories required 1,231 photographic hours.

In Central Canada, five detachments photographed around Coral Rapids and Orillia in Ontario, and Oskelaneo and Ste. Agathe in Quebec. Support was also provided for ground triangulation parties in the Oskelaneo area. It took less than four hours to complete vertical photography for a mosaic of the Ottawa-Hull area.

In the maritime provinces, two detachments photographed around Campbelltown, Newcastle, Chipman, Shediac, Sherbrooke, Bridgewater and Halifax, and in the Annapolis Valley. A stretch of the north shore of the St. Lawrence River from Anticosti Island to the Strait of Belle Island, and the shoreline and islands of the Saint John River from Fredericton to Saint John were photographed for the Hydrographic Service of the Department of Marine. A variety of other tasks included photographing the Cape Breton coal fields, the Halifax–Truro highway and Fort Louisburg.

Improved equipment for cold weather operations made photography possible during the winter of 1931–32. One operation to photograph 2,750 square miles near Moose and Ladder Lakes, Saskatchewan, for the Forest Service was completed in 108 flying hours. Another operation for the Geodetic Survey Branch near Great Bear Lake took 139 hours.

A 17 September telegram from Ottawa to F/L W.M. Emery, CO of No.7 PD, then at Coral Rapids on the Abitibi River in northern Ontario, gives an insight into the personal tribulations faced by the detachment personnel, absent from their families for extended periods. Preceding directions for the next photographic assignment was the brusque message, "Congratulate Wynd birth daughter yesterday all well stop."[114] F/O R.M. Wynd was second pilot of No.7 PD.

The search for economies continued. A review of the previous year's expenditures identified $878 spent by the PDs on boat rentals and motivated S/L

114 Ibid.

Tackaberry to ask if it would be better to purchase boats, and sell them at the end of the season. Ottawa decided to pass the buck to the OIC PDs who were required to "exercise the most economical method: rental, local purchase or supply from base." On 25 November 1931, F/L Jellison, OIC No.11 PD working on the east coast, recommended the purchase of a piece of canvas, 6' x 10', for suspension under engines being serviced: "Loss of tools and engine parts by airmen servicing engines in seaplane aircraft has been a continual source of trouble and expense." He quoted his solution, apparently at his own cost, with a $5 piece of canvas lashed to two wooden 2x4-inch supports.[115]

Winters were devoted to aircraft maintenance, but it was also the silly season for catching up with correspondence and trying to smooth the bureaucratic path for the following year. F/L W.M. Emery, OIC No.7 PD, unburdened himself on 19 November 1931, regarding the perpetual irritant of travel claims. He explained at some length the difficulty of obtaining receipts in sparsely populated territory and when a move had to be made in haste. One scenario illustrated his problem:

> In the case where a Detachment is messing itself, and every article must be itemized when claimed for, in order that receipts may be checked and the exact cost of each meal determined, the detail involved becomes a burden to the Detachment Commander, who must handle this matter himself in order to ensure that the mess is properly and economically run. The matter becomes more complicated when some of the personnel are detailed for part time duty away from the mess, or when the whole Detachment moves its temporary base for a short period.[116]

Emery commented on the inherent inefficiencies of requiring receipts for every expenditure and cited an example of his detachment having to stay overnight waiting for the return of their host in order to obtain accommodation receipts; having the weather close in and necessitating another overnight stay—and having to obtain more receipts. He pointed out that this work fell on the shoulders of the OIC, also the pilot, who was fully occupied, especially when a move was taking place. At the close of the 1931 season, Emery said it took him two weeks of "concentrated effort" to prepare his claim, accounting for 205 separate movements, which, when typed, comprised a volume of 67 pages. "This whole claim could have been completed in a day, had we been able to claim on a flat rate basis." This procedure he discussed with all PD commanders who "heartily endorsed" his recommendation that a flat rate be instituted.

Headquarters rejected Emery's recommendation. He, and others, soon engaged in a running battle with the administrators of Personnel and Administrative Regulations. There is even a recommendation on file that all receipts should be in the handwriting of the originator, on which an unknown author has pencilled, "How is the Auditor to know whose handwriting it is?"

115 Ibid.
116 Ibid.

A 17 June 1931 memo from W/C G.O. Johnson, A/DCGAO, to Winnipeg, lit a fuse with his suggestion that OIC PDs were claiming the maximum permissible allowance, covering their claims with receipts and "pocketing the difference." Apparently there was unrecorded correspondence on the falsification of travel documents in the interim, but S/L N.R. Anderson, CO Winnipeg, shoots in defence of the PDs on 13 January 1932, ". . . it must be presumed that the average individual met with in life is of an honest nature. . . . It is not considered that the Photographic Detachment personnel would go so far as to commit a forgery."[117]

F/O R. Sunnocks, OIC No.6 PD, then at Oskelanéo in northern Quebec, had a more serious administrative problem to resolve on 13 June when approached by the Quebec Fire Protection Service to help find two rangers missing for two days and believed drowned. The local telegraph office was closed and it was not possible to consult headquarters, so Sunnocks complied under the authority given to detachment commanders to deal with exceptional circumstances. He sent A/Sgt. F.J. Ewart in his Fairchild 71 'VX on two trips carrying fire inspectors to Obiduan village, 40 miles distant, to organize the search. Covering the total distance of 160 miles took two hours and ten minutes for which the normal charge would have been $240. The director of Civil Government Air Operations supported Sunnucks's request that no charge should be made, but insisted on him raising a work order with supporting letters from the fire service. He warned Sunnucks that, in future, the spirit and intention of Orders and Instructions must be adhered to, and persons requesting such flights advised that payment must be made at the rates laid down. Sunnucks had hardly been in a position to refuse help in the search for missing rangers, whose bodies were not found until 18 June, and he probably did not appreciate the reprimand from W/C G.O. Johnson, A/DCGAO: "It is understood that at the time the flights were undertaken it was known that the fire rangers had been drowned and the visits of the fire inspectors were only in connection with dragging operations for the bodies. If such were the case, there was no urgency for the saving of life or relief of suffering, such as contemplated by Order and Instruction 30,070 (150/11/1) Para.2 (a) (iii) and therefore, there is no justification for your recommendation that no charge be levied."[118]

From 19 to 26 June 1931, S/L S.L. Stevenson conducted an inspection of PDs operating in the maritime provinces with a particular reference to accommodation, living conditions and costs. He was also required to select a suitable operating base for No.11 PD. He piloted his aircraft to Montreal, Saint John, Moncton, Shediac, Moncton, Enfield, Halifax, Chester, Digby, Saint John, Montreal and back to Ottawa in seven days. His detailed report gives a picture of the accommodation problems faced by the PDs and GPDs.

At Millidgeville, near Saint John, Mrs. White's boarding house offered the only accommodation. Canadian Airway pilots stayed there, at $1 per night

117 NAC RG24 Vol. 4889.
118 NAC RG24 Vol. 4916.

and 50 cents per meal, with reductions for prolonged stays. The next closest suitable accommodation was in Saint John where hotels were $3 to $5 per night. At Shediac, hotel rates were similar and boarding houses charged $8 to $10 per week.

Stevenson found no suitable accommodation at Grand Lake: "Personnel would be required to live under canvas." Government buildings at Dartmouth Air Station still contained stoves, beds, etc., but the services of a cook would be required. At Williams Lake, accommodation was available in private houses at $8–12 per week, but Chester, Nova Scotia, "a popular summer resort," was more expensive. Lovett House, the cheaper of the two hotels in Chester, where F/L Jellison, OIC No.11 PD, was staying, charged $5 per night. The difference in allowances for officers and airmen meant that his non-commissioned crewmen had to stay at a boarding house costing between $15 and $20 per week.[119]

Stevenson considered Chester to be a good base as it offered excellent support facilities, sheltered moorings, and good takeoff and landing areas. On the down side, however, living expenses were excessive, salt water presented corrosion problems, and the harbour was congested with small boats. With the completion of the Bridgewater and Halifax sheets, Stevenson directed Jellison to move to a new base, east of a north-south line through Halifax, to reduce costs and be nearer the area to be photographed. Charlotte Lake was considered the best location, but close to the sea and the perils of salt water, and lacking railway, telegraphic or telephone communication. William Lake, about 15 miles north of Halifax, had railway, telephone and telegraphic connections, so was selected as the new base, with Grand Lake, about 10 miles farther north, as an alternate.

Evidently, consideration was being given in 1931 to reactivating Dartmouth Air Station because Stevenson inspected the facilities. He found the hangar and slipway in poor repair, but the dock, workshops, mess buildings and barracks in reasonable shape apart from leaking roofs. His rough estimate of bringing the facilities to a good state of repair totalled $2,400, $2,000 of which was for painting.

F/L Jellison displayed proper concern for economy with a recommendation to DCGAO on 25 November 1931 to discontinue the practice of shipping barrack room tables more than 200 miles. He calculated the cost of one-way express shipment of two tables from Ottawa to Waverley, Nova Scotia, at $8. Instead, he purchased local materials at a cost of $1.31 to build a table 20' long by 3'6" wide for the purpose of laying out field copies of prints for indexing and error tabulation.

Eastern Lines, a subsidiary of Canadian Airways Limited, operated airmail, freight and passenger services in Eastern Canada, and also took oblique photographs of 171 square miles and verticals of 3,432 square miles along the St. Lawrence River. A second subsidiary, Western Lines, operated

119 Ibid.

similar services in the west and absorbed Commercial Airways' routes along the Mackenzie River. The three companies used 72 aircraft of many different types.

The Ontario Provincial Air Service continued aerial photography, logging 200 hours in 1931. The Compagnie Aerienne Franco-Canadienne vertically photographed 3,400 square miles for the Quebec Department of Land and Forests.

The manufacturing industry continued in a survival mode with fewer orders from the RCAF although Canadian Vickers did provide six Bellanca Pacemakers for photographic work and completed the sole metal-hull conversion of a Vedette. De Havilland built 24 Moth variants, six of them adapted for "blind" flying for instrument training. Curtiss-Reid introduced the Courier, a mail plane version of the Rambler. Ottawa Car delivered six Avro Avians. Boeing had a slow time, but initiated the design of a new all-metal flying boat, the A-213 Totem. Fleet Aircraft delivered 20 Fleet trainers to the RCAF. Fairchild was relatively busy with five new 71B aircraft, assembling the largest aircraft then in Canada, a Junkers JU-52, and marketing the Pitcairn PCA-2 autogyro for which a dealership had been negotiated. Aero Engine of Canada and Canadian Pratt and Whitney continued engine assembly, repair and overhaul, as did Armstrong Siddeley, which also built one new Siskin.

1932-1933

EFFECTS OF The Depression were evident in 1932, remembered by the RCAF as the year of "The Big Cut." The air service bore the brunt of the reduction in military funding, receiving only $1.7 million, a reduction of $3.6 million from the previous year. Flying time was severely restricted, and 78 officers, 100 airmen and 110 civilians were let go. The strength of the RCAF reached a low point of 694 officers and men on 31 March 1933. The trickle of transfers to the RAF, more aware of what the future held, began rising to a flood within a few years. No new aircraft or engines were purchased, even maintenance of the existing fleet was reduced, and training was severely curtailed. One reported positive result of the economies was a reduction in accidents at the seaplane school in Vancouver because only senior, experienced pilots were undertaking flying boat conversion. The air force went into a holding pattern during 1932, except for a new venture in co-operation with the RCMP to intercept rum-runners, and exploring airmail routes.

On 15 April 1932, S/L Tackaberry wrote with obvious regret to Ottawa and Winnipeg Air Stations on behalf of the A/DCGO: "As it is expected that few, if any, Photographic Detachments will be operating as such during the coming season, it is considered advisable to have the equipment of the Detachments returned to stores where this has not already been done."[120] Maintaining some photographic momentum, however, GPDs completed 377 hours of photography in the first year of austerity, mostly in the Vancouver area for the Marine, Mines and Interior departments. Aircraft from Ottawa also undertook small assignments in Ontario, Quebec, New Brunswick and Nova Scotia. On cloudy days, John Carroll, DLS, kept the crews busy with an experimental program of landing on lakes of known altitude, setting his barometer and proceeding to establish the altitude of nearby lakes. When they were too small for landing, the pilot overflew at five feet above the surface while Carroll took his readings. This technique proved valuable in later years to establish elevations and contours in the most inaccessible regions of Canada.

120 NAC RG24 Vol. 4890.

One investment was made in 1932 for potential long-term benefit of the photographic units. S/L Art James was sent to RAF Cranwell to study the Smith's Automatic Pilot's Assister, one of the first autopilots that showed promise of holding an aircraft steady and able to fly a consistent "line," even when haze or smoke obscured the horizon. Sgt. J.D. Hunter followed in his footsteps in 1933 for a more intensive course.

The experience gained during detached photo operations led to other challenging tasks. S/L Roy S. Grandy was in charge of a detachment of two Bellanca Pacemakers, two Vickers Vancouvers and a pair of Fairchild 71s charged with speeding up mail delivery for the Imperial Economic Conference held in Ottawa during the summer of 1932. A RCN launch collected mail from incoming transatlantic liners in the Strait of Belle Island and delivered it to the sheltered waters of Red Bay, Labrador. A Bellanca then started the air relay to Havre St-Pierre on the North Shore, continued by a Vancouver to Rimouski, and a Fairchild to Ottawa via a refuelling stop at St. Hubert. Weather permitting, the delivery was completed in less than 14 1/2 hours. In reverse, mail was delivered to the ships as they departed the Strait of Belle Island. Over the summer, eight of these aircraft-to-ship links were attempted with only partial success, due to inclement weather.

Red Bay, named for its red cliffs, on the rugged south coast of Labrador, one thousand miles east of Montreal, was a lonely spot in 1932. In the latter half of the 16th century, however, it was probably the largest whaling port in the world, where as many as 1,000 Basque sailors from Spain slaughtered thousands of whales. One of their unique *chalupa* rowboats was found in 1978, was restored, and is now the focus of an interpretation centre expected to attract tourism dollars to the village of 250 inhabitants.

Morale remained high in the RCAF, despite all of the restrictions and economies. F/L C.R. Slemon, OIC of a detachment working in the Northwest Territories that summer, commented on the "devotion to duty" shown by his men:

> The personnel of the Detachment successfully overcame the temptation to participate in the mining and prospective activities of the Great Bear Lake area. With sensational finds of silver and other minerals being reported in the near vicinity and high-grade samples common throughout the area, the urge is sometimes great to permit one's interests to be somewhat diverted from Service duties.[121]

Slemon did not intend his comments to be taken seriously and they probably originated around the campfire, when crews passing through the area would bemoan, "We're in the wrong business, guys!" As G/C F.H. Hitchins commented, tongue in cheek, "Greater love hath no airman than this, that he gave up a potential fortune for the sake of the Service."[122]

Before closing down its Vancouver facility due to the market slump, Boeing sold its first, and only, A-213 Totem flying boat to Canadian Airways. The facility was reactivated briefly in 1937 to build Blackburn Sharks for the RCAF.

121 F.H. Hitchins, *Air Board, Canadian Air Force, Royal Canadian Air Force*, op. cit., p. 270.

122 Ibid.

1933-1934

THE RCAF BUDGET was reduced by $53,000 in 1933 to $1.7 million, covering all expenditures. No money was made available for new aircraft although the inventory was rapidly becoming obsolete, exacerbated by reduced use and minimal maintenance. Attrition in the ranks was not replaced, reducing the total permanent strength on 31 March 1934 to 106 officers and 586 airmen. On the credit side, the Non-Permanent Active Air Force was beginning to take shape with 26 officers and eight airmen formed into three squadrons, but with no aircraft. Of great future benefit to Canadian aviation, under the provisions of the National Defence Unemployment Act, clearing and grading commenced on 12 airfield sites across Canada for a trans-Canada air route.

Almost 80 percent of the total 3,491 flying hours in 1933 were carried out from Ottawa where emphasis was on "preventative patrols" over the St. Lawrence and the east coast. Photographic hours increased slightly to 586 during the year—but still only a small fraction of the record 4,821 hours in 1931.

The PD designation was completely discarded in favour of GPD to reflect the more versatile duties expected of a smaller air force. No.1 GPD photographed along the coast of British Columbia, Vancouver Island and the Queen Charlotte Islands for the Department of Marine. No.2, based at Winnipeg, photographed in northwestern Ontario and the Northwest Territories for the Interior and Mines Departments. No.6 from Ottawa was active in northern Ontario and Quebec, while No.7 worked along the Gulf of St. Lawrence and the interior of Quebec, New Brunswick and Nova Scotia.

The fortunes of No.7 GPD for the next four years were typical of the times. It was formed at Ottawa on 4 May 1933 under the command of F/O J.A. MacInnis with Sgt. F.J. Ewart as second pilot and fitter, Cpl. J.E. MacLaughlin and LAC J.A. Wiseman as camera operators and fitters, and LAC A.G. Morrison as rigger. Two Bellanca Pacemakers, 'VJ and 'UX, with Wright J6E engines and floats carried all personnel and equipment, including Fairchild camera systems. Of necessity, spares and emergency equipment were reduced to the minimum.

General transportation duties were undertaken until 31 July when No.7 GPD departed on its photographic mission. The first destination was Rimouski, Quebec, where the aircraft were refuelled. Low fog and rain delayed the aircraft for one day before taking off for Havre St. Pierre on the north shore of the Gulf of St. Lawrence, opposite Anticosti Island, the designated base for Operation Gulf. Continuing bad weather prevented landing at Seven Islands en route to refuel as planned so they flew on to Havre St. Pierre, arriving at noon on 2 August. Fuel not being due for delivery there for another week, the aircraft returned the next day to Seven Islands to refuel at the RCAF cache. The exposed nature of this cache necessitated taxiing for an hour and a half to the shelter of some islands in order to take off with any degree of safety.

Moorings for the aircraft were laid down in a small bay on Esquimo Island, about two miles off Havre St. Pierre. Each aircraft was moored to a 200-pound anchor with 60' of 1 1/4-inch diameter new manila rope rented from the local Hudson's Bay Company store. While sheltered at anchor from the prevailing winds, the aircraft had to be taxied two miles for refuelling at the Hudson's Bay store through high winds, rough water and cross swells, so this was done in relatively calm periods during early morning or late evening. Both aircraft were periodically beached near the Hudson's Bay Post at high water and inspections carried out.

Takeoffs and landings took place in the shelter of Esquimo Island, about 100 miles from the Anticosti site and 50 miles from the North Shore site. Cumulus clouds formed every fine day before noon so photography from 10,000' was limited to periods of two to three hours. Accommodation for all ranks was in a private house a few minutes' walk from the Hudson's Bay store, where a darkroom tent was rented. A motorboat was rented from the village as required. Communication was by the Department of Public Works government telegraph ten-hour service. There was then no road along the North Shore, and the weekly steamship from Quebec City provided the only means of transportation. The detachment's oil and gasoline supply was off-loaded on 6 August, under contract with the Imperial Oil Company.

Problems with the persistently tail-heavy 'VJ could only be corrected by keeping the load well forward. The engine throttle control arm broke on the same aircraft when making a landing on Lake Rimouski, but was soon fixed at a local welding shop. The rocker arm ball bearing race seized on 'VJ at Havre St. Pierre, but a replacement was installed from the spare parts carried.

The Coast Guard Ship *Acadia* anchored in Havre St. Pierre for one day and provided some relief from the GPD's daily routine with an invitation to dine on board. The detachment was well treated by the French-speaking villagers who, it was emphasized, helped in every way possible. The Hudson's Bay people were particularly obliging and charged very nominal rates for services and rentals. The detachment reciprocated by showing scenic photographs of the area, and of the arrival of the Papal Delegate from Rome on the steamship *North Shore*, en route to Quebec City. F/O MacInnis reported that the local people were most interested in the prospects of airmail service as they were cut off from the rest of the world for three months every year, except for the telegraph service.

Vertical photographs were taken for the Hydrographic Service, Department of Marine, of the mainland from Natashquan Point to Walrus Island, and the shoreline of Anticosti Island from Carleton Point east to Heath Point and then west as far as Southwest Point. No.7 GPD completed Operation Gulf on 15 August in 34 hours and 40 minutes' flying time, exposing 11 rolls of film. The two Pacemakers were loaded and departed from Havre St. Pierre on 17 August. The 280-mile transfer to the RCAF base at Shediac on Northumberland Strait, near Moncton, New Brunswick, for Operation Cobequid, was completed the same day.

A permanent buoy was utilized at Shediac by No.7 GPD for one aircraft, but the second buoy—a 700-pound anchor with half-inch steel cable—had to be provided and laid by the visitors. The detachment did have the use of an RCAF motorboat kept at Shediac, and the detachment stationed there was of considerable help. A wharf only 100 yards from the well-sheltered mooring site was convenient for refuelling with oil and gasoline delivered by Imperial Oil Limited. All ranks were accommodated at the Shediac Inn, one-half mile from the wharf, where a storeroom and office were rented for equipment and a darkroom. Communication with headquarters was by Canadian National Telegraph's 12-hour service from Shediac and 24-hour service from Moncton.

All the oil lines were inspected and the oil pump removed and examined without finding the cause for the large amount of oil used by 'VJ on the flight from Havre St. Pierre. Excessive clearance of the scraper rings was evident, however, when the two bottom cylinders were removed, remedied by replacing the lower ring on each cylinder with the rings from the piston carried as a spare. "This lowered the oil consumption considerably." The water rudder control cable, located at the mouth of the exhaust, also had to be repaired after it burned through and broke while taxiing. Additional canvas bumper guards had to be made and installed on the floats of both aircraft for protection when moored at the dock where they were in continual movement due to wave and wind action and the wash from passing boats. The only medical problem noted during the whole season occurred at Shediac, but no details were provided.

Vertical photography from 10,500' of approximately 630 square miles in the Cobequid Mountains, near Truro, Nova Scotia, was completed in 37 flying hours for the Topographical Survey, Department of the Interior. Thirteen rolls of film were exposed during intervals in generally overcast weather some 100 miles from the Shediac base. Operation Cobequid was completed on 31

August and No.7 GPD moved on to Dartmouth on 2 September for Operation Halifax.

The RCAF base at Dartmouth provided sheltered waters and welcome facilities in the way of permanent moorings, a refuelling wharf only 50 yards distant, slipway, hangar, workshop and a fresh water supply for washing down the aircraft. The crews, however, had to travel four miles to find suitable accommodation at the Thorndyke Hotel. A private motor car was rented at six cents per mile, and water transport was by the RCAF motorboat. Royal Canadian Navy 24-hour radio service was used for communication with Ottawa.

Advantage was taken of the Dartmouth facilities and personnel to beach and thoroughly inspect each aircraft in turn. The aircraft were equipped with floats manufactured by different companies—'VJ with Edo X-5280 and 'UX with Short 51-S. The available beaching gear only fitted the Short floats, so planks had to be laid down to give sufficient clearance between the keels and the slipway when beaching 'VJ. The floats and struts of 'VJ were completely cleaned and painted, those of 'UX where necessary. The fabric of both aircraft was opened and the fuselage exposed for corrosion inspection, cleaned and painted. Controls were also checked and the aircraft were both back in service by 21 September. During this maintenance period, one aircraft was kept in flying condition and about 120 square miles in the Halifax area photographed from 10,300' for the Militia Service, Department of National Defence.

The greatly improved reliability of engines had by now removed most of the concerns that had hitherto made it wise for aircraft to fly in pairs. On 22 September 'UX with Sgt. Ewart at the controls and LAC Wiseman doubling as photographer and fitter left Dartmouth for the next assignment at Magog, near Sherbrooke, Quebec. They had to divert to Shediac and spend the night there due to low fog at their destination before completing the journey.

Reports on the Cobequid photographs had been received from Ottawa and F/O MacInnis with camera operator and fitter Cpl. MacLaughlin, and rigger LAC Morrison returned to Shediac in 'VJ on 23 September to fill in the gaps identified in the previous coverage of Cobequid. MacInnis exposed two rolls of film at Cobequid on 1 October in seven hours' flying time and continued directly to home base at Rockcliffe on 4 October.

Sgt. Ewart anchored 'UX in a sheltered bay on Lake Memphrémegog, about two miles from Magog where accommodation was secured and the loan of a private motor car obtained. "Operation Magog," vertical photography from 6,000' of the Canadian shore and outlet of Lake Memphrémegog, and scenic obliques in the vicinity for the Topographical Survey Branch, was completed by 26 September and 'UX returned directly to Rockcliffe.

F/O MacInnis's summary report of No.7 GPD's 1934 season with two Bellanca Pacemakers in Quebec and the Maritime provinces illustrates the typical problems experienced with photographic detachments. Operating from salt water required that all metal parts—controls, cables, pulleys, undercarriage struts and fittings—were kept thoroughly greased. 'VJ had zip fasteners on the underside of the fuselage to facilitate access to these parts, but they were difficult to fasten and the grease apparently deteriorated the fabric to the extent that it failed in the vicinity of the stitching when drawn together by the fastener.

MacInnis considered the general performance of both Wright J6E engines satisfactory, his measure being, "there were no forced landings during the season's operations." He concluded, "the engines functioned satisfactorily, requiring only ordinary maintenance and inspection." The broken throttle control arm and rocker arm ball race previously noted were evidently not considered unusual and were covered by routine defect and failure reports. Inertia starters worked well and care was taken to cover the magnetos with heavy, waterproof canvas after each flight. Engine covers were not, however, used as it was found that they held moisture, causing increased corrosion.

The cylinders and all metal parts of the engines were oiled daily with a spray gun using penetrating oil when available, otherwise a mixture of equal quantities Marvelube engine oil and kerosene. The lower inter-cylinder cowlings were removed to facilitate spraying and regular inspection of lower cylinders, the most vulnerable to corrosion. The exhaust collector ring and stubs were removed every 24 hours of operation, rubbed with emery cloth, cleaned, and painted with a mixture of aluminum powder and varnish. MacInnis noted that this protection stood up very well on the inside of the collector ring, forming a corrosion-resistant barrier. Paint on the hotter outer surface and exhaust stubs, however, soon burned off, but this did not present a problem, as the exposed surfaces could easily be protected with a film of oil.

Floats took a beating from the salt water and rough conditions but, despite some corrosion on the ventral surfaces and around rivet heads, none of the compartments leaked. During refurbishing at Dartmouth, the corrosion was removed and metal parts coated with a mixture of aluminum powder and varnish. Significantly different performance was obtained from the Edo floats on 'VJ and the Short floats on 'UX. When taxiing in rough water or a crosswind, the Edo floats dug into the water more than the Short type, throwing considerably more water in the path of the propeller, resulting in greater erosion of the propeller tips on 'VJ than on 'UX. On takeoff, the Edo floats, however, had superior performance than the Short type, which appeared to skip over the water surface and the aircraft had to be hauled off. In comparison, the Edo floats could be rocked onto the step and would gradually clear the water as speed increased, without any undue pressure on the control column.

The Fairchild cameras were generally reliable and, with regular maintenance, gave little trouble. The most significant problem was keeping sand out of the mechanism on windy beaches such as experienced at Havre St. Pierre. Repairing a loose electrical connection cured an erratic intervalometer on one camera, and the winding handle was accidentally broken off the second camera. Other handles had to be removed from both cameras because the cabin floor restricted movements when large angles of crab were necessary in crosswind conditions.

Although only minimum spares and emergency equipment were carried because of weight limitations, no serious problems resulted. There was need for a second calibrated altimeter and a second spark plug disassembly wrench when the detachment split up at Dartmouth. The persistent tail-heavy condition of 'VJ was investigated by the Test Flight after return to Rockcliffe, but no answer was found to the problem other than to keep the load well forward.

F/O MacInnis had a few suggestions to make following the 1933 operations of No.7 GPD. He recommended a spray gun able to deliver oil to the back of the cylinders, "where corrosion appears to be worst and this part of the engine is almost impossible to reach with the common type of Flit gun." He suggested a spray guard for Edo floats, similar to that used on some Fairchild floats, to reduce spray onto the propeller. He felt provision should be made to clean both sides of the camera window, especially when operating in salt water, as "so much oil and grease is applied to the engine and exposed metal parts to prevent corrosion when the aircraft is moored, that it requires nearly a half-hour flying before the engine stops throwing oil." Other suggestions, evidently learned the hard way, were: the engraving of drift markings on the cabin floor window to aid in following a photo line; an improved engine oil cooler; parachute storage, with rapid release clips, on the cabin roof; inclusion of fresh water in emergency kits; redesign of the electrical control box, as the switch was installed in a cup-shaped receptacle that collected moisture and fed it into the mechanism; and recessing projecting oil plugs on mooring buoys, or locating them on top, to "lessen the possibility of puncturing the metal of the float when the buoy is caught by hand and pulled around to the nose of the float while mooring the aircraft."[123]

F/O MacInnis's report on the activities of No.7 GPD concluded with his summary: "Twenty-six film magazines were exposed. Total flying time for 'VJ was 67.10 hours and 'UX 70.25 hours, with an average gasoline consumption of 13.7 gallons per hour and oil consumption of 0.48 gallons per hour."

As a result of recommendations from pilots attending the photographic course at Ottawa, several aids to accurate pilotage were developed by the Aeronautical Engineering Division. The most useful helped the pilot maintain the aircraft in a stable attitude. A version of the familiar fore-and-aft level, it consisted of a vertical glass tube mounted on the instrument panel and connected by metal tubing to a vessel about four feet behind the panel; for each degree of pitch, the liquid in the glass moved about 1 1/2 inches, thus helping

[123] NAC RG24 Vol. 4917.

the pilot maintain a very accurate fore-and-aft attitude in reasonably calm weather. A similar system was used to indicate lateral attitude.

It was F/Sgt. H.J. Winny's turn in 1931 to fall foul of the perpetual problems resulting from differing, and often inadequate, travel allowances for officers and men. He flew from Vancouver to Prince Rupert with S/L A.B. Shearer. They stayed overnight at the Prince Rupert Hotel and were each charged $4, whereas only $1 was allowed for NCOs. Shearer managed to get one dollar refunded from the hotel to each pilot, and he moved to a cheaper inside room to await the arrival of Commodore W. Hose, director of Naval Service. Winny returned to Vancouver to fight the battle over the excess $2 with the financial superintendent in Ottawa. After the intervention of Shearer, his claim was paid in full because it was proved that no cheaper accommodation was available.

On the commercial front, Robert Bernard Cornelius Noordyn brought his extensive experience with Fokker, Bellanca and Pitcairn to Montreal and opened an office to design the first real bush plane, the Norseman.

1934-1935

THE PURSE STRINGS were loosened a little for 1934–35. Parliament appeared to have an enhanced appreciation of the benefits offered by aviation and increased the appropriation for air services by $565,000 to $2.3 million. Permanent RCAF strength was increased during the year by 110, bringing the total to 118 officers and 676 airmen. Construction and the hiring of local staff was authorized to improve the facilities at Dartmouth, intended to be a permanent station for No.5 Flying Boat Squadron.

While aircraft assigned to civil government operations had hitherto been given civil registrations it was becoming increasingly difficult to maintain this distinction, which also inhibited flexibility of their use. It was decided that all aircraft operated by the RCAF should be included in its numerical series registry.

Preventive patrols continued to be a primary occupation of the RCAF during 1934, but photography almost doubled from the previous year to 1,085 hours. Nos. 2 and 3 GPDs from Winnipeg did most of the work, covering sections of northwest Ontario, the Rice and Gods Lakes areas of Manitoba, and large tracts around Great Bear and Great Slave Lakes and the Nahanni River in the Northwest Territories.

No.2 GPD with a pair of Bellancas under the command of F/L C.L.Trecarten and with Eric Fry as navigator progressed north, on the heels of the retreating winter. From Cameron Bay on Great Bear Lake, they photographed a large expanse between Great Bear Lake and the Yellowknife River. They also identified a route for a road between the lake and the MacKenzie River to enable traffic heading for Eldorado Mine to bypass the difficult rapids on the Great Bear River.

The detachment was then ordered to the Rocky Mountains on the border of the Northwest Territories and the Yukon to photograph around the headwaters of the South Nahanni River. This particular assignment had a somewhat romantic and legendary background originating with the murder in 1906 of two McLeod brothers near, appropriately named, Dead Man's Valley.

Their partner was located several years later in Vancouver where he had deposited a large amount of gold, starting a minor gold rush to the area and several more mysterious disappearances. Many claims were staked, none of which could be precisely located until topographical maps were available. Until then, fortune seekers had to rely on sketch maps prepared from information gleaned from Indians, trappers and prospectors. Trecarten himself successfully panned for gold on non-photographic days.

Nos. 6 and 7 GPDs based in Ottawa operated around Peterborough and Haileybury in Ontario, along the west shore of James Bay and around Eastmain River, Chibougamou and Kakabonga Lakes in Quebec. A detachment of No.5 Squadron sent to Shediac photographed the shoreline of the Bay of Chaleur. This detachment had an unusual assignment on August 24 to rendezvous with the SS *Champlain* off Wood Island, circle, and drop flowers of welcome to the French delegation arriving to celebrate the 400th anniversary of Jacques Cartier's landing in Canada.The crews of No.7 GPD thought they had, for a welcome change, found comfortable quarters for the 1934 season. The OIC, F/O R.C. Hawtrey, noted in his 3 August 1934 memorandum to W/C A.E. Godfrey, Rockcliffe CO, that Chibougamau Consolidated Gold Fields Inc. had built five log cabins on their property at Meril Island, adjoining the RCAF fuel cache at Lac Dore, some 300 miles north of Montreal . The company's representative, a Mr. Carpenter, had offered use of the buildings free of charge. Hawtrey proposed using the office building as the OIC's office and orderly room, the dining cabin as a mess, the

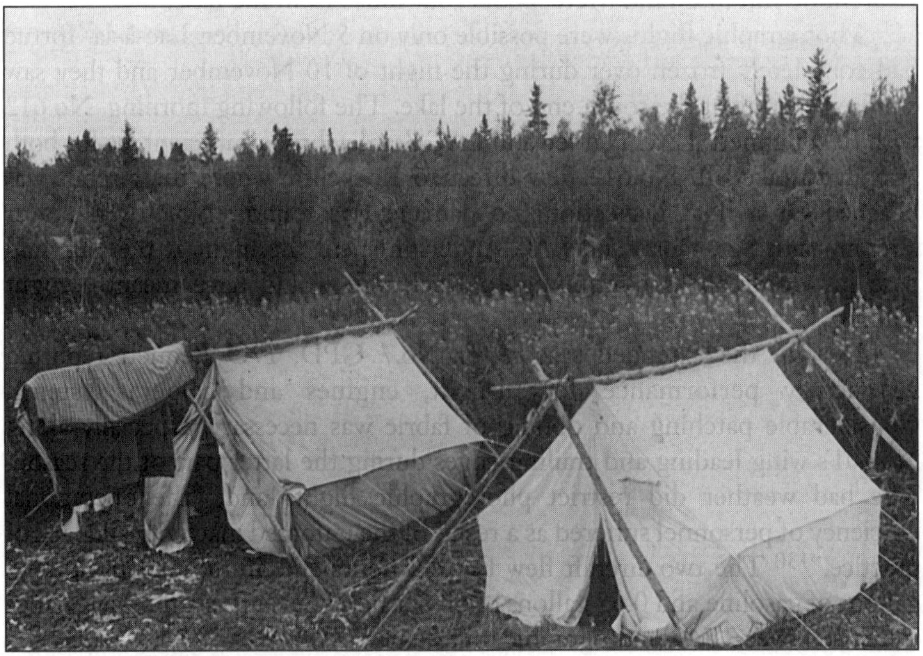

The photo crews generally had to "make do" with whatever accommodation they could find. This is the camp of No.3 GPD by the Taltson River on the southeast side of Great Slave Lake, Northwest Territories, 1934.
DNDPU PMR 74-484

bunkhouse as sleeping quarters and draughting room, and the store cabin for supplies. The blacksmith's cabin was not to be used. All was contingent on No.7 GPD accepting responsibility for fire hazard, which was considered small due to the existence of Pyrene fire extinguishers and the "extreme dampness of the district due to weather conditions"—which would, no doubt, make the shelter doubly welcome. Godfrey supported the request, but it was denied by W/C G.O. Johnson, A/DCGAO, because "the Department cannot accept the responsibility for fire hazard." No.7 GPD went back to its tents.[124]

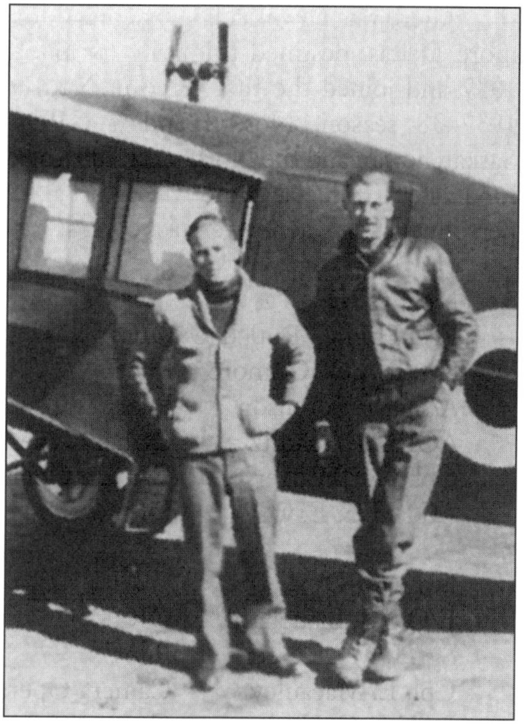

Pilot Sgt. Jack D. Hunter completed a course on "Smith's Automatic Pilot's Assister," an early version of the automatic pilot, at RAF Cranwell in the United Kingdom in 1933. He returned to conduct trials at Ottawa in 1934 with Pacemaker G-CYUY, Serial 602, fitted with the Assister. The stable camera base enabled photographs to be taken under conditions of poor visibility caused by smoke or haze, hitherto requiring impossibly slow shutter speeds. The sheer size, weight and complexity of the Assister's gyros and plumbing, however, severely limited its use. Hunter is on the right in this 1934 photograph with photographer LAC J.A. Wiseman, who became a pilot before retiring as a Wing Commander. Hunter was not particularly tall and their relative size indicates why Wiseman was known as "Squirt" throughout his career.
PHOTO COURTESY CAHS JOURNAL

Sgt. Hunter was called to Rockcliffe to conduct tests of infrared film and a recording barograph over Ottawa and Montreal from 18,000'—without oxygen, which Hunter recalled as "a tiring experience."[125] The infrared film improved penetration of haze, but the emulsion speed was too slow to give acceptable results with the fast exposure necessary to overcome aircraft movement. The ability of infrared film to differentiate topographical features was also explored in anticipation of improved emulsion speed. The recording barograph was synchronized with the camera to graphically record the height at which the photograph was taken. In this way, the scale of each photograph could be readily determined. Work also continued on the design and testing of a three-camera mount for multiple oblique photography.

Sgt. Hunter complemented his autopilot work with testing a crude, but effective, "blind landing" system in a Vedette. He trailed a weight beneath the flying boat from a cable during a long, power-on approach. When the weight touched the water, it switched on a sensitive altimeter that was graduated from 80 to zero feet. At five feet, the power was cut and, all being well, the hull settled smoothly on the surface.

Increased funding allowed winter operations to resume in January 1935. Two Fairchild 71s on skis, under the command of F/O L.E. Wray, with pilot Sgt. W.F. Tourgis, fitters Cpl. L.P. Parmenter and LAC J. Fortey, and surveyors J.L. Rannie and F.P. Steers from the Geodetic Survey Branch, set out from Lac du Bonnet for aerial reconnaissance of two areas north and west of

124 Ibid.
125 J.D. Hunter, "Flying in the Prewar RCAF," *CAHS Journal* 29, no. 1 (spring 1991): 20.

Lake Superior. Wray was ordered not to exceed 120 flying hours or a total cost of $2,500. Arrangements were made for the detachment to lodge at Wagaming, north of Lake Superior, where an unemployment relief camp had been established to clear land for an airfield. Flights were also made from refuelling bases at Kenora, Collins Lake, Port Arthur, Sioux Lookout and Nipigon River. The detachment had to arrange its own food supplies. As the Wasp engines consumed up to 30 gallons of gasoline per hour, the original allocation of 2,000 gallons for 120 hours' flying was insufficient and 1,000 additional gallons had to be authorized. However, Wray evidently practised economy as the total consumption of gasoline was only 2,466 gallons, prompting a request from Ottawa for a final accounting as early as possible, "as the balance is urgently required." Operation Kenora was completed on 24 February and the air force crews flew back to Lac du Bonnet while the two surveyors returned to Ottawa by train.[126]

While at Port Arthur, F/O Wray made an "emergency mercy flight," at the request of the Indian agent of the Department of Indian Affairs, to pick up an Indian woman and two children at the reserve on MacIntyre Bay, Lake Nipigon, and fly them to Port Arthur for urgent medical treatment.

Mutual congratulations on Operation Kenora were extended. The director of the Geodetic Survey Branch expressed his "sincere appreciation for the type of equipment and personnel which were supplied." The senior air officer, A/C G.M. Croil, added his words of appreciation for the crews, "for the manner in which this operation has been successfully completed," and added, "the success of the operation was due in a great measure to the co-operation of Mr. Rannie and Mr. Steers, not only whilst flying, but also in the servicing and maintaining of the aircraft on the ground, which during winter flying operations involves considerable manual labour."[127]

Winter flying still required extraordinary effort in 1935; the crews suffered from the cold, and maintenance was both difficult and time-consuming. The two Fairchild 71s, Serial 629 and 'XB, used for Operation Kenora, were thoroughly checked at Lac du Bonnet and test flown before setting out. Draughts in the cabin were eliminated "as far as possible" by placing felt strips around the door frames, packing the sliding window grooves, and closing openings in and around the instrument panel and firewall. Local magnetic disturbances at Lac du Bonnet necessitated re-calibrating the compasses at Kenora. The aircraft were outdoors for the four-week duration of the operation.

"For future reference," F/O Wray detailed the procedures followed in the bitter cold after a long, tiring day during Operation Kenora.[128] At the end of each day, the oil was immediately drained, taking care to prevent any ice that might have formed in the tank drain from mixing with the hot oil. Sometimes ice blocked the drain and had to be removed with a screwdriver. A marginal note was added to the report: "The Fairchild 71 is a most difficult type for winter flying as the oil tanks are much exposed and have a great deal of surface

126 NAC RG24, Vol. 4917.
127 Ibid.
128 Ibid.

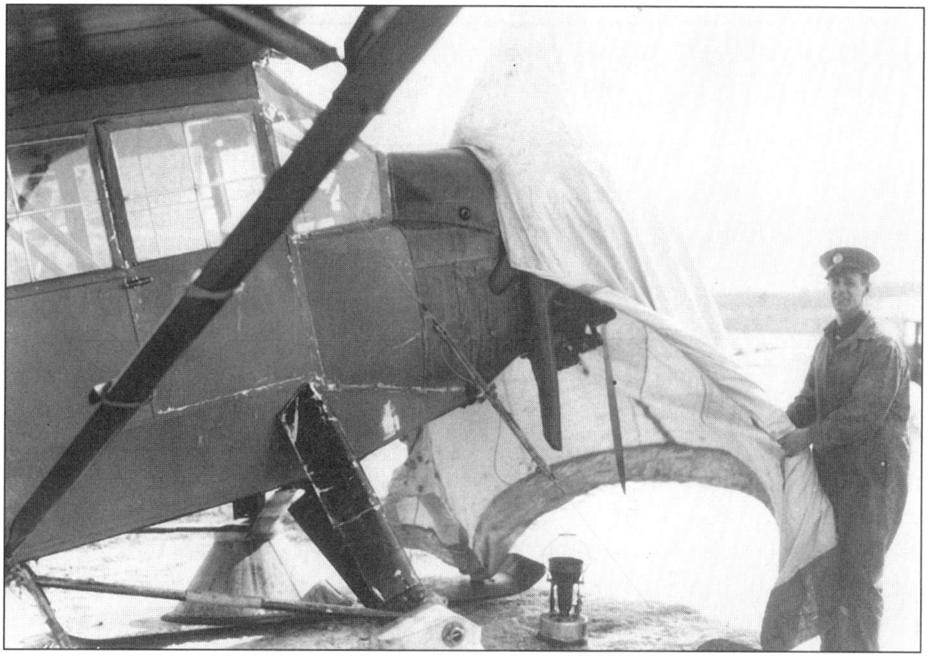

Warming the engine of a Fairchild FC-2, Rockcliffe, 19 March 1929.
Until 1938, standard procedure in cold weather was to drain off all the oil after a flight, keep it warm overnight and replace it the following morning, after heating the engine with a blowtorch or heater—a time-consuming and dangerous procedure. Thomas W. Siers of Western Canada Airways perfected a system, originally conceived by Weldon Worth of the US Army Air Corps, to satisfy Canadian aviation needs. The Worth Oil Dilution System introduced gasoline to the engine oil system during shutdown, thinning the oil so it no longer solidified, even in the coldest conditions. The cold engine could easily be turned over the following morning and started with adequate lubrication. The gasoline would then burn off and the oil return to its normal consistency. Tommy Siers received Canada's premier aviation award, the McKee Trophy, for this and other aviation innovations.
NAC PA62629

for their capacity." It was found that erecting the engine-heating tent over the engine each night, banked with snow, provided better protection than just the engine cover. Bungee cords had to be released and the skis placed on poles or boards to prevent them freezing to the ground. Finally, controls were lashed, and elevator and rudder chocks tied on. All this done, the crews could then look after their own welfare.

In preparing for flight, two fire pots were placed in each engine tent, with an attendant and a fire extinguisher nearby. This precaution was necessary as strong winds sometimes tore the tent loose from the snow banking and blew the canvas against the flame. The time required to heat the engine varied with the outside temperature and the wind velocity. As a rough guide, when the propeller ceased "sweating" the frost appeared to be out of the engine. If inside facilities were available, the oil was heated on a wood stove and stirred frequently, until it was "too warm to touch with the fingers." If there were no inside facilities, the oil was heated in a container suspended under the tent by ropes slung from the propeller shaft, taking care that the fire pot did not touch the bottom of the container, "so as to avoid uneven heating and lumpiness."

After the engine and oil had been heated and replaced, the subsequent "drill" had to be carefully executed to avoid cooling and repeating the warming procedure. The tent was removed, the engine primed and the propeller turned several times by hand before using the electric starter. Warming up after a satisfactory start took about 25 minutes' running to reach the prescribed oil temperature. If the engine had to be idled, a second warming-up period would be necessary, "for unless approximately 1,000 rpm could be maintained, the slipstream cooled the engine much more in comparison than the internal heat of the engine was able to maintain in oil temperature."

The Fairchild skis with "oil-draulic" legs had one severe disadvantage: they were extremely difficult to free when stuck in the snow. The usual method of rocking on the wing struts did not work because the oil-draulic leg absorbed the alternating loads. Without two men doing "very hard manual labour," Wray said that on many occasions he would not have been able to move the aircraft. In every other respect, however, the ski was satisfactory, its rugged construction and shock-absorbing legs enabling it to withstand the heavy loads landing on and taking off from rough, frozen surfaces. As the Fairchild 71 on skis was difficult to turn on snow in limited areas, manual assistance on the rear of the fuselage was necessary, exposing the pushers to the slipstream, which in a few seconds could cause frostbite. To avoid this, the detachment devised a method of attaching a rope to the tail skid or sternpost and pulling on it out of the slipstream blast.

In concluding his report, F/O Wray put in a plea, obviously from personal experience, that some means be devised to heat the space in front of the pilot's seat. "Efforts to block drafts were of little effect, and in extended flights at low temperatures, the pilot's legs became so numb that his ability to control the aircraft was affected."

Although the airmen were still far from comfortable in the cabin aircraft now in general use, the cameras could be mounted permanently and protected from the elements. For oblique photography, three cameras were mounted on a frame, the aircraft set in "flying trim" on the ground, and the cameras aimed to achieve overlapping views, left, centre and right, through fuselage apertures, and locked in position. At photographic altitude of 5,000' to 8,000', the three cameras made simultaneous exposures at about three-mile intervals along flight lines about eight miles apart. Accurate determination of ground reference points, as always, was necessary and, during 1934–35, surveyors John Carroll and Eric Fry were transported for this purpose. From the ground they would make stellar observations with a British Admiralty Pattern prismatic astrolabe, chronometer and chronograph, and radio for time signals. However, it was necessary to rely on sun observations, with some consequent deterioration in precision, for the many photo flights made north of the Arctic Circle, where the sun shines permanently during most of the summer.

The introduction of improved optics and wide-angle lenses gave vertical photography greatly increased coverage and this technique, with its added advantage of stereoscopic interpretation, largely displaced the more complicated oblique photography. The ability to plot contours from vertical photographs was of great value in the preliminary planning of roads and railways, and in identifying the land to be inundated by the construction of hydroelectric dams.

1935 - 1936

THE RCAF emerged slowly from the Depression doldrums and Parliament agreed that some catching up was necessary after the previous few lean years. The aviation appropriation totalled $4.3 million for 1935–36, almost twice that of the previous year. The permanent force increased to 1,024, bringing the total personnel to more than before "The Big Cut" in 1933. Wholesale reorganization accompanied the improved financial state of the air force and 28 aircraft were ordered for 1936 delivery. Five more squadrons were organized on paper, in addition to the two already operating with aircraft, one at Vancouver and one at Dartmouth.

Photography took a back seat to other activities during the 1935 season although Ottawa aircraft accumulated 754 hours of photography. They operated in Ontario at Sault Ste. Marie, Sudbury, Haileybury and Peterborough; in Quebec at the lower Gatineau Valley, Senneterre, Lac Wasanipi, Lake Chibougamau, Joliette and Baie Lavalliere; and in the Maritimes at Saint John, Charlottetown, Ship Harbour and the coast of Cape Breton Island.

Since 1932, experiments had been conducted with infrared film in the hope of penetrating the haze or smoke that plagued photo operations during most of the season. Little success was achieved until the introduction of new dyes sensitive to practically all visible light except infrared made the manufacture of plates commercially viable. Experiments during 1933 with the new film plates and an infrared filter showed favourable comparison with regular panchromatic film at the centre of the plate, but there was insufficient contrast between land and water and a progressive decrease in clarity away from the centre of the plate. In 1934, more infrared photographs were taken with Ilford film, the best available, but with limited shelf life and requiring an exposure of 28 seconds at F64. This was considerably slower than the 1/50 second at a full lens aperture of F4, considered the slowest possible to overcome aircraft vibration and movement. The conclusion remained that, despite its attractions, infrared film was not yet suitable for aerial photography.

In 1934, Newfoundland requested the Department of the Interior to quote on a triangulation survey covering a large area. The estimate of $10,500 for two Fairchild 71s over 60 days was not accepted, possibly because Newfoundland already had two de Havilland Fox Moth seaplanes on hire from Imperial Airways. In the spring of 1935, the request was renewed with the comment, "Our officers would much prefer having the work done by RCAF planes and personnel, as they have full knowledge of the type of work." The RCAF again examined the project and confirmed the previous estimate with a qualification: "The question of undertaking this work on payment of the out-of-pocket costs only depends on the political expediency of such a concession." This meant referral to the chief of general staff, Maj. Gen. A.G. McNaughton, who sent it to the Inter-Departmental Committee on Air Surveys and Base Maps, with the further provisos that no aircraft additional to those already allocated (eight actively engaged and two on reserve, but only eight crews) could be made available, and the air force vote should be reimbursed by the actual cost of the flying less salaries of the personnel. The committee reported back, pointing out that, "should the geodetic reconnaissance in Newfoundland be undertaken in lieu of air photography in Canada, it will, of course, be impossible for the remaining three detachments to complete the air photography program." There the matter rested.

Late in January, 1936, the Photographic Section was redesignated as the RCAF Photographic Establishment and made a long-anticipated move from the Jackson Building in Ottawa to new quarters at Rockcliffe. The three-storey "White House" was constructed as an unemployment relief project. Probably designed by DND's Directorate of Engineering Services, it was described by the Federal Heritage Review Office as "a modern classical building with a stylised Gothic decorative element." The Photo Establishment—one officer and 15 men under the command of S/L Edward Rosser Owen—occupied only the ground floor. The basement and other floors were occupied by the Geographical Section of the General Staff (GSGS) under the command of Maj. E.L.M. Burns. To make room for an intensive aerial mapping program the GSGS moved out in 1945 and the RCAF occupied the whole building.

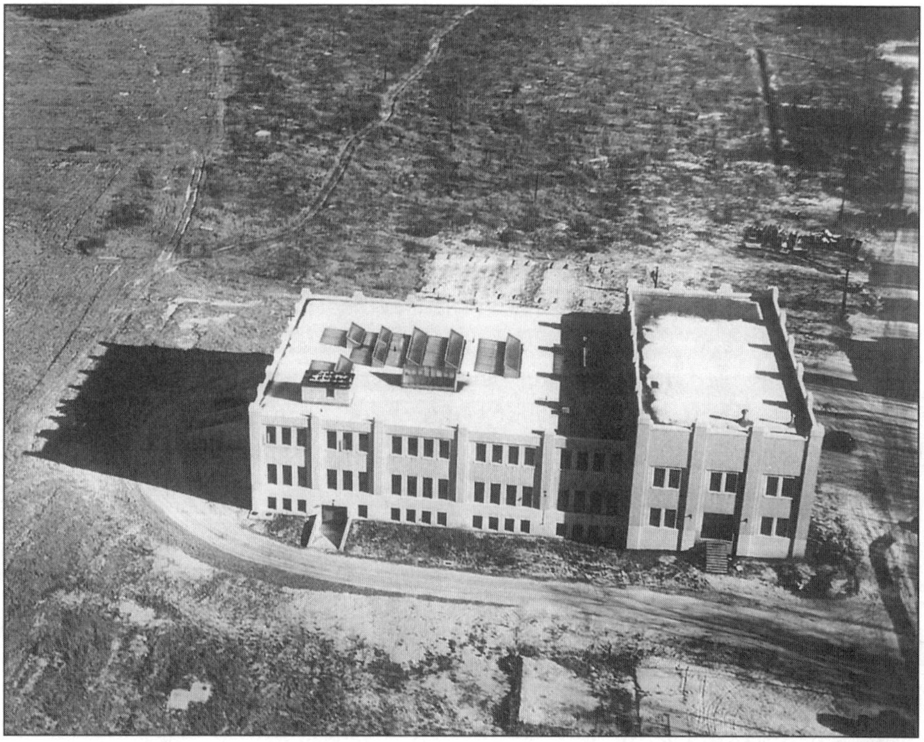

Landscaping remained to be completed at the White House in November 1937. *NAPL A5863-37*

1936-1937

AVIATION RECEIVED an increase in 1936 appropriations of $2.5 million to a total of $6.8 million, most of which was directed to the purchase of new aircraft and building new facilities for the RCAF. Manufacturing companies were so busy at this time, however, that orders for new equipment could not be completed and almost one million dollars remained unspent at the end of the fiscal year.

New aircraft continued to be introduced, with the Fairchilds on floats, skis or wheels gradually replacing the old flying boats. The all-metal Northrop Delta had attracted the attention of the RCAF, both as a photographic aircraft and as an opportunity to introduce stressed metal skin techniques to the Canadian manufacturing industry. Three, manufactured by Canadian Vickers with engineering support from Northrop, were ordered for delivery in 1936. Pilots, used to the excellent field of view from the high-wing monoplanes with highly visible wing struts then in use, remained somewhat suspicious of the low-wing Delta with no external—and visible—bracing. A need for additional windows in the cockpit sides and floor was evident to improve visibility when following photographic "lines." Manufacturing deficiencies and these modifications, which included the installation of under-wing hard points for bomb racks or guns, delayed effective deployment of the first three Deltas until 1937.

Bellanca Pacemakers were used almost exclusively for photography and were distributed in 1936–37 as follows:

Station	GPD	OIC	Aircraft
Winnipeg	No.2	F/O Wray	604 and 605
Winnipeg	No.3	F/O Rutledge	607 and 608
Ottawa	No.6	F/O Carscallen	603 and 611
Ottawa	No.7	F/O Bennett	601 and 612
Ottawa	GSGS	Sgt. Ewart	602

Pacemakers 610 at Winnipeg and 606 at Ottawa remained in immediate reserve.

No.7 (GP) Squadron at Ottawa had a total establishment of seven Pacemakers, which completed 1,358 hours on photographic assignments. They participated in wide-ranging operations, included vertical photography from bases at Trout Lake near North Bay, Ontario; at Rouyn, Bark Lake and Grand Mère, Quebec; obliques from Lake Waswanipi in northern Quebec; and both verticals and obliques from Cat Lake in northern Ontario.

On 17 August 1936, LAC Fortey and pilot F/L S.W. Coleman in a Fairchild 71 were returning to Fort Reliance from a photo mission and delivery of engine parts to a stranded Pacemaker at Hunger Lake, southeast of Bathurst Inlet. In rain and low cloud they searched in vain for the landmarks Fortey had photographed five years previously. Out of fuel they put down on Point Lake, 140 miles west of their intended route. They were, fortunately, near trees and were able to cut poles for a tent carried against such emergencies, and to improvise a stove from a gasoline drum. They subsisted on meagre emergency rations, berries, and two ground squirrels they managed to snare until found 23 days later, as ice was beginning to form on the lake, on the hunch of Canadian Airways' veteran bush pilot Matt Berry. This was one of longest and largest air searches in Northern Canada, with seven RCAF and four civilian aircraft accumulating 802 flying hours.

F/O W.E. Bennett commanded No.7 GPD in 1936 with Sgt. J.E. Doan as pilot and fitter, Cpl. J.M.M. Vinet and LAC E.G. Doran as camera operators and fitters, and LAC H. Stanley as camera operator and rigger. A civilian cook, M. Genest, was carried on the strength. LAC N.E. Harvey replaced LAC Doran on September 22, and LAC H.F. Anderson replaced Cpl. Vinet on October 30. They flew two Bellanca Pacemakers with Wright J6E engines; No.601 had Short Type 5100 floats and No.612 had Edo Type X floats. Four Fairchild cameras were carried in each aircraft, three for oblique and one for vertical photography. One additional camera was operated by hand for the oblique pictorial work. The two aircraft were fully loaded with six crew members, cameras and accessories, four 8x10-foot tents and a mess tent, camping gear, aircraft and engine spares, including one complete cylinder, and emergency kits. The continuing saga of No.7 GPD demonstrated that, even with vastly improved equipment, the crews still had the same trials and tribulations to contend with as had their predecessors during the 1920s.

Arriving at the fuel cache on Trout Lake, north of North Bay, on 14 May, No.7 GPD found good accommodation at a hunting lodge, Camp Champlain, where the aircraft could be secured on the beach. Cloudy weather delayed scheduled photography of surface conditions after a recent earthquake in the Lake Timiskaming and Kipawa areas. Visitors, however, provided some relief from the boredom of waiting for improved conditions. On 21 May, F/O Miles arrived in a Fairchild transporting A.M. Narraway, controller of surveys, and Dr. E.A. Hodgson, chief of the Dominion Observatory's Seismology Division, who gave instructions for "illustrative obliques" to be taken of Timiskaming, North Bay and Trout Lake. On 12 June, the detachment completed photography in the area for the Dominion Observatory, packed up, and crossed the provincial boundary to Rouyn, Quebec.

General Airways' base at Lake Osisko was utilized and accommodation found at the nearby Elwyn's Hotel. Bennett was apprehensive about the two 400-foot-high smoke stacks at Noranda Mines, "a hazard to be avoided in conditions of poor visibility." No.7 GPD took vertical photographs and illustrative obliques of mining operations in the Rouyn area for the geological survey on 16 June, and then moved northeast to Lake Waswanipi on 18 June.[129]

No.7 GPD had used the cabins of an abandoned fishing camp at Peninsular Bay on Lake Waswanipi the previous season. Tents were pitched inside the largest cabin while the other cabins were repaired and made weatherproof. The addition of a cooking stove leased from Mr. MacLeod at the nearby Hudson's Bay post made them reasonably comfortable. This was fortunate because wet weather delayed flying until 5 July. The Hudson's Bay post had satisfied some needs, including a rental canoe, but they made frequent trips to Seneterre, 90 miles east, for supplies and, when the weather improved, to collect film and deliver it after exposure. Canadian Airlines brought in five hundred gallons of fuel. Advantage of frequent poor weather was taken to beach the aircraft for airframe and engine inspection, resulting in a push rod of No.612's engine being replaced.

Still dogged by rain and poor visibility, some obliques were taken for the geological survey before S/L Morfee, with LAC Denny, arrived on 12 July in Fairchild No.641 from Ottawa to inspect the detachment. The same day, provincial land surveyor Perry and two men arrived by outboard-powered canoe. Perry's party was taken by air to a series of observation points before they left Waswanipi by canoe for Lake Doré on 14 August. Considerable prospecting was evident in the area, four different canoe parties calling in on the detachment while it was in residence. All the commercial aircraft operating from Rouyn, Senetterre and Sioux Lookout were also noted as working to full schedules, "with more work than they could handle."

Rain ceased for a few days, but photography continued to be handicapped, this time, surprisingly, by haze and smoke from forest fires. James Burns, general merchant and long-term resident of Senetterre stated that the weather experienced that summer was the worst he remembered; rain fell on sixty percent of the days the detachment spent there. En route to Senetterre on 1 September, bad weather forced down F/O Bennett and Cpl. Vinet in No.612 at Lake Kamshigama. Headquarters decided that enough was enough and that No.7 GPD should transfer to Bark Lake, Quebec, via Senetterre, on 5 and 6 September. Sgt. Doan and Cpl. Vinet returned to Lake Osisko in No.601 to photograph gaps found in the previous coverage, and to take obliques of the

129 Ibid.

experimental farm at La Ferme, but were delayed by smoke from Noranda Mines and did not reach Bark Lake until 16 September.

"Very good" accommodation with storage and office space was provided at the Canadian International Paper depot on Bark Lake. Limited supplies were available at the Hudson's Bay store. Film was flown 30 miles south-southeast to Lacroix Depot and transported from there 25 miles by road to the railhead at Maniwaki. Vertical photographs were taken in the area and both aircraft returned to Ottawa on 19 September.The engine of No.612 was changed before the detachment set out for the Ontario Provincial Air Services base at Sioux Lookout via Haileybury and Longlac on 25 September. Movement to Cat Lake, 120 miles north, was completed on 28 September. Mr. MacCauley, the Hudson's Bay factor, made the loft over his store available for sleeping accommodation. He greatly assisted the detachment, installing a raft at the edge of deep water, some 70 yards from the store, for mooring the aircraft, and a dock at a sandy point about 600 yards east where the aircraft could be beached.

Winter was beginning to make itself felt. Ice formed in shallow water on 2 October, and on 4 October, "Geese were observed going south." It was necessary to drain oil from the engines at night time and store it at room temperature to be sure of starting the engines in the morning. Bellanca 612 with F/O Bennett and LAC Harvey had a forced landing at Lake Kapikik en route to Sioux Lookout on 6 October due to low oil pressure. Transporting surveyor Eric Fry of the Topographical and Air Survey Bureau to Spirit Lake was interrupted when, for a welcome change, good photographic weather was encountered en route. A hasty visit to Cat Lake to exchange Fry for cameras and film enabled photo lines to be run before Sgt. Doan took Fry to Spirit Lake in No.601. The next few fine days were busy with photography and taking the exposed film to Sioux Lookout. Fry was moved from Spirit to Cedar Lake and, on 12 October, to Sandy Lake where F/O Bennett decided to drain the oil and stay overnight because he was apprehensive that he might not be able to return to the rapidly freezing lake. His concerns were justified the following morning when it was necessary to break through about 30 yards of ice to reach open water before returning Fry to Cat Lake. From the air, they observed that Cedar Lake was completely frozen over. The same day, they took Fry to Sioux Lookout and started evacuating Cat Lake, fortunately for LAC Stanley who was admitted to Sioux Lookout General Hospital with a high temperature and what turned out to be tonsillitis.

Pushing their luck somewhat in the face of inclement weather and advancing winter, No.7 GPD, less LAC Stanley, departed Sioux Lookout for Rouyn, via a refuelling stop at Longlac, on 19 October. The two aircraft lost contact in cloud. F/O Bennett and LAC Harvey in No.612 force-landed at Opazatika Lake because of rain and sleet and a 400-foot cloud base. Sgt. Doan and Cpl. Vinet in No.601 continued to the General Airways base at Lake

Osisko, joined by Bennett the following day. Due to lack of beaching space, No.612 was moored to the dock with almost disastrous results. Heavy snow during the night of 20 October weighed down the aircraft until the float heels rested on the bottom of the lake. Salvage was effected with the aid of General Airways personnel and equipment, and the only damage sustained was a small puncture in one float, easily repaired. Two commercial aircraft moored to buoys were submerged to the engines by the 12-inch snowfall.

By 24 October, Rouyn Lake and other smaller bodies of water had frozen over with ice forming at the north end of Lake Osisko. On 26 October, commercial aircraft were unable to take off from Osisko because heavy ice formed on propellers and undercarriage while taxiing, and the lake had frozen over during the night. General Airways aircraft broke an east-west runway on 27 October and No.7 GPD made a hasty exit to photograph gaps in the previous lines before flying direct to Ottawa.

The two Pacemakers were removed from the water at Rockcliffe, but soon put back as the detachment was ordered on 31 October to proceed to the Canadian Airways base at Lac-à-la-Tortue, north of Three Rivers, to photograph projected ground survey routes. Bad weather delayed departure until 4 November. With experience at Osisko fresh in mind, F/O Bennett refused offered moorings at Lac-à-la-Tortue and beached the aircraft. Mr. Allard, the Canadian Airways caretaker, built them a service dock and provided a boat and storage space for their equipment. Bennett commented on the great assistance this man gave, "and in this French-speaking community, his co-operation was particularly welcome." They found accommodation at the Laurentide Inn in Grand Mère, about four miles from the base.

Photographic flights were possible only on 5 November. Lac-à-la-Tortue had completely frozen over during the night of 10 November and they saw children skating at the south end of the lake. The following morning, No.612 with F/O Bennett, LAC Harvey and LAC Stanley broke sufficient ice for both aircraft to take off. No.612 flew direct to Rockcliffe where the aircraft was beached for a float inspection (no damage was found). No.601, a slower aircraft, with Sgt. Doan and LAC Anderson spent the night at the Fairchild base in Montreal, as continuing to Rockcliffe would have meant a night landing.

Despite the extended season for No.7 GPD, F/O Bennett reported satisfactory performance for aircraft, engines and camera systems. Considerable patching and doping of fabric was necessary, especially along No.601's wing leading and trailing edges during the latter part of the season. The bad weather did restrict photographic flights and, "the operational efficiency of personnel suffered as a result of the enforced inactivity and lack of practice."[130] The two aircraft flew for 332 hours and 50 minutes using 16.5 gallons of gasoline and 0.65 gallons of oil per hour. Seventy-six film magazines were used over 3,549 photographic miles.

130 Ibid.

The four Pacemakers at Winnipeg were joined by Serial 610 in June after completion of an overhaul at MacDonald Brothers. All came under the control of No.8 (General Purpose) Squadron at Ottawa. Areas photographed included Fort Hope and Osnaburgh House in Ontario, north of Great Slave Lake, Northwest Territories, and the right-of-way for the road joining Lake Winnipegosis to The Pas, Manitoba.

The RCAF no longer relied on civilian surveyors to navigate its photographic aircraft although Geological Survey retained overall responsibility for maps and mapping. Surveyors in general were enthusiastic supporters of aerial photography. C.S. Macdonald, DLS, president of the Canadian Institute of Surveying, commented in his presidential address in 1936 on technical advances made in aerial photography since he navigated for Mawdesley on the epic 1930 flight to the Northwest Territories:

> Altitude for obliques has increased from 5,000 feet to 8,000 feet; aircraft range has increased from 230 to 500 miles; and techniques have improved as has quality of photographs. All resulting in economies as spacing of flight lines has increased from six miles to eight or ten miles and intervals between photographs increased from two to three miles—and simultaneously for each set versus the previous necessity to fix each photograph of a set of three (when taken individually).

1937-1938

PARLIAMENT WAS MORE concerned with national defence in 1937 than at any time since 1918. The myth of that conflict being the "war to end all wars" had been scattered on the winds of news from Europe, Africa and Asia. The hard truth emerged that the RCAF, with its concentration on civil tasks, was ineffective as a fighting force. Corrective measures included more than doubling the RCAF estimates to $11.7 million. This was an even larger increase than apparent on the surface as the $1.6 million cost of civil aviation, previously borne by the RCAF, had been transferred to the Department of Transport. During the year, the permanent strength doubled to 178 officers and 1,523 airmen, and the non-permanent force increased by 40 percent to 80 officers and 729 airmen.

A special branch was formed at Ottawa to deal with armaments. It is a commentary on the air force's concentration on civil operations that the air force section of the 1937–38 annual departmental report to Parliament makes direct reference to weapons for the first time since 1919. Also marking an important change in air force operations was transfer of the Victoria Beach station in Manitoba, an air force base for 17 years, to the Department of Mines and Resources. The RCMP took over the responsibility for preventive patrols, and the RCAF no longer flew forest patrols in Manitoba and Saskatchewan. Only photography and transportation remained in the RCAF's roster of "civil" tasks.

Moves were made in 1937 to locate airfields on each coast where reliance had hitherto been placed on waterborne aircraft. On the west coast, two aircraft photographed possible sites for land defences on behalf of the Geographical Section of the army. The existing base at Jericho Beach was to be expanded and an "aerodrome" built at Patricia Bay on Vancouver Island. Alliford Bay in the Queen Charlotte Islands and Prince Rupert in northern British Columbia would be developed as advanced seaplane bases. A decision to concentrate on land-based aircraft for the defence of the east coast saw aerodromes built at Dartmouth, Yarmouth, Sydney and Truro.

There was now much less demand for civil aircraft in the RCAF, but four more Deltas, destined primarily for photographic duties, were delivered in 1937 and joined the first three in No.8 (GP) Squadron at Ottawa. For the 1937–38 season, Nos. 2 and 3 GPDs were designated for Manitoba, Saskatchewan and the Northwest Territories; No.6 to the west coast; and No.7 for Eastern Canada. Although, as usual, changes were made later, the initial personnel assignments were as follows:

No.2 GPD

F/O H.H.C. Rutledge	OIC
Sgt. R.A.W. Gilmore	Pilot
A/Sgt. P. Sorenson	Pilot, Transportation Aircraft
AC1 G.F. Schultz	Camera Operator/Fitter
Cpl. G.F. Walker	Camera Operator/Fitter
AC1 H.C. Edwards	Fifth Man

No.3 GPD

F/O S.S. Blanchard	OIC
Sgt. J.W. McNee	Pilot
Cpl. E. Macauley	Camera Operator/Fitter
LAC D.C. Hudson	Camera Operator/Fitter
AC1 V.A. Margetts	Fifth Man

No.6 GPD

F/O H.M. Carscallen	OIC
Sgt. R. Thomas	Pilot
LAC N.E. Harvey	Camera Operator/Fitter
LAC W.H. Lord	Camera Operator/Fitter
LAC H.F. Henderson	Rigger

No.7 GPD

F/O C.H. Porter	OIC
Sgt. J.E. Doan	Pilot
LAC E.G. Doran	Camera Operator/Fitter
LAC J.M. Dexter	Camera Operator/Fitter
LAC H. Stanley	Rigger

Aerial mapping remained high on the government's list of priorities for the RCAF. The Interdepartmental Committee on Air Surveys and Maps, under chairman Dr. Camsell of the Department of Mines and Surveys, with influential A.M. Narraway as secretary, identified an ambitious program for 1937. Poor weather hampered operations, but 67,000 square miles were photographed, plus an additional 67,850 square miles for the Geographical

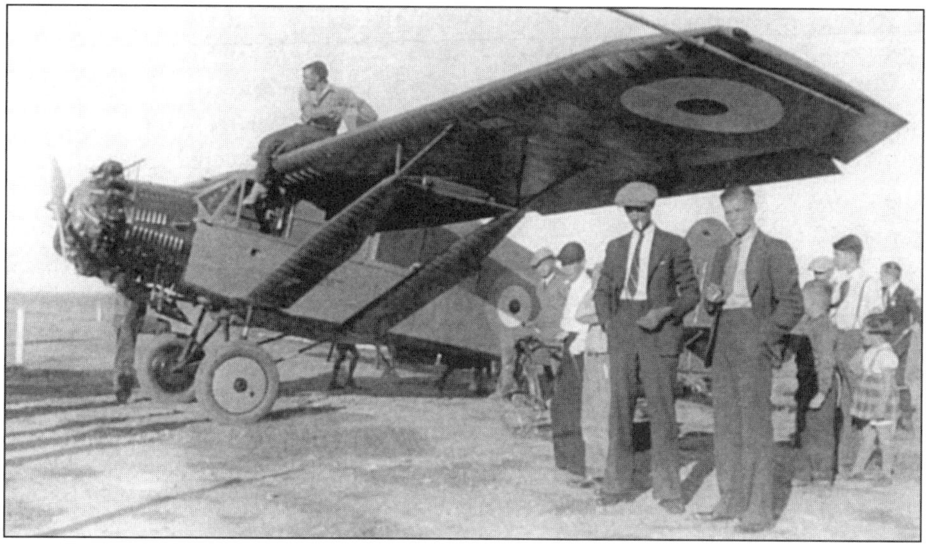

The largest operation in 1937 was on behalf of the Department of Agriculture, covering prairie drought areas, including the Frenchman and Souris watersheds and Youngstown in Alberta, and Regina in Saskatchewan. Bellanca Pacemakers were based at Calgary Municipal Airport in October under the command of F/L H.H.C. Rutledge. One of the photographers was LAC J.A. "Squirt" Wiseman who played an increasingly important role in aerial photography during World War II.

Facing the camera wearing a cap is W.F. Field, a leading member of the Calgary Aero Club. He enlisted in the RCAF as a W/O pilot and was killed in a flying accident at MacGregor, Manitoba, in January 1942.
RON CAMPBELL

Section of the army. This was only half the anticipated work, but, completed in 1,807 photographic hours, compared favourably with the 76,000 square miles in 4,821 hours accomplished in 1931. It was an increase of about 30 percent from 1936. Ottawa-based aircraft took virtually all the photographs and the Photographic Establishment processed 448 rolls of film.

In western Ontario, a long strip was photographed along the route of the proposed trans-Canada airway for which no detailed maps existed. Other operations covered parts of northern Quebec, Manitoba and the Northwest Territories. By this time, greater reliance was being placed on commercial air transport to support ground survey teams. During the 1937–1938 winter, a Manitoba government team surveying the northern stretch of the Manitoba–Ontario boundary was the first to rely completely on air support. Thanks to a combination of aerial photography and air transport, the team completed its work three weeks ahead of schedule in mid-March 1938. In contrast with previous years when the journey in and out would have consumed weeks of arduous travel, the men were back in Winnipeg within a few hours of finishing work. This survey left 168 miles of border to be surveyed, a task not completed until after World War II.

Fairchild Super 71 Serial 666 initiated a mystery, still unsolved, in August 1937 when it crashed near Morrison Lake, Manitoba. F/O W.C.H. Porter, OIC No.7 GPD, and observer LAC E.G. Doran had refuelled at Cormorant Lake en route to Lac du Bonnet, but subsequently disappeared. Visibility in the area was poor due to smoke from forest fires and the burned wreckage containing evidence of only one body was not discovered until 9 August. On 3 November, however, Porter's body was found on the shore of Morrison Lake, about a quarter mile from the crash site. The aircraft was flying too low to permit the use of a parachute and no evidence of one was found. The court of enquiry concluded that the primary cause of the accident and how the pilot left the aircraft in flight were unknown.

Despite the mystery, books had to be balanced: Porter had received an advance of $500 for travel and subsistence, and $200 for incidental expenses. Some receipts, $120 in traveller's cheques and $13 in cash were recovered from the remains, but in December, the accounts department noted that $193.39 remained outstanding. W/C A.B. Shearer, CO Ottawa, responded to the effect that Porter must have incurred considerable expense during the flight from Twin Lakes to Edmonton and in the repair of 666 at Edmonton. "In view of the indeterminate conditions surrounding the Outstanding Balance of the Advance it is strongly recommended that the balance be written off."[131] It was.

NAC RG24, Vol. 4890.

1938-1939

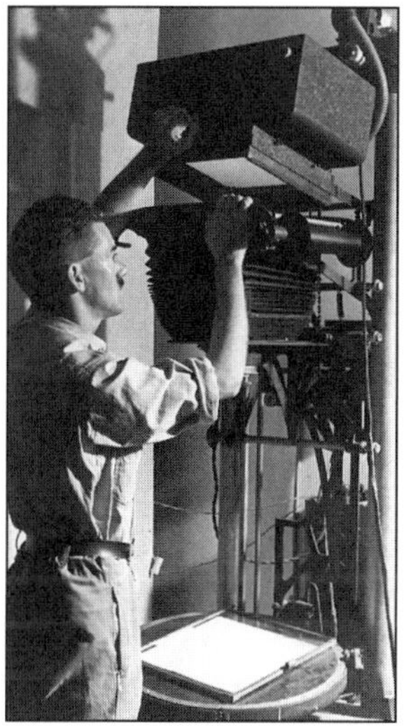

Cpl. E.B. Hacket operating a new enlarger at the White House, 1938.
DNDPU PL5791

THE RCAF ordered fewer aircraft in 1937 than anticipated, resulting in a reduction of almost $1 million in the 1938 appropriation to $10.7 million. However, supplementary votes for bombs, ammunition and other equipment restored $884,900. The strength continued to grow: the permanent force to 261 officers and 1,930 airmen, and non-permanent (now called auxiliary) to 99 officers and 867 airmen. In addition, the reserve of officers numbered 155. As part of the reorganization associated with the expansion, a photography branch was established within the Air Staff Division of Air Force Headquarters.

No.8 (GP) Squadron at Ottawa Four acquired more Deltas in the fall of 1938, with a little apprehension as an unexpected tendency to spin had been experienced. Wind tunnel model tests by the National Research Council resulted in a recommendation that the rear fuselage be strengthened and an anti-spin parachute installed.

Photography continued at about the same level as in 1937 with 1,774 flying hours, mostly with aircraft from No.8 (GP) Squadron at Ottawa. The introduction of a more sensitive film extended the "photographic day." In the Great Slave and Athabaska Lakes areas, for example, the film could be used as early as 0500 hours during June and 0600 during July.

No.6 GPD, with two Deltas, spent six months working along the coast of British Columbia. Nos. 2 and 3 GPDs with four Deltas left early in May on wheels for photography of the drought areas on the prairies until ice broke up in the north when they transferred to floats and headed to the Northwest Territories. One of the new Noorduyn Norseman accompanied the detachment north for transport support. No.7 GPD with two Bellancas went first to Truro, Nova Scotia, for an unidentified "high-priority task," and then to Blue Sea Lake, Bark Lake and Grand Mère in Quebec. Aircraft from Dartmouth, Trenton and Vancouver also undertook local photography.

In approximately the same number of flying hours as the previous year, an additional 11,250 square miles were photographed using a total of 591 rolls of film. With faster film and the faster Delta aircraft, the detachments were pleased to be back in Ottawa by 1 October, at least one month earlier than normal.

Fairchild cameras gave reasonably good service, but were considered large, heavy and unwieldy. During 1938, the National Research Council examined the problem and developed a smaller camera with a four-inch-square negative that it anticipated would be ready for test in 1939. A Ross camera with a special wide-angle lens was given preliminary trials during the year. In order to overcome the delays inherent in shipping film back to Ottawa for processing, a photographic trailer was designed and several units ordered to equip detachments in 1939.

A sign of the times appeared in February when A/C G.M. Croil advised the CO at Ottawa that no suitable replacement for F/L Carscallen was available other than an NCO pilot, "due to expansion of the RCAF."[132] Another sign was No.3 GPD being forced to operate all the season without an airframe mechanic because one was not available. Although no serious problem resulted in this instance, G/C N.R. Anderson, CO Ottawa, later explained to the secretary, DND, the perils of continuing this practice. Experience, he stated, had shown that the Wright Cyclone in the new Northrop Delta aircraft demanded more of the fitter's time than the Wright Whirlwind in the Bellanca Pacemaker it replaced. The greater speed of the Delta also enabled more rolls of film to be exposed during a photographic flight, increasing the fitter's workload, as he was also the photographer. The increased operating altitude of the Delta resulted in greater fatigue for all the aircrew, and the 35-gallons-per-hour fuel consumption of the Delta imposed a refuelling task for two men of two hours per mission. The end result was that the fitter had to work for 16 hours per day, including a very tiring six-hour flight. Because photographic weather occurred only spasmodically, the equipment had to be readied the previous night to take full advantage of opportunities, often on several successive days. Anderson argued that holding the fitter responsible for maintenance and inspection of the airframe in addition to his duties listed, especially as no senior NCO was available to supervise the work as required by Administrative Order A.27/5, to be unreasonable. He noted that several Delta aircraft had been found wanting on inspection at the end of the season, and recommended that they be inspected in accordance with AO A.27/5—which specified an airframe mechanic. Three carpenters were transferred in November to No.8 (GP) Squadron at Ottawa and placed on a camera operator's course with the objective of reducing the mechanic's workload.

On 23 June 1938, W/C A.B. Shearer, CO Ottawa Station, in a memorandum to the secretary, DND, tackled the thorny and perpetual problem of travelling expenses. He responded to a rejected claim by F/Sgt. A. Fleming, NCO-in-charge of No.1 Detachment, for overnight accommodation in Montreal at $1.50 per airman, fifty cents over the allowable amount. No

132 Ibid.

doubt drawing on his own experience as a photo pilot, Shearer endorsed the opinion of the OIC No.8 (GP) Squadron that stringent adherence to financial instructions and regulations was causing the detachment commanders and personnel enough worry, loss of rest and mental strain to affect the health of the personnel and the excellent work of the detachments. He tabled the problem:

> The photographic detachment commanders and their personnel are employed away from their permanent stations for a period of six months in the year. During this period they are working at a high nervous tension, as their one objective is to produce photographs of a high quality in the least possible time. Owing to the fact that they are working under service regulations, they are required to submit many more returns and keep a closer accounting of their equipment than the work itself requires.
>
> Also, of four detachments who are supposed to be commanded by Flight Lieutenants, No.8 (GP) Squadron has two detachments commanded by Flying Officers and two detachments commanded by Flight Sergeants. These junior officers and NCOs have been given the responsibilities which it is considered should be given to Flight Lieutenants and, therefore, it is submitted that they should be treated with leniency and consideration.
>
> It is not thought to be conducive to the good health of the detachment commander and his personnel for them to have to scout around a large city to find a hotel which will provide the airman with an inferior bed for $1.00 per night. Further, it should be admitted that a pilot or camera operator who has been employed on photography at 10,000 feet, or higher, for several hours requires as good a night's sleep as can be obtained.
>
> One photographic pilot could not be employed on photographic work this year on account of his medical condition, and another is of a doubtful medical standard as he has lost over twenty pounds since his employment on detached operations.
>
> If it were not that the time factor is of such vital importance in photographic work, no hardship would be experienced in thoroughly complying with the regulations.
>
> It is recommended that special consideration be given to photographic detachments in the approving of claims, or alternatively, that the allowance for a bed for an airman employed on photographic operations be increased to $1.50 per night.[133]

133 Ibid.

Lt.-Col. A.R. Mortimer, assistant director of Pay Services, conceded that special consideration should be given in such cases as referred to, but stopped short of recommending any change to the regulations. The new Ottawa Station CO, G/C N.R. Anderson, continued his long-standing support of the PDs and, after interviewing the returning detachments, took up the cudgels again on 21 October. He stated that on occasion it was impossible to obtain a bed for $1 per night, even after canvassing door-to-door. As an example, he describes the experience of a detachment arriving at Lethbridge, Alberta, to undertake a two-day photography mission:

> The detachment arrives in the afternoon and arrangements are made for hangar space. The Imperial Oil [agent] is phoned to send out ten drums of gasoline, by truck, and the work of unloading baggage and equipment from the aircraft commences. Once the aircraft are completely cleaned out, the camera gear must be set up in them. The gasoline arrives and is pumped into the aircraft by a manually operated wobble pump. This takes at least two and a half hours. The detachment commander has in the meantime seen the radio and weather agents and the airport manager and has gone by taxi into town to enquire at the freight and/or express depots regarding equipment shipped from the previous base, and film expected from Ottawa. He returns to aerodrome with the film in time to sign for the gasoline bought from the Imperial Oil agent and supervise the storage of the aircraft in the hangar for the night. There is no place to load film at the airport, so this will have to be done in a hotel cupboard at night. Personal gear, office equipment, magazines and film are laid aside ready to go into town. Two taxis are called and, after some delay, arrive. In the meantime, a fitter has found that one of the tail wheel oleo legs has lost pressure. It must be pumped up that night in preparation for photography the following morning. Enquiries have been made regarding the cheapest hotel available. The radio operator is sent there with the airmen's personal baggage to commence work on the magazines and film to be loaded. The detachment commander returns to town to obtain a room for himself, set up his office and commence work on flight reports, etc. The remainder of the detachment eventually arrives in from the airport, have their dinner, and commence to load magazines. The time is between 2000 and 2200 hours. The loading of magazines takes one hour. Reports must now be made out for rolls of exposed film which have been unloaded. It is necessary for the pilots to work in collaboration with the camera operators in making out these reports so the crews work together until midnight, or later, completing film reports so the film may be sent to Ottawa the following day. The detachment commander then returns to his hotel,

> sending a telegram to his Squadron on the way to notify his Officer Commanding that the detachment has changed base and is ready to continue photography the following day. Before he turns in he requests that the night clerk call him at 0430 hours for a look at the weather.[134]

Anderson continued at some length to make his case for a 50-cent increase in the overnight allowance for airmen. He emphasized that detachment commanders did not wilfully place their personnel in expensive quarters, but little opportunity existed for accommodation bargain hunting. He noted that all the men from detachments operating in the prairies and Northwest Territories had lost weight by the end of the season and one corporal had an illness directly attributable to a decline in his state of health. He argued that photographic operations incurred longer hours, more exacting work with more costly mistakes, and greater drain on physical and nervous energy than any other type of air force work, often demanding 15- to 20-hour workdays for several consecutive days. Anderson repeated his recommendation that the dollar per night limit be raised: "Otherwise, if regulations are rigidly enforced, it is obvious that the health of the personnel and the quality of the work turned in by them will suffer out of all proportion to the slight extra expenditure involved."[135]

Endorsed all the way up the line to the senior air officer, A/V/M G.M. Croil, and the deputy minister, the Pay and Allowance Regulations were at last amended to increase the airmen's overnight allowance by fifty cents. But not without a last word: "Claims would, of course, be required to be supported by receipted vouchers."

G/C Anderson also stepped forward in July to support the supplementation of No.1 GPD's $200 standing advance with another $200

134 Ibid.
135 Ibid.

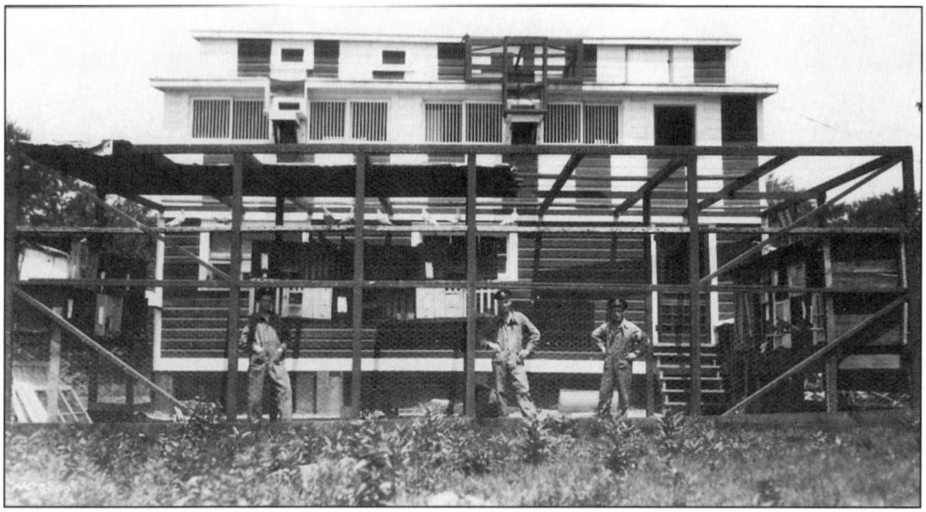

Pigeons continued to play an important part in RCAF communications until well into World War II. Although radio had been developed and was in general use, it was vulnerable to eaves-droppers and direction finders while pigeons offered a secure means of communication. This "high-rise" pigeon cote was operated by a special "Pigeon Division" at Rockcliffe in 1938.
NAC PA62607

because of the high expenses incurred at Halifax. The detachment was commanded by an NCO, F/Sgt. Fleming, however, so he requested that the additional advance go via his commanding officer, Anderson, in the form of two cheques, "for issue as required." Accounts were clearly uncomfortable about the additional advance to an NCO and queried its status in August, before Anderson had even forwarded the cheques to Fleming.

In a postscript to this incident, a 22 April 1939 memo authorizes advances of $1,000 each direct to F/L S.S. Blanchard, OIC No.3 GPD, and F/S R.I. Thomas, NCO-in-charge No.6 GPD, both at Shediac, New Brunswick. F/Sgt. Fleming, by then NCO-in-charge No.1 GPD at Camp Borden, where expenses would be considerably less, however, had his $400 advance sent in four equal cheques via his CO.

1939-1940

WITH WAR a virtual certainty, $60 million was appropriated for DND in 1939. The RCAF received almost one-half, an increase of $18.2 million from 1938. The stated objective was to build up to a permanent force of 525 officers and 4,500 airmen, and an auxiliary of 220 officers and 2,014 airmen.

Serious consideration was given to ceasing all photographic work. Military needs were pressing and the demands made on both men and machines preparing for war seriously diluted resources. Maintenance and repair would be required for photographic aircraft dispersed as far north as Aklavik and Hudson Bay, and impossible to mobilize quickly for the defence of Canada. Dr. Camsell, chairman of the Interdepartmental Committee on Air Surveys and Maps, was advised in April that aerial photography would have to be abandoned on 1 December 1939.

Operations planned for the Northwest Territories were cancelled, but maps of the Maritimes were urgently needed for defence purposes. A detachment of two Deltas and one Norseman with veteran photo pilot W/C G.E. Brookes aboard left Rockcliffe on 14 July and spent one month inspecting and photographing Mary Harbour, Cartwright, Hopedale, Table Harbour, Rigolet, Pottles Bay, Halton Harbour, Makkovik, Canairektok Bay, Nain, Davis Inlet, Hebron and Port Burwell. Another Delta kept busy at Petawawa, Camp Borden and Toronto for the Army Geographical Section before flying west to complete other tasks at Regina, Moose Jaw, Saskatoon and North Battleford. On this survey work, a special six-inch wide-angle Ross lens was tested and found to be superior to those in use.

During an abbreviated 1939 season, the detachments from No.8 GPS photographed 25,100 square miles, consuming 237 rolls of film during 424 flying hours. The new, small camera developed by the National Research Council satisfactorily passed ground tests, but air tests, scheduled to commence on 24 August, could not be undertaken due to lack of a suitable aircraft, and the project was shelved.

Most of the Deltas were used on an intensive program to photograph the east coast during the 1939 season. The future role of this aircraft was in doubt, although one more had been added to the fleet in January 1939. The high performance Deltas had no real military application—too fast for army co-operation, too slow for combat, unsuitable for over-water operations, too small for realistic transport duties, and too complex for training purposes.

Friction continued between the incompatible field detachments and the accounting department over financial advances made both ways—from the government, and from the private pocket of the OICs of PDs. S/L K.M. Guthrie, Rockcliffe CO, with long experience on photo flights, responded in March 1939 to a query from the accounting section regarding the previous year's transactions by No.3 GPD. He pointed out the impossibility of searching approved, closed and destroyed accounts and went on the attack: "It was often necessary for the detachment commander to use his personal funds to finance his detachment. This was directly attributable to the almost always prolonged delay in payment of detachment claims. The officers commanding northern photographic detachments have had to make numerous contributions of this nature during the course of the past operational seasons." He detailed the accepted end-of-season balancing of accounts, usually ending with the officer in a credit position— "money . . . which had, in effect, been loaned to the Dominion Government, without interest, to enable it to carry out photographic operations, in spite of the restrictions imposed upon public servants by the regulations governing the handling of Public Funds."[136] The government had changed its ways very little since the 1840s when the first director of the Geological Survey, Sir William Logan, often put up his own cash to finance the initial surveys of Canada East and Canada West.

The international situation became ominous. On 24 August 1939, the chief of the Air Staff instructed the officer commanding No.8 GPS that all aircraft engaged on photography must "return to Ottawa forthwith."[137]

Over the previous two decades, the Canadian air force had turned its hand to a wide variety of civil tasks—forest patrol, firefighting, transport of Indian Treaty payments, immigration patrols, customs patrols, preventive patrols, spore collection, airmail, pilot training and passenger route development—but only photography remained a constant factor. This era ended in September 1939. Aerial photography of Canada did not regain its old momentum until a new generation of aircraft and crews returned from the battlefields six years later to take up the baton.

137 F.H. Hitchins, op. cit., p. 385.

The importance of early photographs steadily increases over the years. As memories wane and records disappear, historians, town planners, civil engineers and the general public make increasing use of the six million aerial photographs, particularly the 20,000 oblique views, taken by the air force prior to 1939 and readily accessible at the National Air Photo Library. One such photograph was taken from an RCAF aircraft flying over Ottawa on 12 August 1925. The crew would be surprised at the changes evident 73 years later on 26 May 1998 when Marshall*Maruska Aerial Images of Chelsea, Quebec, "matched" their photograph with a modern version.

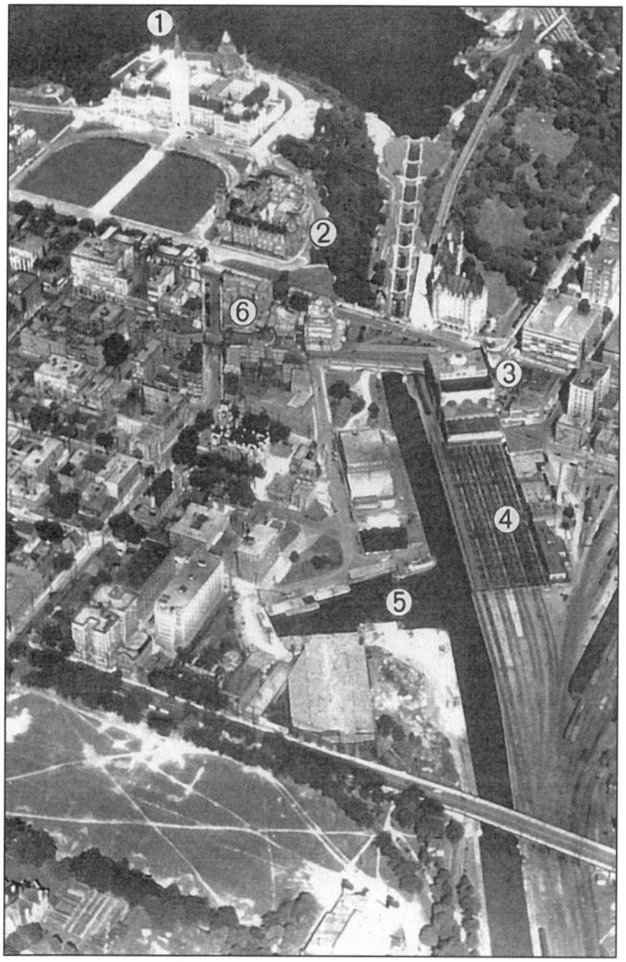

Looking northwest along the Rideau Canal to Parliament Hill, Ottawa, 12 August 1925. *NAPL HA41-6*

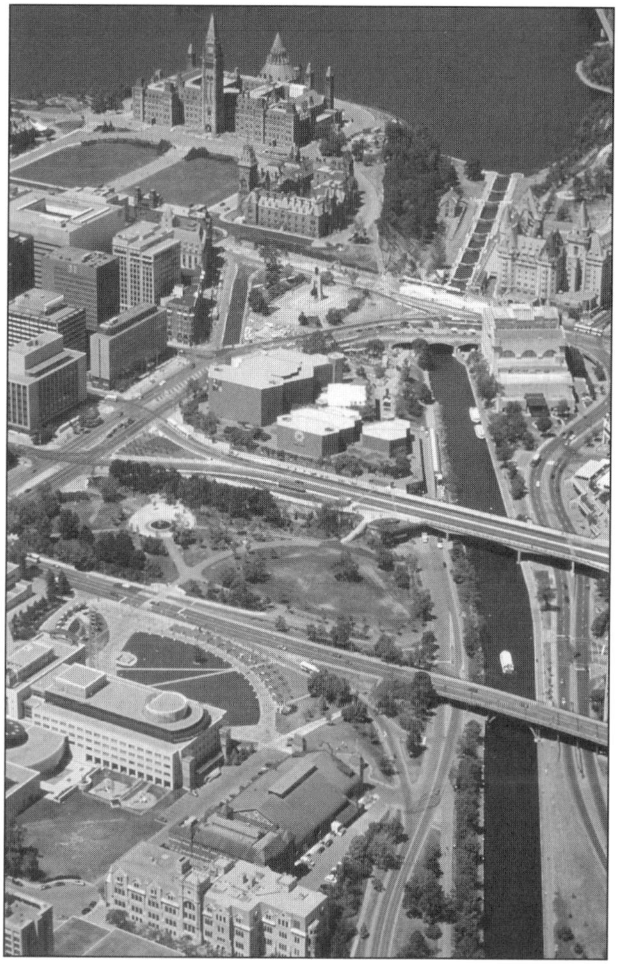

The same view, 26 May 1998. *MARSHALL*MARUSKA AERIAL IMAGES*

In 1925, the Peace Tower on Parliament Hill (1) had just been rebuilt after the disastrous 1916 fire and its new sandstone gleamed in the sunlight; the east wing of the Chateau Laurier Hotel (2) had not yet been built; the railway station (3) is now the Federal Conference Centre destined to become the Canadian Sports Hall of Fame; the railway tracks (4), previously a turning basin for Rideau Canal boats, have been replaced by the Westin Hotel; foundations for the MacKenzie King Bridge replaced the turning basin (5) in the early 1950s and the National Arts Centre was built to the north; and the area that became Confederation Square in 1939 (6) was still occupied by the Post Office.

Part II

THE MEN

THE SELECTION of heroes to profile was difficult, and I may be correctly accused of grave omissions, for which I apologize. My objective was to describe the lives of men representing the widest possible range of trades and professions involved in aerial photography. Decisions had to be made, in many cases, on the basis of my ability to obtain information from the individual, his family or his friends—and permission to publish.

Among the many heroes not profiled are high-spirited B.G. Carr-Harris, who died in a 1942 crash; RNAS veteran and bush pilot pioneer Stuart Graham, whose experiences have been well documented by his son Robert; B.D. Hobbs, who sank a German U-boat and shot down a Zeppelin in World War I, and led the epic expedition to Reindeer Lake in 1924; J.D. Hunter, who went from carpenter to photo pilot and superintendent of operations for the transport department; J.E. Jellison, acknowledged by A/M C.R. Dunlap as "tops in the field";[1] C.C. MacLaurin, who played a key role in establishing the Canadian Air Board and launched the base at Jericho Beach, but died in an accident there in 1922; R.E. McBurney, who held senior positions with the famed Canadian 6 Group of Bomber Command and whose invaluable advice from his years as a photo pilot is reflected throughout this book; R.A. Logan, a surveyor and photo pilot, who went on to head the Irish Air Force and become a colonel in the US Army; E.L. MacLeod, a World War I aviator acknowledged to be the RCAF's best flying boat pilot; C.F. "Black Mike" McEwen, who had 22 victories in the First World War and commanded 6 Group in the Second; J.H. Tudhope, World War I "ace," who played a leading role in establishing Trans-Canada Airlines; C.C. Walker, who went from chemical warfare in World War I to become an aerial photographer and authority on Genghis Khan; and J.A. "Squirt" Wiseman, who joined the RCAF as an AC2 labourer and went to become an expert on northern operations and command 413 Photo Squadron.

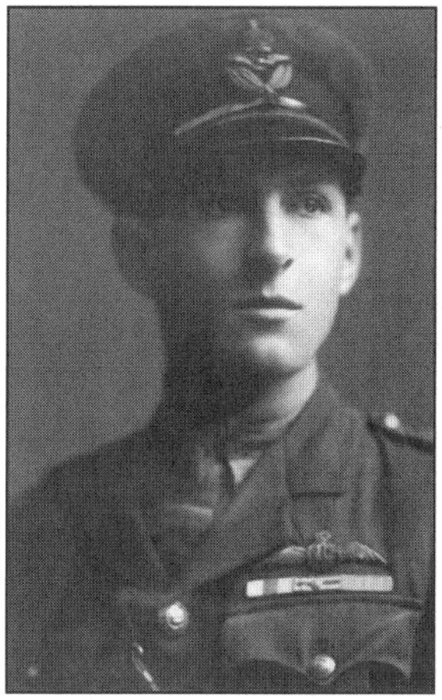

Lt. C.M. McEwen, 1918.
DNDPU RE14938

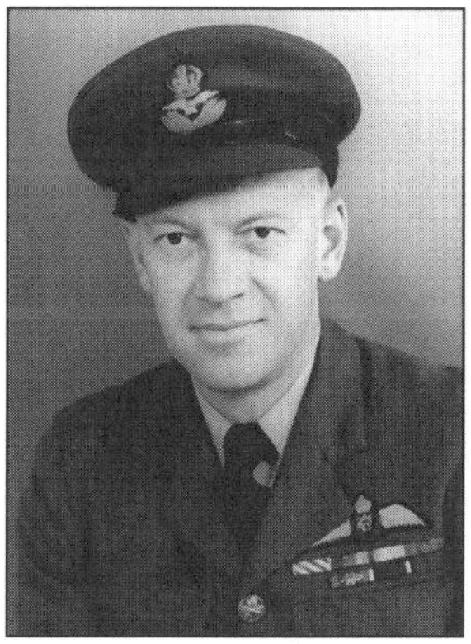

S/L J.A. Wiseman, 1946.
DNDPU RE4185

1 C.R. Dunlap, Lecture to the Canadian Aviation Historical Society 3 June 1972 (Willowdale, Ont.: CAHS, 1972): 14.

David James Robertson Cairns

BORN ON 3 October 1892 in Edinburgh, Scotland, where his father was a gunpowder works foreman, Jimmy Cairns played one of the most influential roles in developing Canadian aerial photography. He attended George Herriot's School for Boys, art college and university in Edinburgh, specializing in chemistry, photography, English and elocution. Evidently of an artistic inclination, he enjoyed photography, elocution, sketching and painting.

Cairns emigrated to British Columbia before World War I. He worked in the United Photographic Stores and a commercial art studio for a total of three years before advancing his photographic knowledge the hard way as a self-employed demonstrator and travelling salesman for photographic materials. In the expectation and hope of being sent overseas, he joined the 44th Wing, RFC at Victoria, BC as a second-class airman on 1 September 1917. His skills were such, however, that he was retained in Victoria with the rank of first-class airman to instruct in the use of aerial cameras, a job which gave him a taste for flying and an appreciation of what aerial photography could accomplish. He was demobilized on 28 January 1919, joined Gordons Ltd., a photographic store in Victoria, and supplemented his income by teaching elocution. He married Elsie Samson in Victoria on 21 July 1920.

Very keen on participating in the emerging science of aerial photography, in November 1920, Cairns applied for the position of civilian photographer for the CAB at Victoria Beach, Manitoba. He was accepted at a salary of $1,800 per annum, and reported in April 1921 to Dartmouth for "training on a special camera to be used for northern Ontario operations." Shortly afterwards, Victoria Beach Station Superintendent Basil D. Hobbs noted that he seemed "absolutely under the command of his wife" and recommended that he be dismissed "if he allows this to interfere with his work." Evidently the problem was satisfactorily resolved, as the following year Hobbs strongly recommended him for a commission to the newly created position of photographic officer. Cairns was commissioned a pilot officer in the RCAF on the day of its formation, 1 April 1924.[2]

2 NAC Personnel File (applies to subsequent quotations not otherwise cited).

Photographer P/O D.J.R. Cairns had to face 60-mph winds while operating his Fairchild camera from the exposed front cockpit of a photographic flying boat. S/L B.D. Hobbs, on left, in Viking G-CYET with Robert D. Davidson, DLS, at Rabbit River, Manitoba, July 1924. *NAC REINDEER LAKE REPORT*

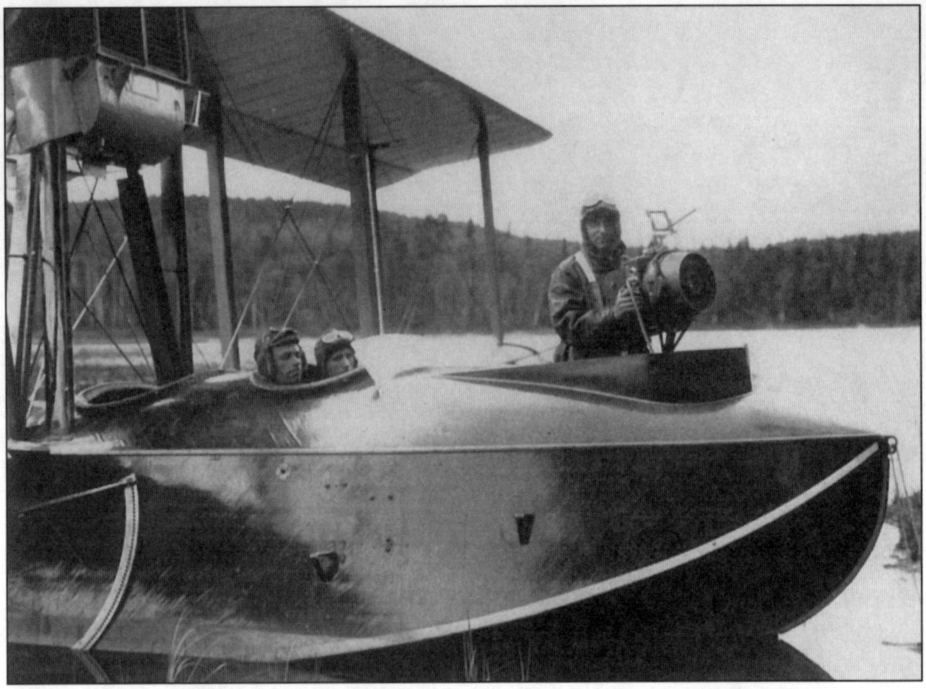

Again evidencing his support, Basil Hobbs selected P/O Cairns along with Cpl. Alex Milne, mechanic, and Robert D. Davidson, DLS, as Viking crew members for the epic 1924 survey flight to virtually unknown country along the Reindeer and Churchill Rivers in northern Manitoba. The crew had to depend on its own resources, except for fuel caches laid down the previous winter, and flew 2,810 miles in one month. Cairns took about 1,700 photographs covering 15,000 square miles. Hobbs had this to say at the end of the remarkable venture:

> During the whole operation, the conduct of the crew was beyond praise. . . . To Flying Officer Cairns fell the heavy work of operating the camera. The physical labour entailed in this when everything works smoothly is sufficient to try the strength of anyone, and when it is considered that on the majority of flights made he had to work the camera by hand, the strain imposed by this task can well be imagined.[3]

The photographs taken made a perfect mosaic of the Churchill River-Reindeer Lake region, and proved to all the skeptics that northern survey by aerial photography was feasible—and, perhaps most significantly, economical. This expedition was the precursor of thousands of such flights, which eventually resulted in a complete and accurate map of Canada.

In common with other officers of the period, Cairns had to be proficient in drill and attended the Royal School of Infantry in 1924 where he did "quite well, but needed more practice in developing a stronger word of command and

3 B.D. Hobbs, "Report on Reindeer Lake and Churchill River Photographic Operations 1924," NAC Accession No. 912040 (Winnipeg: n.p., 1924).

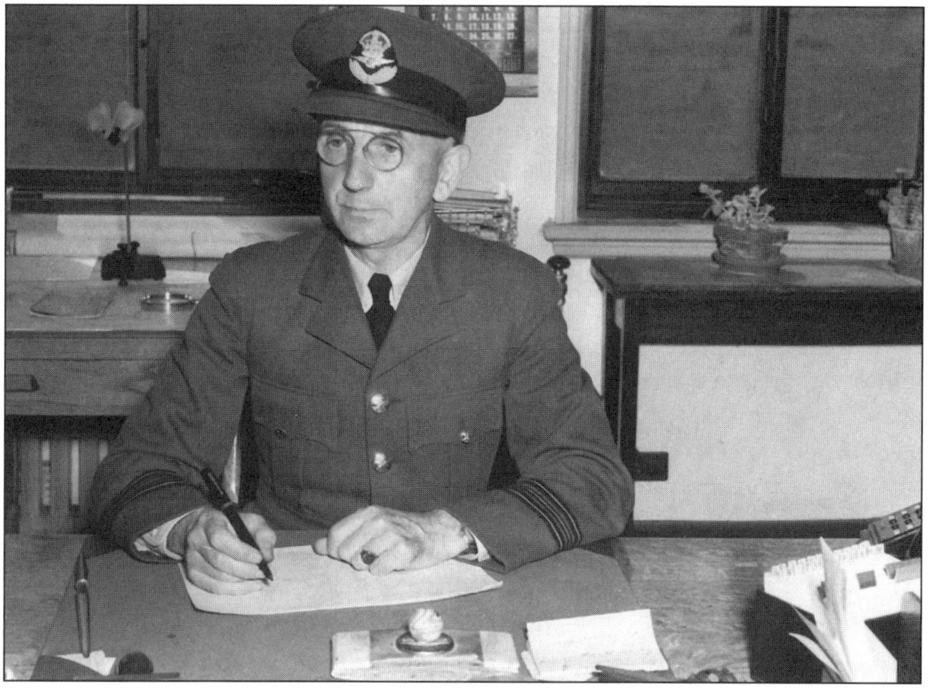

S/L D.J.R. Cairns, CO RCAF Photographic School, Rockcliffe, 7 August 1940. *DNDPU PL-830*

in the use of the sword." Cairns progressed steadily to flying officer in 1925 and flight lieutenant in 1928, spending many hours in the air and developing photographic equipment and techniques. He visited the Fairchild and Kodak plants on several occasions, co-operating on camera improvements and inspiring letters of appreciation from both companies. Even through the austere Depression years he received travel authority to continue the visits, perhaps influenced, as W/C G.O. Johnson reported, by the fact that he saved the RCAF over $500 in 1930 alone by obtaining replacement parts "free of charge."

F/L Cairns was devastated in 1934 when removed from the general list and aircrew status because he had reached the age limit for his rank—with consequent loss of pay. He was reinstated after successfully pleading that he had two boys "to be given an education necessary to give them a fair chance in life to meet the competition which is prevalent today in the business world," and hinting that special circumstances existed regarding his own abilities.

Cairns's appraisals from 1921 followed a consistent theme of high technical competence, keenness and reliability, but this was not achieved without overcoming a serious handicap. In 1937, Cairns reluctantly revealed why he was having difficulty in passing promotion exams: "I have the misfortune to suffer from a form of temporary amnesia or mental aberration brought on by nervous strain when studying." His commanding officer at the Trenton School of Army Co-operation, S/L T.A. Lawrence, was sympathetic, noted his distinguished service record and recommended accelerated promotion. Although Cairns commanded the school in Lawrence's absences, he did not become a squadron leader until 1 January 1940, by which time he commanded the School of Photography at Rockcliffe.

Hundreds of officers, airmen and airwomen graduated from the Photo School during World War II. The basic course lasted 12 weeks with an additional six weeks for those destined to become instructors at the ten training schools for air observers under the British Commonwealth Air Training Plan. In addition to taking photographs, processing and interpreting them and making maps, instruction was given in infrared techniques and maintaining the camera gun used at the air gunnery schools. A key training device, the camera gun took photographs on 16 mm film instead of firing bullets. Back on the ground, the results projected on a screen gave indisputable proof of the student's accuracy. In a fireproof vault were stored more than one million aerial photo negatives accumulated to that time.

An indication of Cairns's keenness and initiative is illustrated by the sole reprimand in his service record. As part of a plan to transfer him from photographic duties, he was removed in 1935 from the Associate Committee on Survey Research subcommittees on Infra-Red Photography and Air Camera Auxiliaries. He attended their meetings anyway and was rebuked for doing so without authorization.

Based at Winnipeg for most of his photographic career, administrative and instructional duties subsequently took Cairns to Summerside, PEI, and to Ottawa, Trenton and Mount Hope, Ontario. In 1942, he was taken off the flying list, promoted wing commander and sent to command No.1 Manning Depot at Trenton. With victory in sight and training winding down, it was decreed that Cairns could not be "advantageously employed at his present rank." He was released from active service on 3 March 1945, retired effective 23 September and went to live in Toronto where he died on 5 November 1976. A lasting tribute to his memory is Cairns Lake, east of Pelican Narrows, Manitoba, named in his honour in 1974.

Jimmy Cairns initiated a remarkable tradition. He presented pilots' wings to his son Norman who flew Spitfires in Britain during World War II and returned to Canada to fly with No.413 Photo Squadron from Rockcliffe, moving on to more exciting work with Vampires, Sabres, CF-104s, CF-5s, CF-100s and, on exchange with the USAF, A-7s, F-105s and T-33s. Norman, in turn, presented wings to his son Brett in 1978. Brett went on to fly CF-5s and RF-4s on exchange with the USAF, and CF-18s with No.416 (CF-18) Squadron, which he commanded. As of 1999, Brett had risen to colonel and was serving as deputy commander of Alaska NORAD Region. The Cairns family lays claim to a longer continuous record of service than the RCAF itself, which did not acquire that title until 1924, three years after Jimmy joined its predecessor, the Canadian Air Board.

Thomas Frederick Cooper

TOMMY COOPER was born on 14 August 1891, the eldest of three brothers, in Birmingham, England, where his toolmaker father worked for a baby carriage manufacturer. He was destined for a career in music, starting cornet lessons at the age of eight, and was soon performing solos, standing on a wooden soapbox, with a Salvation Army band. Music remained his lifelong passion, but circumstances directed him to a military career and he served in uniform for over 40 years.

After demobilization from the RAF, Cooper worked for several automobile manufacturers, notably BSA in Birmingham. He had met Lily Maud Wray before he went to France and she remembered that they used to take rides into the countryside on his motorcycle and he would often stop and tinker with the engine if it "didn't sound right."[4] They were married on 31 July 1920 and emigrated to Canada in April 1921.

.After short stints as an independent mechanic and working for the Ford Motor Company in Winnipeg, he returned fulltime to his love of music in April 1922 and signed up as a bandsman with the Princess Patricia's Canadian Light Infantry.

Moving again from the army to the air force, Cooper transferred to the Canadian Air Force on 15 April 1923 as an aero-engine mechanic at Winnipeg. Versatility was the order of the day in the embryo service and he spent the summers maintaining aircraft and flying on forestry patrols and photographic surveys from Victoria Beach and Lac du Bonnet. Winters were devoted to repair and maintenance of the aircraft stored in Winnipeg.

Cooper had fond memories of his early years at Winnipeg when the establishment numbered only about 30 men. As he wrote in *Roundel*, Jan–Feb 1959, "We were a happy gang, despite the fact that we were handicapped by lack of tools and equipment."[5] Social life revolved around a seven-piece orchestra led by violinist Cpl. Jack Henderson with F/L G.V. Walsh at the piano. Other regular members were F/L L.F. Stevenson–saxophone, Sgt. G. Moon–trombone, Cpl. T. Terry–drums, "Lofty" Horton–guitar, and Sgt.

4 Irene Cooper, personal communication.

5 T.F. Cooper, "Trials and Tribulations of the Ground Crew," *Roundel* 11, no. 1 (1959): 8–10.

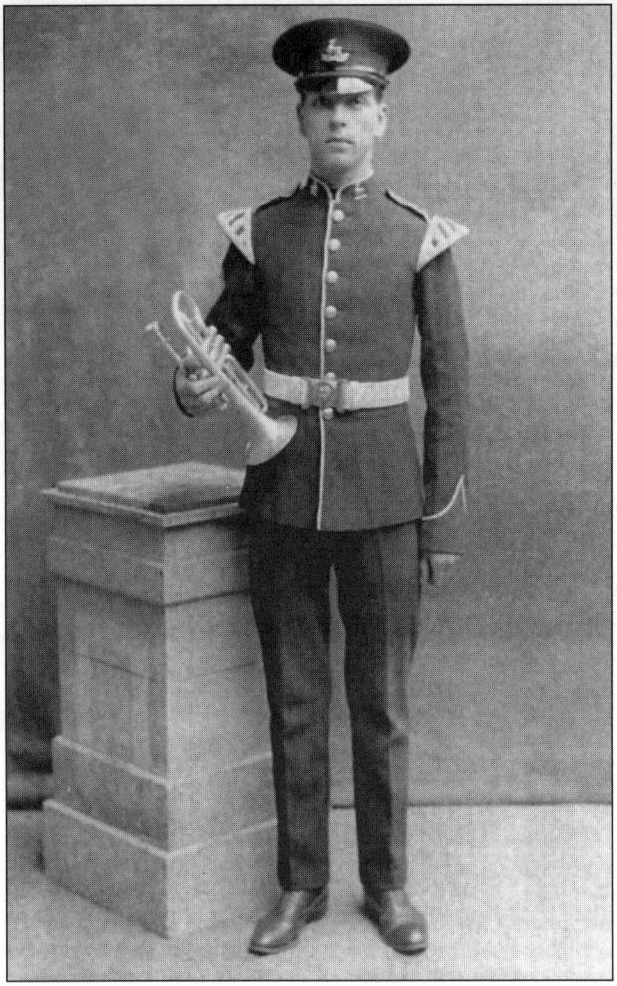

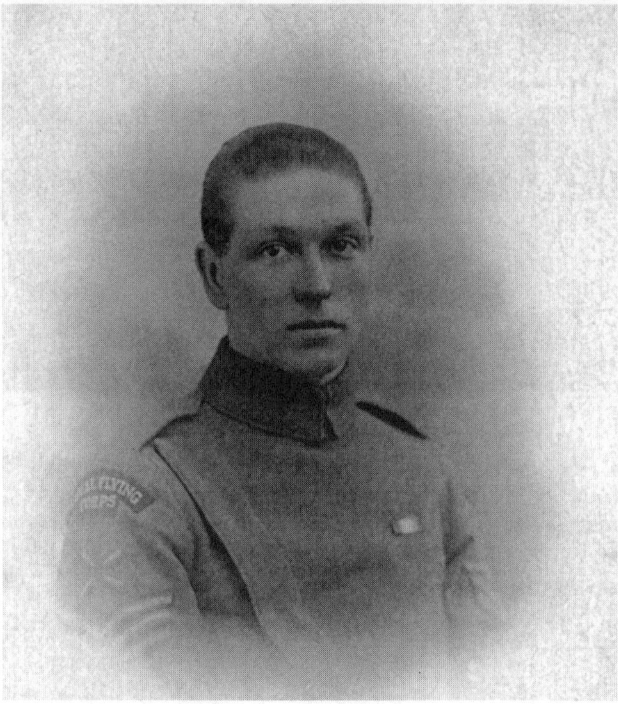

Left: Tommy Cooper joined the 5th Battalion, Royal Warwickshire Regiment, Territorial Army, as a bandsman at the age of 18. *IRENE COOPER*

Above: He successfully completed his seven-year engineering apprenticeship with the Wolsley Tool & Motor Company before being shipped in the infantry to France. He managed to transfer to the RFC in December 1915, and remained as a sergeant mechanic with the RAF at Henlow in England until April 1919. *IRENE COOPER*

Tommy Cooper–trumpet. Their uniform was a Norfolk-style jacket with cloth-belted waist; cap and lapel badges were silver wings with "CAF" in the centre and the motto *Sic iltur Ad Astra* below. Cooper recalled, "It was dynamite to be near a railway depot as you were always asked the time of departure for trains."

The speed of Cooper's rapid rise up the promotional ladder—from airman second class, the lowest rung, in April 1923 to acting sergeant on 1 February 1924—indicated his ability. Perhaps a more significant measure of his worth in those austere days was being placed on the permanent force and paid at his acting rank. On 1 April 1924 he remustered into the new RCAF as a sergeant fitter. In 1925, F/O (later A/M) Roy Slemon summed up his technical performance as "Very good," identifying his other interests by adding that he was also a good rifle shot, good cornet player and a keen drill instructor.[6] Cooper's 1938 assessment was typical of those throughout his

6 NAC Personnel File (applies to subsequent quotations not otherwise cited).

career: "A zealous efficient and thoroughly reliable NCO. Excellent with men and a first class machinist. A bandmaster of considerable merit and ability."

The possibility of a transfer raised its ugly head in 1926. The Coopers did not want to move their two young daughters, Dorothy and Irene, around the country so Tommy "bought himself out" of the RCAF. For the next three years he sustained his family by playing with local orchestras and dance bands and accompanying silent movies. He also worked part time as a mechanic at Athey's Garage. Despite his small income and the demands of two children, he managed to salt away enough dollars to demonstrate his affection for Lily by delighting her with a fur coat at Christmas in 1928.

In 1929 he re-enlisted in the RCAF on the promise that he would regain his former rank of flight sergeant, but the Depression intervened and he was not promoted until 1938, after his posting to Trenton. In the meantime, he remained a sergeant based at Winnipeg and its sub-stations.

In charge at Victoria Beach was F/L H. "Gus" Edwards who, according Cooper, performed an act hardly in keeping with the dignity of his later exalted rank of air vice-marshal. As S/L G.O. Johnson taxied away from the Victoria Beach after inspecting the base, Cooper in his best uniform "walked" the wing to the end of the dock. As the tailplane passed, he leaned forward to make sure everything was clear. Edwards could not resist the temptation and applied his foot to the obvious target, "thereby demonstrating a sense of humour not fully appreciated by me at the time."[7]

Cooper usually flew with F/L A.D. Ross, who called him "Plugs" and considered himself lucky, as he reported in *Roundel*, "to have one of about the only six people in Canada who really knew how to get the best out of the Rolls-Royce Eagle IX engines which constituted the inadequate power plant of the Viking."[8] They were practising step landings at Cormorant Lake, Manitoba, with Viking IV, C-GYEV, on 3 October 1930 when the aircraft caught a wave and dived under water. The only injury resulting was a broken finger for Cooper, but he soon recovered. An emergency landing with F/Sgt. Bob Marshall in another Viking caused no damage, but the pair had to survive for eight days in the bush until they were found, "a little worse for wear, but still in good spirits."

Cooper took six weeks' leave over Christmas, 1930, to visit his family in England. Always keen, he obtained permission to visit the Armstrong Siddeley factory at Coventry to "gain the latest information on Lynx engines." Cooper made one of the first direct flights from Winnipeg to Vancouver with pilot Elmer G. Fullerton in a de Havilland Gypsy Moth in 1933. His daughter Irene, by then a commercial pilot with instructor rating, retraced the route 21 years later in her Aeronca Chief.

Transferred to the Technical Training School at Camp Borden in 1935, Cooper became assistant bandmaster of the station band. After moving to Trenton the following year as an aero-engine instructor, his love of music

7 T.F. Cooper, op. cit., p. 19.
8 E.P. Wood, "Northern Skytrails," *Roundel* 1, no. 3 (January 1949): 26.

inspired him to organize the RCAF Trenton Band. Retired F/L Chris Lockyer has vivid memories of those days. Newly graduated from high school in 1938 and with no job in sight, but an accomplished musician, he was pleasantly surprised to be propositioned by Cooper to join the RCAF. Plans were afoot for the band of the junior service to play an important role in the 1939 visit of King George VI and Queen Elizabeth. Although there was no official trade of RCAF bandsman in those days, Cooper succeeded in recruiting young Lockyer, and in organising a volunteer band. The band toured Ontario with the king and queen in 1939 and performed for the governor general, political leaders and visitors from around the world. Lockyer remembers this period as one of the happiest in his long career:

> Both in concert and on the parade square this band in blue barathea with lots of gold braid certainly made its impact, thanks to Tommy Cooper, who was like a father to us. His dedication, counselling, understanding and humour made us like a family . . . and we made a tremendous impact on the people of Ontario. I was proud to be one of Tommy's boys.[9] .

Lockyer's father Sidney was bandmaster of the Hastings-Prince Edward Regimental Band; common interests led to a lifetime friendship with Cooper and an illustration of the regard in which he was held. A few days after Sidney died in 1948, Cooper visited Chris Lockyer who commented: "To this very day, I'm grateful for Tommy to have taken the time to come from Winnipeg to Picton to pay his respects to my father and to visit his surrogate son of air force days."[10]

After the Trenton Band was disbanded in 1939 as the volunteer bandsmen were called to more urgent duties, Cooper played a leading role in establishing the RCAF Central Band. These musical distractions generated some criticism that he was not adequately attending to his engine fitter duties on Fairy Battles. Nevertheless, he was awarded an additional task as drill instructor. In all probability he welcomed the opportunity to exercise his skills on the parade square for the edification of raw recruits flooding in as part of the build-up in response to signals of imminent war in Europe. He was promoted warrant officer, second class, in August 1940 and warrant officer, first class, three months later.

Cooper's organizational abilities were recognized in May 1941 with a commission and promotion to temporary flying officer at No.8 Repair Depot, Winnipeg. Rapid promotion brought him to the rank of wing commander and command of the depot in 1943. His talent for fixing problems was in constant demand at the busy base. For example, a USAAF P-47 Thunderbolt arrived during the 1943–44 winter for testing on skis, but it was a warm winter and the runway was dry. Taxiing on the asphalt ruined the USAAF skis, so Cooper

9 Chris Lockyer, personal communication.
10 Ibid.

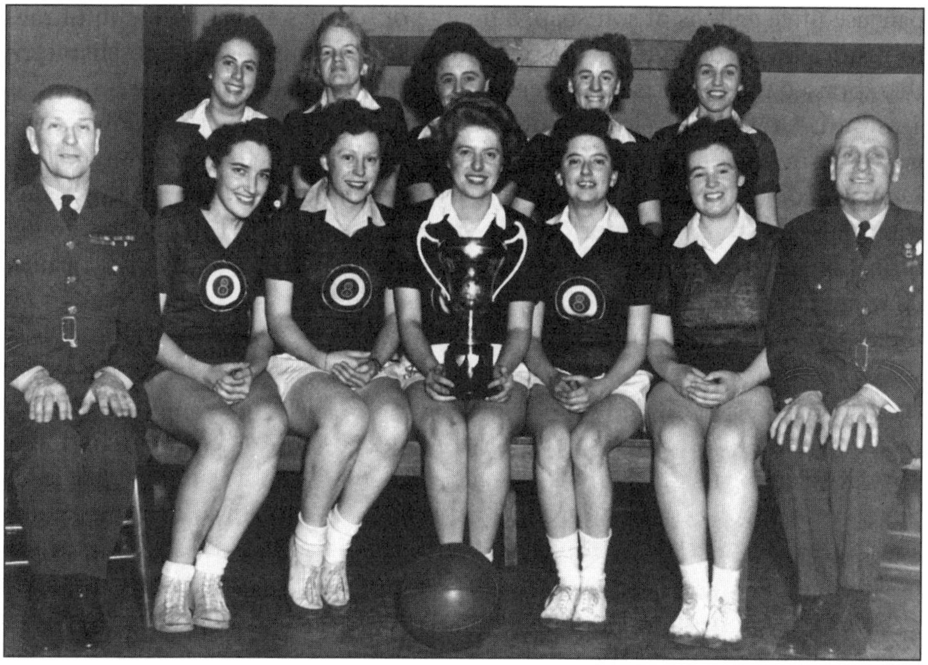

Despite increasing technical responsibilities, Tommy Cooper found time to encourage many popular sporting events. Here, he is on the left with No.8 RD Women's Division Basketball Team in 1945; F/L Harry Jenkins is on the right. He also organized No.8 RD Airwomen's Trumpet and Drum Band. *WESTERN CANADA AVIATION MUSEUM*

designed another set that 8 RD built in short order. Ground crew bulldozed snow onto the runway, but directional control was difficult due to severe engine torque in combination with meagre vertical tail surfaces of the P-47. Cooper added strakes to the skis, which permitted straight-ahead taxiing, but the project was abandoned soon afterwards and the P-47 flew home on wheels.

In recognition of his service at Winnipeg, Cooper received the OBE in 1944:

> This officer's services as chief engineering officer have been exceptional. He has executed his ordinary duties and the special work given to him with skill and dexterity. He has developed numerous modifications and repair schemes for aircraft which have resulted in increased aircraft serviceability at reduced costs. Giving unstintingly of his time and with concentrated effort, this officer has built up an exceptionally well organized technical branch at a Repair Depot.

> Cooper also possessed the 1914–15 Star, British War Medal (1914–1920), Victory Medal (1914–1919), Canadian Volunteer Service Medal, Canada Medal, King George VI Coronation Medal, Queen Elizabeth II Coronation Medal, Long Service Medal (RAF), Territorial Efficiency Medal, and Canadian Forces Decoration. He was very proud of the distinction, held by few, of having served on the honour guard for three reigning monarchs—Edward VIII, George V and George VI, the latter two at their coronations. Cooper Lake, near his old base at Cormorant Lake, Manitoba, was named in his honour.

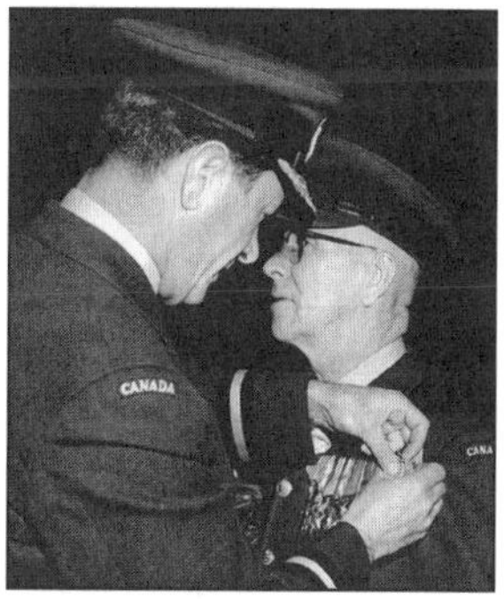

Air Commodore H.H.C. Rutledge presenting W/C Cooper with a bar to his CD at Winnipeg. *E. RAYNER*

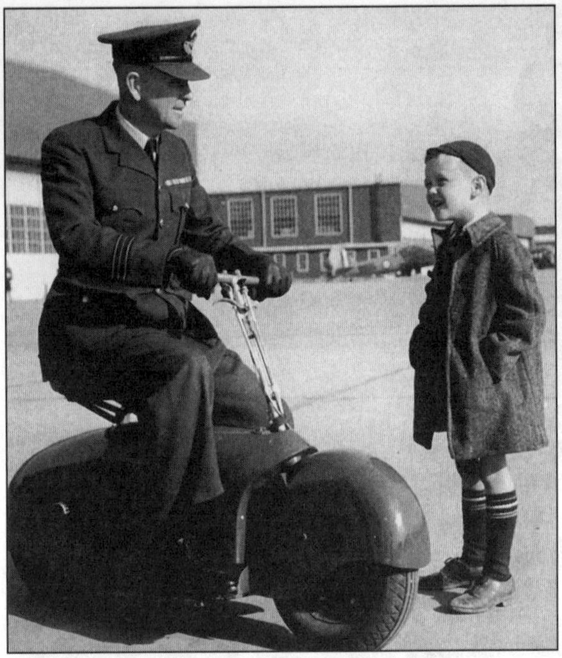

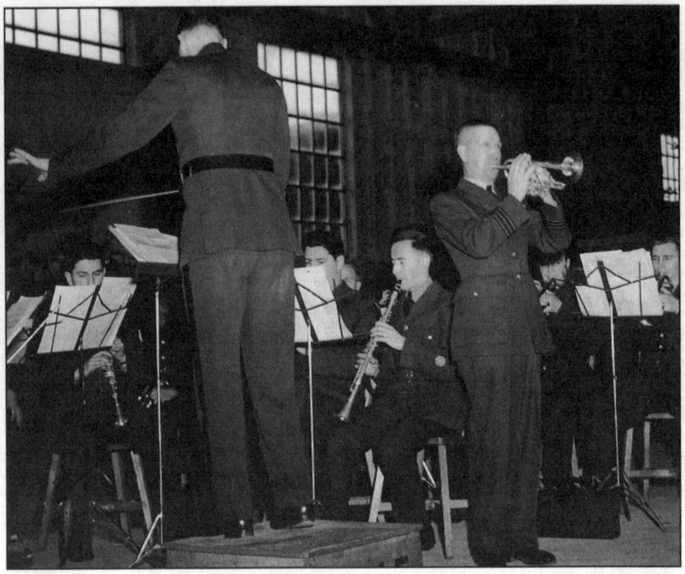

Tommy Cooper could always be relied on to participate in any musical event, even when he was commanding No.8 Repair Depot.
IRENE COOPER

Tommy's daughter Irene remembers him as a very active and interested man who never quit. "He was always a handyman and could fix almost anything. In his retirement he took up furniture making and, when he was over 80, got in trouble when his doctor found him up on his roof making repairs."[11] Throughout his life, he demonstrated his "hands-on" attitude with the introduction of a number of innovations, including a motorized scooter used for transport around the wide-spread Winnipeg base. He always had a good rapport with children.
IRENE COOPER

The wind-down of aircrew training in 1945 resulted in a decrease in the workload at No.8 RD, but the unit had to be maintained in an active status pending conclusion of the war with Japan. In order to ameliorate the Winnipeg unemployment problem, compounded by cancellation of orders at MacDonald Bros. for Avro Ansons, No.8 RD was charged with converting five B-24 Consolidated Liberator bombers to transport configuration. This was a "do-it-yourself" project with no drawings or directions from the manufacturer. The practical Cooper rose to the challenge, masterminding what amounted to a complete rebuild of the aircraft. Norman Malaney in an unpublished paper on the B-24 conversion quotes Cooper: "When we got them, I had to completely redesign the whole hull because the keel had to come out. Do you know what sort of job that would be? It was like taking the backbone out of a fish. We designed all that and built new frames. The bomb bays were taken out, flooring put in. You could convert them to litter carriers, mail carriers or passengers. You could perform the conversion in about 25 minutes."[12]

Continuing in uniform after his retirement from the RCAF in 1946, Cooper organized the seven Winnipeg air cadet squadrons into No.3 Wing. He commanded the wing until 1957, having obtained a one-year extension past his official separation date at 65 years of age. Retired RCAF and Canadian Airlines pilot Bill Davidson remembers being inspected by reviewing officer W/C Cooper when he was a new air cadet in 1951. Davidson, somewhat lacking in height, was looked down on and in a stern voice told, "Cadet, when you get home tonight you will have to ask your mother to put some manure in your boots, so you will grow a bit taller."[13] Davidson will not say if he followed

11 Irene Cooper, personal communication.
12 N. Malayney, *The B-24 Liberator Conversion Program at No. 8 Repair Depot, RCAF Station Winnipeg, 1945* (Pittsburgh: unpub., 1999).
13 Bill Davidson, personal communication.

the instructions, but admits that he is still a little "height challenged." He remembers Cooper as being both popular and highly respected by all ranks.

His enthusiasm for drill and a smart turnout is still recognised by the Winnipeg air cadet squadrons that compete annually for the Tommy Cooper Drill Trophy. Ted Rayner, one of "Tommy's boys," recalls with considerable pride that he was not only a drummer in 3 Wing band formed by Tommy, but was the flight sergeant in charge of 220 Squadron drill team that won the competition for the third successive year, thereby gaining permanent possession of the trophy. True to character, Cooper replaced it.

W/C Walter A. Mildren, an ex-RAF observer who trained in Western Canada and emigrated to Winnipeg in 1947, took over command of 3 Wing on Cooper's retirement in 1957. He offered an opinion shared by everyone who knew Cooper: "He was a fine man and a good friend. He was devoted to the welfare of young people and did everything he could to interest them in music and a healthy, productive life. He was absolutely honest and had not one devious bone in his body."[14]

Although never a military pilot, Cooper obtained a private pilot's licence in 1953 and purchased an ex-RCAF Tiger Moth. He was a director of the Winnipeg Flying Club for many years. From 1952 to 1959, he was a civilian inspector in the RCAF Aircraft Inspection Division at Standard Aero Engines

14 Walter Mildren, personal communication.

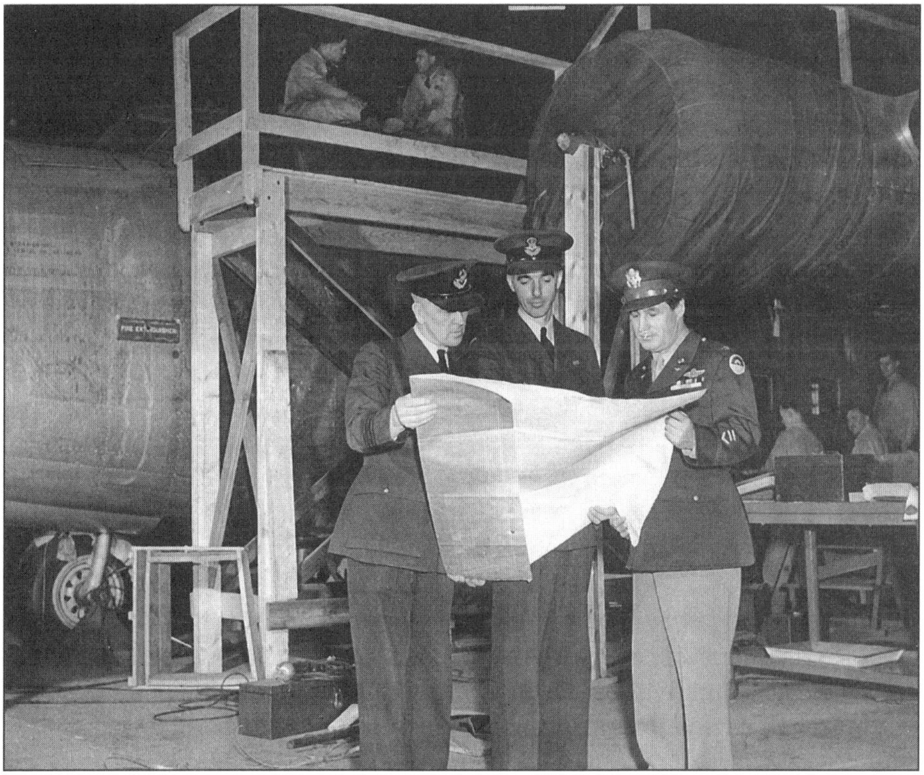

Cooper discusses Liberator modifications with S/L Sharp, RCAF, and Col. Hodgson, USAAF. To Cooper's great disappointment, his B-24 transports were never used and were sold for scrap. *IRENE COOPER*

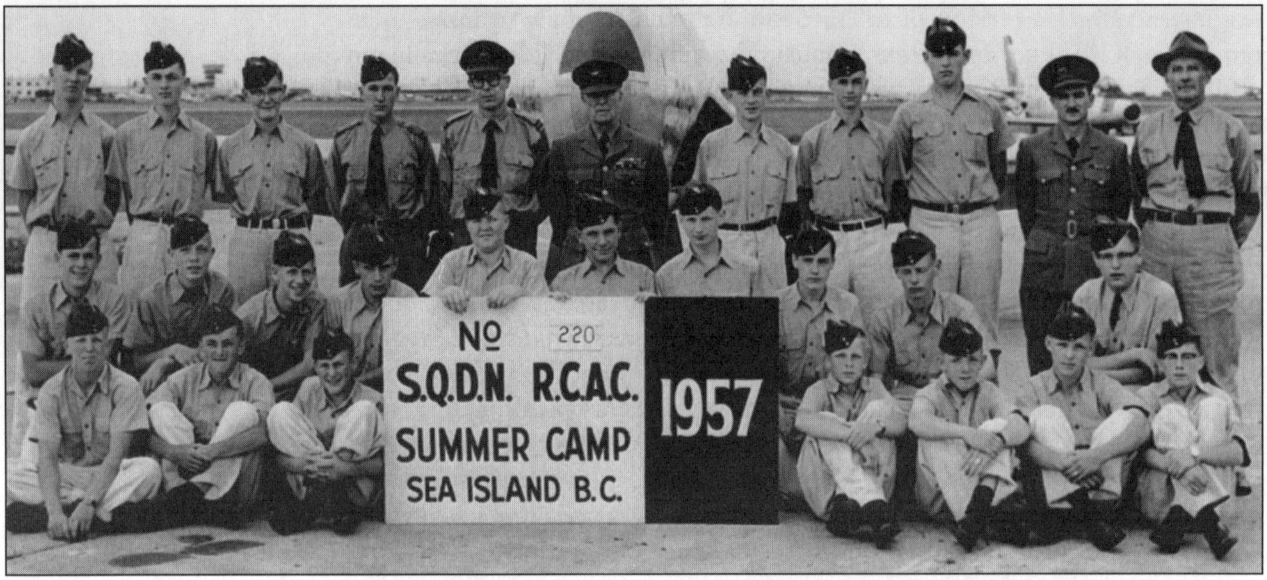

W/C Cooper attending his last air cadet camp in 1957. He sponsored Officer Cadet Ted Rayner, standing, fourth from left, who eventually became a captain and commanded No.6 Cadet Squadron of 3 Wing.
E. RAYNER

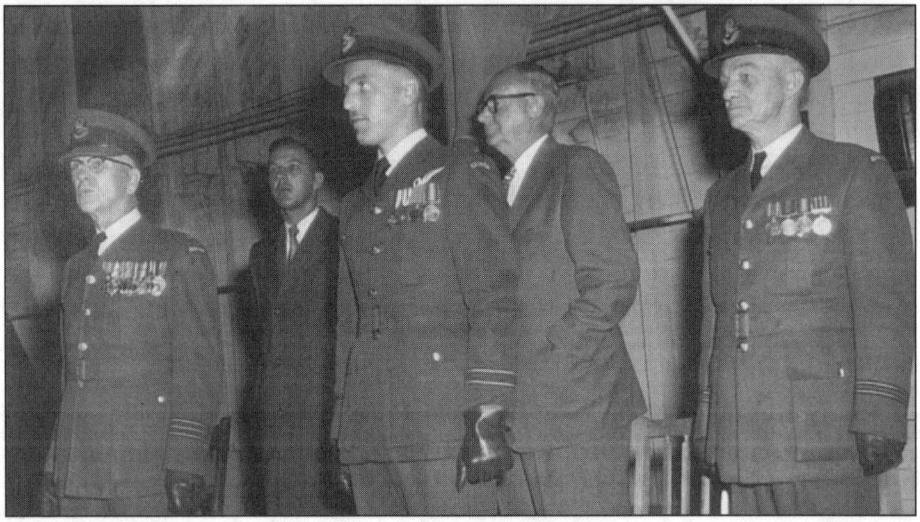

Handing over ceremony, 3 Wing, Winnipeg, 1957. (L to R) W/C T.F. Cooper, Bill Ellis, S/L W.A. Mildren (new CO), Harold Steele, F/L Art Jowett.
W.A. MILDREN

in Winnipeg, which gave him a small pension, but planted the seeds for an unwelcome problem many years later.

Tommy Cooper's real retirement years were spent at White Rock, BC, where he married Eileen Hannah Bunyan in 1970, his first wife Lily having passed away in 1961. Barely managing on a small pension, he was astounded to learn in 1976 that he should never have been receiving a second government pension for his years at Standard Aero in addition to his RCAF pension. Reimbursement was demanded for 16 years overpayment. Fortunately, common sense prevailed and it was admitted that he was an innocent party and that repayment would cause hardship, and the "debt" was forgiven.

Tommy Cooper died on 22 November 1977.

Stewart Clarke Dearaway

STU DEARAWAY was born at High River, south of Calgary, Alberta, on 2 February 1905. As an impressionable teenager in 1921, he watched the construction of an airfield near his home town as the embryo Canadian Air Force transferred its western base about 100 miles east from Morley to escape the prevalent turbulent winds in the lee of the Rockies. Open-cockpit D.H.4 biplanes flying from the base on forest fire patrol and photo missions fired the youth's imagination. He left his job as projectionist at the High River movie theatre to enlist in the Canadian Air Board on the bottom rung as an airman fitter, motor transport, in 1923.

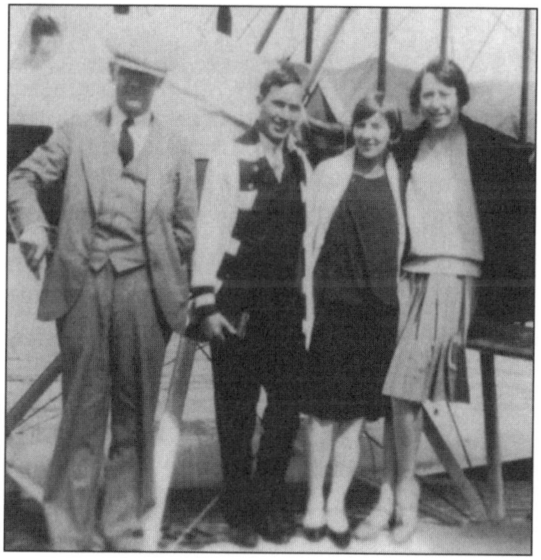

Stu Dearaway married his hometown sweetheart Myrtle Florence Wakeford in Calgary on 14 March 1928 before he was posted to Rockcliffe Air Station, near Ottawa. The young couple are in the centre with friends in front of a de Havilland Moth at St. Donat, Quebec in 1928. *PERRY DEARAWAY*

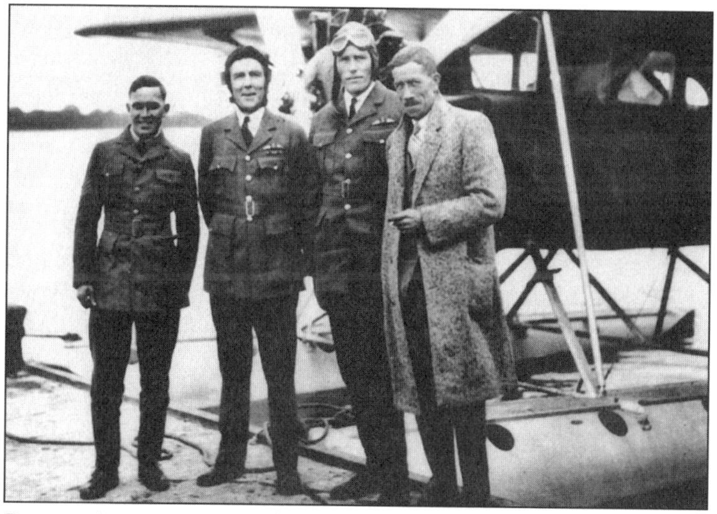

Dearaway's name went into the history books in 1930 when he was both mechanic and photographer on the epic three-month, 13,000-mile expedition to inspect fuel caches and photograph the main water and air routes in the Northwest Territories. (L to R) A/Sgt S.C. Dearaway, F/L F.J. Mawdesley, F/Sgt. H.J. Winny and C.S. Macdonald, DLS, on their return to Rockcliffe, 1 October 1930. The previous night was spent at Sudbury where they shaved and showered, and the three servicemen changed into uniform, discarding their well-used civilian clothes. *DNDPU PMR94-486*

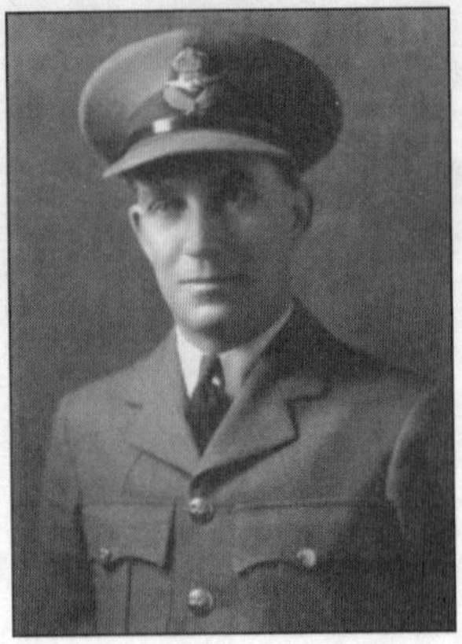

Stu Dearaway soon after receiving his commission in 1940.
PERRY DEARAWAY

The CAB was reorganized as the RCAF on 1 April 1924 and Dearaway, along with other airmen, was released. As the new organization became established, he was rehired on 3 October as an AC2 motor transport driver. In December 1927, he and his friend Harry Diller were posted from High River to Camp Borden, Ontario, on a photographic course. They both followed much the same path as versatile mechanics and photographers for the next decade.

For the remainder of his career, Dearaway and his wife, eventually joined by their son Douglas who now lives in California, moved to various stations in Eastern Canada and the Maritimes. By 1936, they were at Trenton where Dearaway was stationed until 1940 when he received his commission and moved to Toronto. After six months he was posted to Montreal. In January 1943, he went to No.9 B&GTS, Mont Joli, Quebec, as chief engineering officer with the rank of squadron leader, soon to be promoted wing commander.

In January 1946, W/C Dearaway was awarded the MBE, the citation summing up the previous few years of his career:

> This officer is an exceptionally fine Chief Engineering Officer, who has done a particularly excellent job in organizing Central Maintenance. Due to his outstanding ability and extremely high sense of devotion to duty, he is directly responsible for the efficient and smooth running Maintenance Unit, and for the present high state of serviceability which continues to exist at this unit. By his zeal and devotion to duty he has always set an excellent example.

Never one to be idle, after retirement from the RCAF on 13 November 1945 Dearaway became an inspector for Bristol Aircraft in Vancouver. He was a busy volunteer with the RCAF Pre-War Club and his widow proudly displays a certificate giving him life membership in recognition for his many years of service. After Stu's second retirement in 1966, the Dearaways spent the winters in Hawaii and the summers travelling extensively around the world.

Stu Dearaway died in Vancouver General Hospital on 22 November 1991.

Harry Sylvester Diller

HARRY DILLER was born in Anthony, Kansas, on 17 October 1891. His father, a Mennonite minister, and mother separated when he was four years old, and he was taken in by an uncle who expected him to labour as a field hand for his keep. He managed to complete Grade VI before quitting school to work full-time at a variety of jobs including delivering telegrams for Western Union and picking cotton. By the time he was 21 he had made his way north to Alberta where he worked as a coal miner, logger and ranch hand. His subsequent military career was surprising in view of his Amish and Mennonite background. Members of his family were killed during an Indian raid in 1757, primarily due to their pacifist views and refusal to defend themselves.

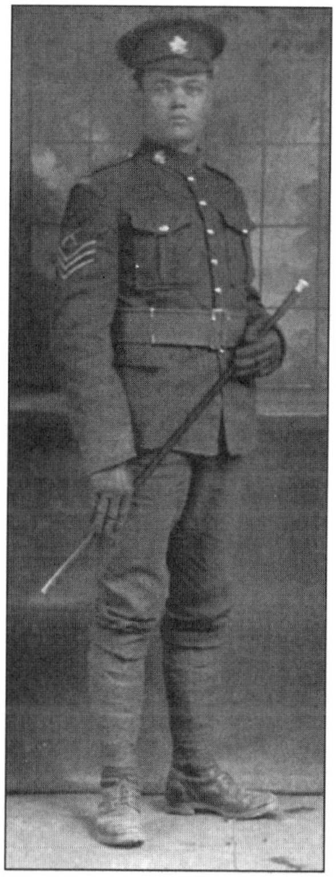

Sgt. Harry Diller, 137th Battalion CEF, England, 1916.
GORDON DILLER

Although still a US citizen, Diller enlisted as a private in the 137th Battalion of the Canadian Expeditionary Force at High River in January 1916 and was promoted corporal in May. He sailed from Halifax for Britain on the SS *Olympic* on 22 August 1916 and was promoted sergeant in September. The heavy demand for replacement troops in France had him in the front line with the 50th Battalion in February 1917.

At the Battle of Vimy Ridge, considered by many to have been the "Crucible of Canada," Sgt. Harry Diller was hit in the calf during the attack on Hill 145. Unable to keep up with the advance, he crawled from shell hole to shell hole under heavy fire, dressing wounds and attending to the needs of other soldiers until rescued the following day. His self sacrifice and endurance were considered "most praiseworthy" and earned him the Military Medal. Gangrene, a common fate of many wounded in the incredibly unsanitary conditions of the trenches, soon made its presence felt and Diller was quickly repatriated via a field hospital to Cherry Hinton Hospital at Cambridge, England. It was only his own strenuous objections that saved his leg from amputation. He did recover, but was left with a huge cavity in his right calf and constant pain to remind him of Vimy Ridge.

Despite his wounds, Diller was put to work as a physical training and bayonet fighting instructor in England from November 1917 until he sailed

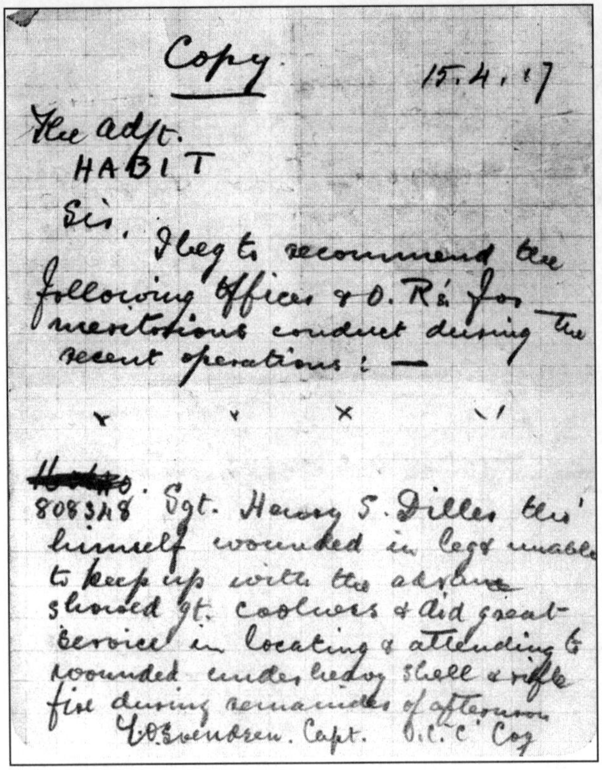

Copy 15.4.17

The adjt.
HABIT

Sir,
I beg to recommend the following Officer & O.R's for meritorious conduct during the recent operations: —

x x x x

808348 Sgt. Harry S. Diller tho' himself wounded in legs unable to keep up with the advance showed gt. coolness & did great service in locating & attending to wounded under heavy shell & rifle fire during remainder of afternoon

[illegible] Capt. O.C. 'C' Coy

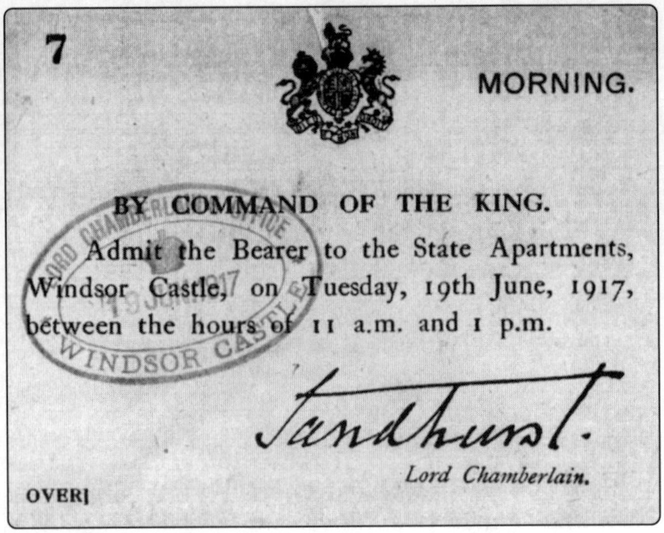

7

MORNING.

BY COMMAND OF THE KING.

Admit the Bearer to the State Apartments, Windsor Castle, on Tuesday, 19th June, 1917, between the hours of 11 a.m. and 1 p.m.

Sandhurst.

Lord Chamberlain.

OVER]

Above: Harry Diller received his Military Medal from King George V at Windsor Castle, 19 June 1917. *GORDON DILLER*

Left: Copy of Harry Diller's field commendation at Vimy for the Military Medal, April 1917. Brig. Gen. Edward Hilliam, commanding the 10th Canadian Infantry Brigade, sombrely endorsed the recommendation: "I thank you for the good service you have rendered to date, and trust that you will be spared to continue the good record you have set for yourself." *GORDON DILLER*

for Halifax on the SS *Cassandra* on 22 February 1919. He was demobilized at Winnipeg on 31 March. After visiting his parents in Kansas, he returned to High River where he renewed acquaintance with Edith Naomi Charles. They were married on 29 December 1920.

With no requirement for his military skills, limited formal education and no civilian trade, Diller had little choice but to return to ranch work. Recruitment advertisements by the newly formed RCAF, however, offered the opportunity to learn a trade and he enlisted as an AC2 fitter assistant on 24 April 1924. He always had an instinctive knowledge of machinery and knew how to drive, so his first job was "mobile equipment driver." In a time when airmen were expected to be versatile and few formal training courses existed, Diller had to learn on the job. He was promoted AC1 in July 1924 and qualified as a fitter, aero engines in October. He became an LAC in May 1925 and a corporal in August 1926.

Diller was posted to Rockcliffe in December 1927 and promoted sergeant the following year. His skills expanded to include fitter, photographer and crewman on a variety of aircraft. Perhaps falling into the familiarity trap, he was in the habit of leaving undone the safety strap in the cramped front cockpit of the Vedette, as it restricted his movement while operating the camera. However, on one memorable occasion, only a death grip on his camera

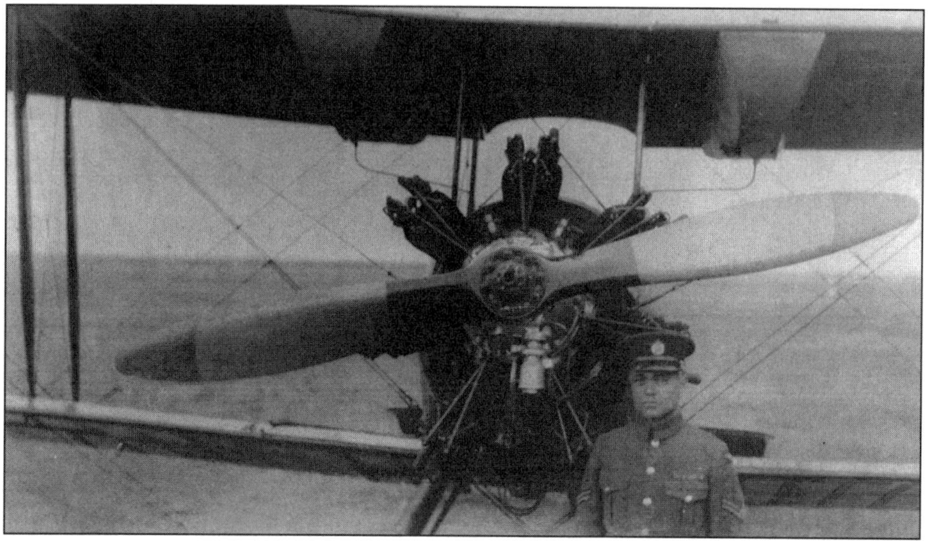

Cpl. Harry Diller with Avro 504N, High River, 1927.
GORDON DILLER

prevented a long fall when his aircraft hit turbulence, popping him right out of the cockpit. Afterwards, Diller always made a fetish of fastening the safety strap.

While stationed at Rockcliffe, he worked as a photographer from 1928 to 1930 with detachments operating from St. Donat, Huntsville, Trout Mills, Remi Lake and Trout Lake, accumulating 380 hours in Vedettes, Fairchilds and Pacemakers. Harry's wife and two young children joined him on these detached missions, living in tents except for one season when his son Gordon remembers that they were lucky enough to move into a rented cabin.

Despite his excellent assessments for duties performed over the previous five years, he was sent for formal training in 1930. He attended the advanced camera operator course at Rockcliffe and the engine fitter course at Pratt & Whitney Canada, Montreal. These were followed by an engine course at Armstrong Siddeley Motors, Ottawa, in 1931. Most likely, the engine training was intended to qualify him for a base maintenance job. He had requested excuse from detached duty because of his wife's illness and their children's medical problems. This request evidently did not affect his career progression as he was promoted flight sergeant in April 1931.

In 1936, Diller went to Camp Borden as senior NCO of the Armaments Flights, moving to Trenton Armaments Training School in 1937. Always a hands-on operator, he obtained permission to move himself from Barrie and rented a truck, submitting an expense claim for $58, saving the service $13.10. In 1938, he was promoted warrant officer, second class, and the following year, warrant officer, first class. During 1938, he took his wife and children to meet his parents in Kansas, his first visit since 1919.

At the outbreak of hostilities in 1939, Diller was posted to No.4 Command in Regina to assist in investigating the increasing number of aircraft accidents resulting from the RCAF's rapid expansion. Commissioned in May

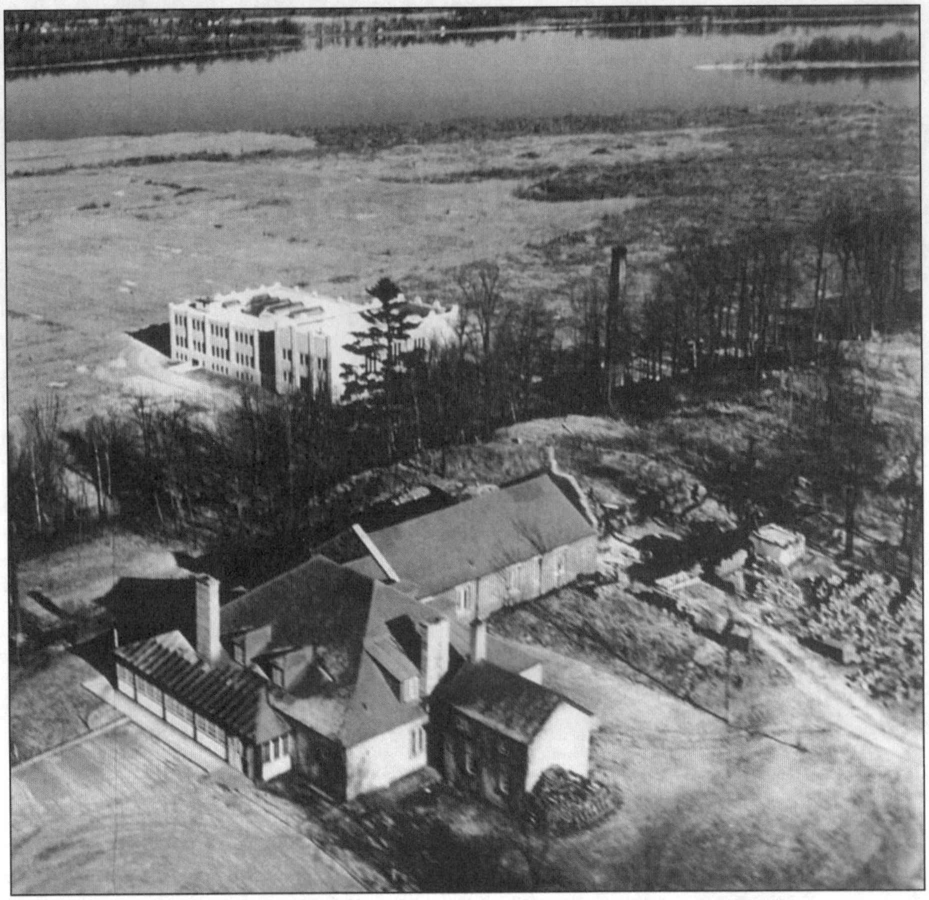

As the Depression deepened and the RCAF budget was progressively cut, there was a reduced requirement for aircrew and Diller, assessed as a "good all-round tradesman," was assigned as "foreman" from 1931 to 1936 to assist with the relief construction projects designed to create jobs.[15] Major projects he was involved in included the new Photo Establishment "White House" (centre) and the Officers' Mess (foreground) at Rockcliffe. As an NCO, he could not benefit from the magnificent Mess, but made up for this later under the invitation of his son Gordon who served in the RCAF from 1944 to 1974. Photo November 1937.
NAPL A5863-34

1941, he was transferred to No.11 Aircraft Inspection Depot at Cartierville, Quebec, responsible for RCAF inspection, test flight and acceptance of Harvard and Norseman production aircraft and various other types after repair. In July 1942, he obtained a compassionate posting as engineering officer at No.6 Repair Depot at Trenton, his home base where his ailing wife was confined to bed and in need of his support. His final posting was in August 1943 to Mountain View as OC Aircraft Repair Squadron. He was released from the RCAF in March 1945.

Edith and Harry Diller returned to the Calgary area, settling at High River in 1950. Receipt of his pension, amounting to $1,616 per year after 24 years of service, was delayed while an order-in-council was approved to change the name on his cheques from "Henry" to the correct "Harry."

An indication of Diller's commitment to the RCAF was the fact that throughout his career he did his best to hide his wounded leg, although it gave him continual problems, for fear he would be considered unfit for service and discharged. He did not even apply for a disability pension until ten years after he retired. It was fortunate, therefore, that he was "handy" and able to earn

15 NAC Personnel File.

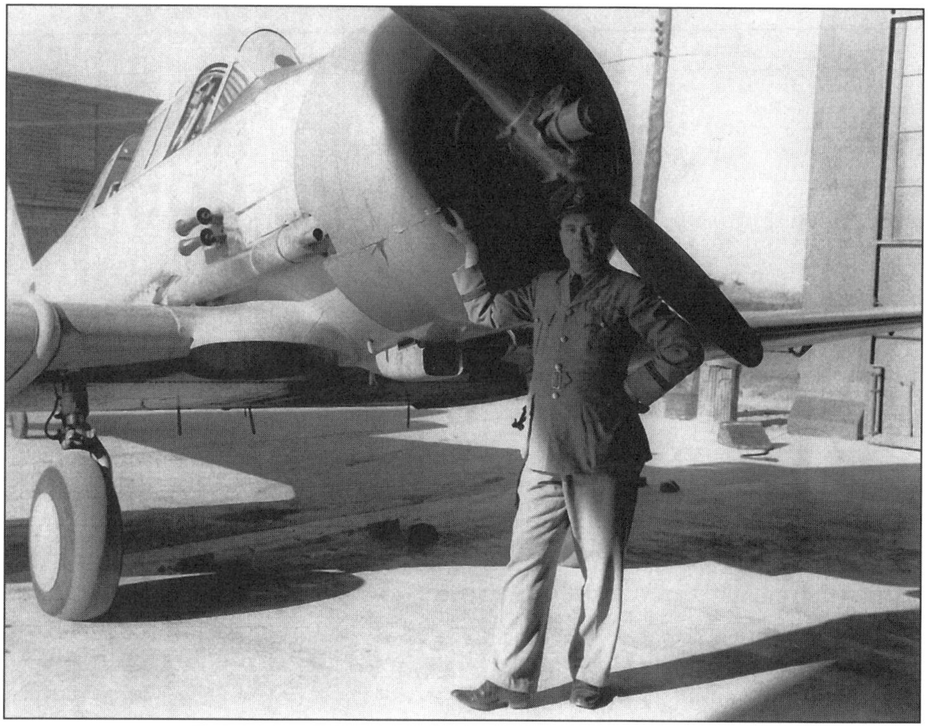

F/L H.S. Diller with North American Harvard, Mountain View, Ontario, September 1944. *GORDON DILLER*

extra income as a cabinetmaker, finish carpenter, handyman and saw sharpener. When Edith passed away in 1963, he turned to a nomadic lifestyle travelling throughout North America in his recreational vehicle, accompanied by only his dog, looking for the best fishing holes. He settled at White Rock, BC, in 1970 where he died on 26 January 1976.

Gordon Diller remembers Harry Diller:

> My father presented a stern, somewhat gruff demeanour which sometimes intimidated those who did not know him. Under this hard shell was a kind, loving, very emotional and sentimental man. He was a man of few words and little humour who kept his feelings pretty well bottled up. He was a good family man, but the lack of example in his own life sometimes made the situation difficult. He was an intelligent man and his lack of formal education never seemed to be a hindrance to achieving his goals. A perfectionist, he believed that "a job worth doing was worth doing right." He was naturalized as a Canadian citizen on joining the RCAF in 1924 and was a strong supporter of his adopted country. He was a very active member of the Canadian Legion, rarely missing a meeting or a Remembrance Day Parade. Conversely, I don't think he ever forgave the RCAF for his sudden retirement at the end of the war—along with many other "old sweats." He rarely spoke of his World War I experiences.[16]

16 Gordon Diller, personal communication.

Clarence Rupert Dunlap

"LARRY" DUNLAP was born in Sydney Mines, Nova Scotia, on 1 January 1908. His father Frank and mother Flora Leleah (Whitman) were both descendants of early settlers in, respectively, New Hampshire and Massachusetts who moved to Nova Scotia in 1761. He attended Nova Scotia Technical College and Acadia University, obtaining a BSc in Electrical Engineering in 1928. His interest in flying was kindled at an early age by the aviation achievements of Alexander Graham Bell, Casey Baldwin and John McCurdy at nearby Baddeck and, in 1918, by the US Navy HS-2L flying boats based at North Sydney. The navy base was within walking distance of his home and he recalled: "On the same perch from which, up to that time, I had watched fishing schooners sail in and out of Sydney Harbour, I now had a even more exciting picture, that of flying boats cavorting about the sky."[17]

After a brief period working as an electrical engineer for Canadian Westinghouse, his ambition was realized in 1928 when he was accepted for pilot training in the RCAF. He was one of the "new bloods" recruited during the 1920s to reinforce the World War I pilots who had nurtured both the military and civil branches of the fledgling air force. From the lowest RCAF commissioned rank he rose to the highest—air marshal and chief of the air staff.

Larry Dunlap is proud of page one of his logbook, dated August 1928, countersigned by F/L C.M. "Black Mike" McEwen and S/L A.B. "Bun" Shearer, both of whom rose to become air vice-marshals. On completion of pilot's training on war surplus Avro 504N biplanes at Camp Borden, Ontario, Dunlap spent six weeks during the spring of 1929 at Jericho Beach, Vancouver, on the seaplane course before being posted to Ottawa for transportation and ferry duties. He was promoted flying officer in July 1929, after his piloting and navigational skills had been put to the test with the delivery of a new Vedette, G-CYZB, 1,600 miles from Rockcliffe to Cormorant Lake in northern Manitoba. To that time his longest flight had been 100 miles from Rockcliffe to the Canadian Vickers plant in Montreal. He recalled that his briefing was

17 C.R. Dunlap, op. cit.

typically terse: "Have faith in your compass and clock . . . don't expect the lakes to look like those shown on the maps . . . don't be surprised if the fuel caches are inconspicuous, for many of them are made that way intentionally . . . make sure you deliver this aircraft safely and in good condition, otherwise your future career might be in jeopardy."[18] He was handed a roll of maps marked with the fuel cache sites. Dunlap and crewman AC2 Ordidge who, with only three months' RCAF service, was even greener than his pilot, set course for North Bay, the first of nine fuel caches along their route. Fuel caches were laid down, usually by ground or water transportation, across Canada and used by both military and civil aircraft, most of which lacked the range to fly non-stop between bases.

Progress was frustratingly slow. The Vedette cruised at about 70 mph, but prevailing westerly winds reduced its progress over the ground to about half this speed, barely exceeding that of a slow freight train. Taking into account the fact that endurance of the Vedette was little more than three hours, it is not surprising that it took four days to reach Berens River in Manitoba. Encouraged by a fine, clear day, Dunlap took off in good spirits for the final two legs of the journey. Alas, on reaching the midpoint of the Lake Winnipeg crossing, carburettor icing occurred of such severity that the whole engine vibrated alarmingly. Losing power, the aircraft descended towards waves far too rough to permit a safe landing. Fortunately, with only a few hundred feet remaining, warmer air cleared ice from the carburettor and the Vedette survived to reach the next fuel cache at Grand Rapids.

Disaster almost struck again on the last leg when the oil pressure suddenly dropped to zero. Dunlap had to put down in a hurry on Moose Lake, only 50 miles from his destination. In virtually uninhabited territory, he considered himself fortunate to see smoke rising from a distant point of land. A canoe put out and a couple of timber cruisers severely chastised him for landing so far from their camp. They were expecting a mail delivery and, after the confusion was sorted out, paddled back to their camp and returned in a couple of hours with a 2 hp outboard and towed the powerless Vedette to anchorage in the lee of a point of land.

The next day, one of the foresters and AC2 Ordidge canoed about 25 miles to a Hudson's Bay post where they were fortunate to find a five-gallon tin of oil, one that had been in stock for many years. Meanwhile, Dunlap repaired the broken oil line with layers of string and glue. After two days at Moose Lake, the apprehensive pair set out for Cormorant Lake. The repair held and the engine ran smoothly, although the oil temperature was a little higher than normal.

Arriving proudly at Cormorant Lake, but with a new, white Vedette smothered in sticky black oil with a quarter-inch layer of insects, they were greeted by the CO, the famously eccentric F/L F.J. Mawdesley: "What do you mean by delivering a new aircraft in such a filthy condition?"[19] The chastened

18 Ibid.
19 Ibid.

pair used five gallons of soft soap, a bundle of brushes and a mountain of rags to restore the Vedette to its pristine condition. Mawdesley, true to character, then praised their efforts and offered his pet de Havilland Moth for a flight as a reward. Dunlap only found out later that this generosity was Mawdesley's way of testing the new pilot's ability to land a seaplane on glassy water at dusk. Dunlap set out on his return journey to Rockcliffe by train and, so impressed was Mawdesley with his performance that he insisted on seeing the young officer on his way. As the train pulled out, it was chased by a lumbering Varuna, its wings almost touching the trees on either side of the tracks as its keel kissed the top of the railway car in a gesture of farewell.

The next five years were devoted to photo survey operations in many areas of Canada, ranging from Nova Scotia to British Columbia. Dunlap recalled fond memories of four years with his "favourite steed," Fairchild FC-2W 'XQ, in British Columbia, "generally under canvas, at such places as Salmon Arm and Williams Lake in the interior; Nimpkish Lake, Sproat lake and Queens Cove on Vancouver Island; Rivers Inlet and Prince Rupert in the northern coastal areas; and to climax it all a period of several months at Queen Charlotte City."

Dunlap gave an insight into the problems of navigation on the west coast and the need for initiative—and a little luck—when describing being forced down by fog at the entrance to Vancouver harbour. "All I needed was a ten-foot ceiling for I was close enough to taxi the rest of the distance." The fog, however, completely enveloped him so he had to resort to paddling on a

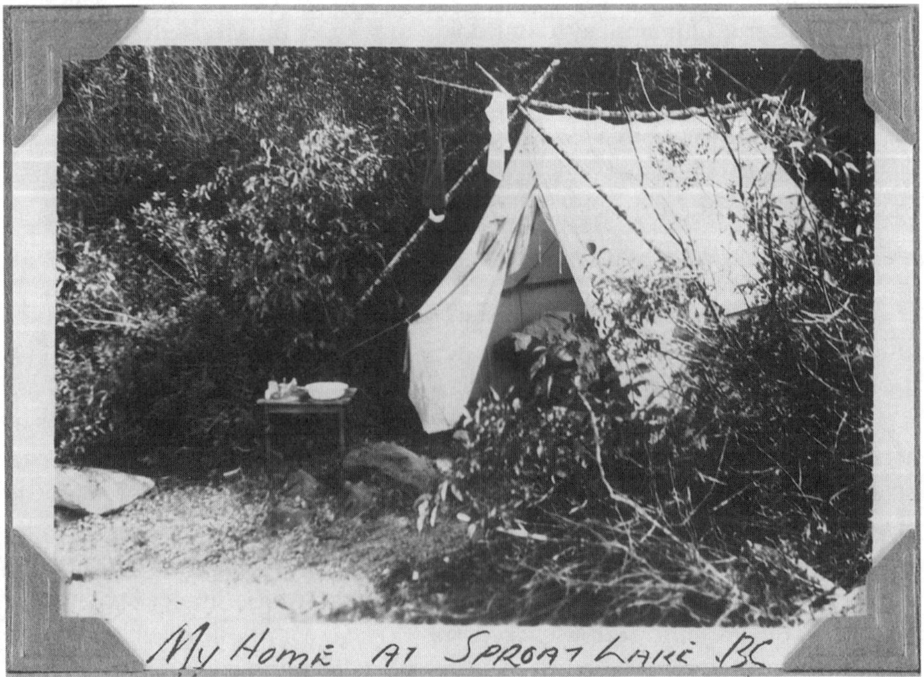

F/O Larry Dunlap's "home" at Sproat Lake, Vancouver Island, 1931.
NAC PA202941/C.R. DUNLAP COLLECTION

compass course for Jericho Beach through the main shipping channel, to an orchestration of ships' foghorns. Assisted by a running tide, he practically ran into a navigation buoy off Spanish Banks and, giving the credit to luck, found his mooring buoy about 400 yards away at Jericho Beach. "Having tied up securely, we bellowed to shore to the astonishment of the base personnel who thought they were hearing from Morley's ghost."

It was time to move on from being a photo pilot and how better to do it than to serve in 1933 as chief instructor of the largest photo survey course ever to be held at Rockcliffe. Graduating from this course were more than a dozen well-known pilots among whom Dunlap readily recalled Wray, McNab, Gordon, Plant, Hawtrey, Coleman, Bennett and Trecarten. Also under training at this time was a large group of "survey camera operators."

In the spring of 1934, Dunlap was assigned to the Air Armament and Bombing School at Camp Borden. Soon his professional and personal lives took major changes in direction. In 1935 he was promoted flight lieutenant and, while on the specialist course at the RAF Air Armament School, Eastchurch, England, he married Hester Catherine, daughter of Dr. and Mrs. E.A. Cleveland of Vancouver. Upon returning to Canada in 1936, he served as chief instructor at the Camp Borden Air Armament School. From 1937 to 1939 he was back at RAF Eastchurch to attend the advanced course in Armament Engineering and Ballistics.

At the outbreak of war he was a squadron leader serving as director of armament at RCAFHQ in Ottawa, a position he held for the next three years. He was promoted group captain in 1942 and spent most of the year commanding the station at Mountain View, Ontario. Late in 1942, he was posted overseas to command a 6 Group bomber station at Leeming, Yorkshire.

Just prior to the Sicilian campaign, in May 1943, G/C Dunlap went to Tunisia in command of No.331 Wing RCAF, comprised of Nos. 420, 424 and 425 Squadrons with 60 Wellington bombers. He directed the wing operations over Sicily, Italy and Sardinia, receiving special commendations from Britain's A/C/M Tedder and the United States' General Doolittle for his unit's notable success.

No.139 Wing later moved to Belgium, continuing daylight raids in support of ground forces, and attacks against V1 flying bomb and V2 rocket sites. Dunlap flew 35 operational sorties and was awarded the Croix de Guerre with Gold Star in recognition of the wing's role in the liberation of France. He received the US Silver Star for his support of Allied forces, the commendation reading:

> Whilst in command of the wing, this officer has carried out a large number of operations with great success. The general standard of bombing, keenness and efficiency has reached a high order under his inspiring leadership. He has operated against many heavily defended targets and has set a very fine example.

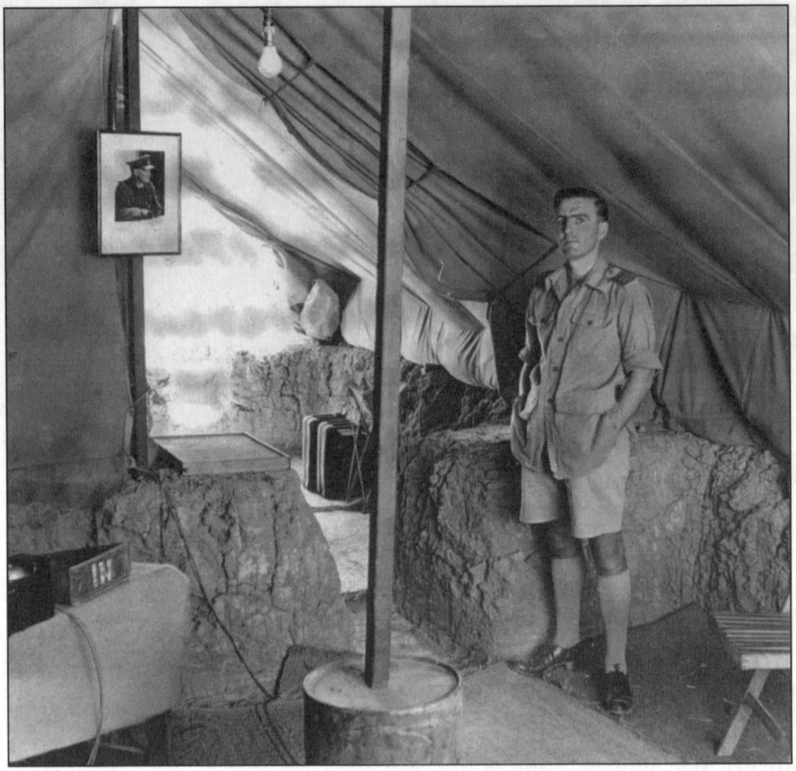

Although in more comfortable conditions than during his photo pilot days, Dunlap was still under canvas in 1943 at Zina, 40 miles southwest of Kairouan, Tunisia. The photograph on the tent pole is of Air Marshal Gus Edwards who was deeply involved in the formative stages of 331 Wing. He insisted that it be 100 percent RCAF and that it be only temporarily on loan to the North African Strategic Air Force. *NAC PA178453/C.R. DUNLAP COLLECTION*

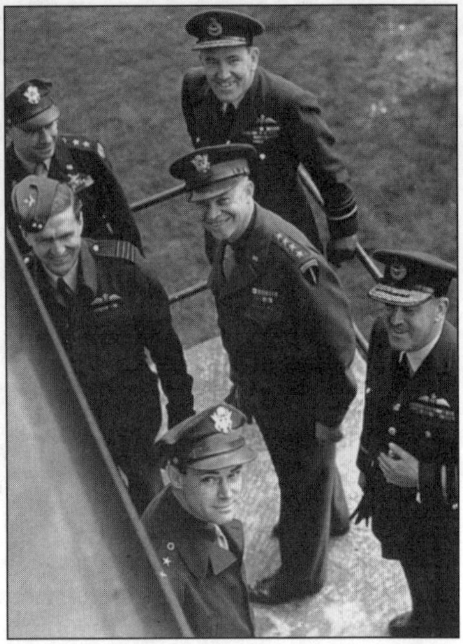

Returning to the UK in November, 1943, G/C Dunlap took command of No.139 Wing (Nos. 98, 180 and 320 Squadrons) of 2 Group, 2nd Tactical Air Force, at Dunsfold, Surrey. Photograph taken from the Dunsfold control tower during a pre-D-Day visit 18 April 1944. Clockwise from bottom left, Vandenburg, USAAF; Dunlap, RCAF; Brereton, USAAF; Coningham, RAF; Supreme Commander Dwight Eisenhower; Leigh-Mallory, RAF. *C.R. DUNLAP*

For the first few months of 1945, Dunlap, now an air commodore, returned to 6 Group as operational commander of a force of 80 Lancasters at Middleton St. George and Croft, Yorkshire. In May 1945, Dunlap returned to Canada and, while serving in the Air Staff Division at AFHQ, he represented the RCAF at the Bikini nuclear tests in the Pacific Ocean. Next, he studied at the US National War College, following which, in 1948, he was appointed deputy air member for Air Plans at RCAFHQ with the rank of air vice-marshal. From 1949 he had senior commands in Canada and was commandant of the National Defence College at Kingston, Ontario, from 1951 to 1954. He then became vice chief of air staff, a position he held for the next four years.

In September 1962, A/M Dunlap became chief of the air staff at RCAFHQ. Two years later he moved to Colorado Springs as deputy commander-in-chief of NORAD, where he served until the end of his active service career.

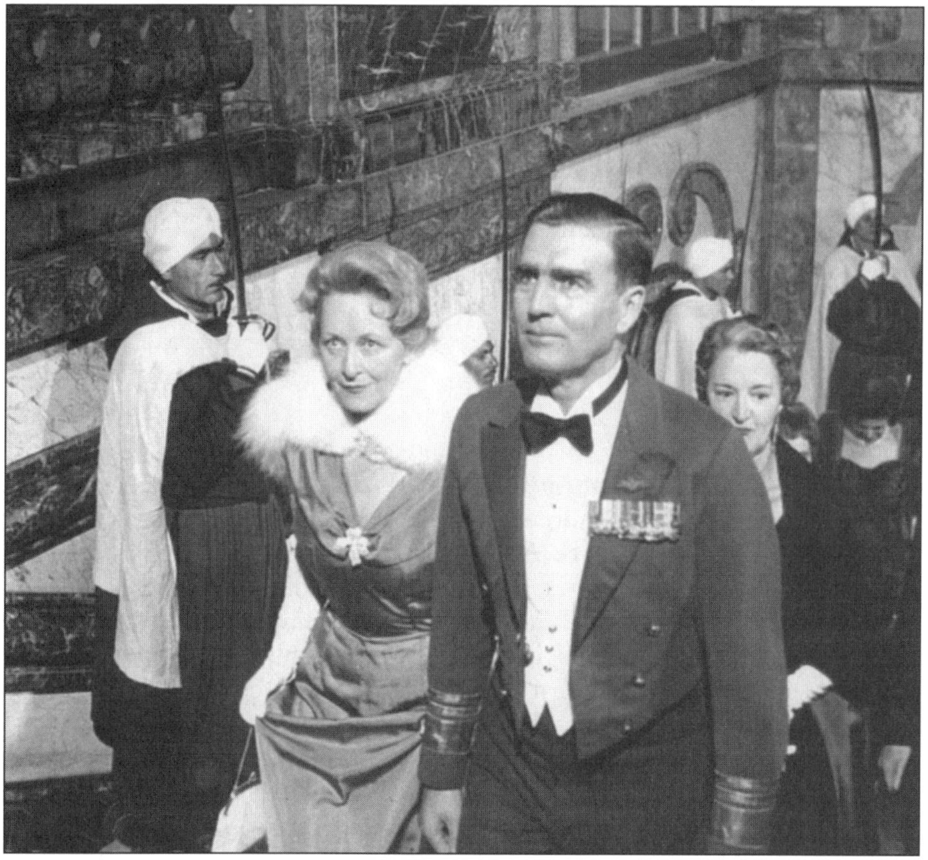

Larry Dunlap was promoted air marshal in 1958 and sent to Supreme Headquarters, Allied Powers Europe (SHAPE) as deputy chief of staff (operations).

A/M and Mrs. Dunlap at a SHAPE Charity Ball, Palace of Versailles, Paris, 1959. *C.R. DUNLAP*

Air Marshal Dunlap, CBE, CD, DCL, DEng, BSc, officially retired in July 1968, but not to a life of leisure. For the next 11 years he worked almost full time on a voluntary basis helping to establish the new National Aviation Museum. Moving from his museum endeavours in 1979, he went to live in Victoria, BC where he plays a lot of golf, but still keeps a watchful eye on aviation activities throughout Canada.

Kenneth MacGregor Guthrie

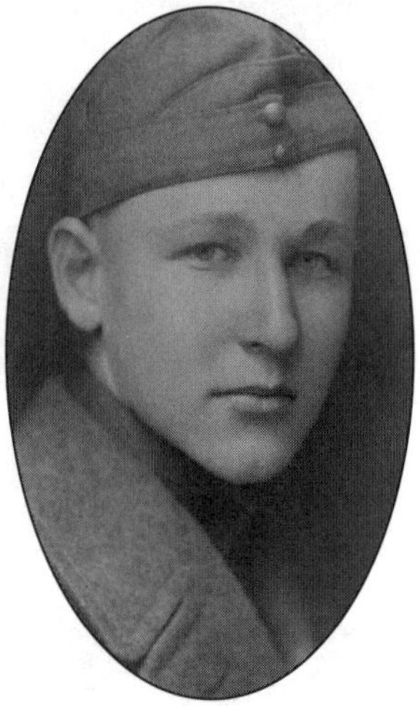

Awarded his pilot's wings and RAF commission in May 1918 after training at Randolph Field, Texas, 2Lt. Guthrie was flying antisubmarine patrols as an observer in Short Seaplanes over the English Channel when the Armistice was signed in November.
ANN POLLAK

KEN GUTHRIE was born on 9 August 1900 in Guelph, Ontario, but the family soon moved to Baltimore where his father was minister of the First Presbyterian Church on Park Avenue. Summers were spent with his four siblings at the family summer home on Farm Point, Wakefield, Quebec. Educated by his mother until they returned to Ottawa in 1912, he attended public school and Lisgar Collegiate in Ottawa, but only graduated from grade X before enlisting in the RFC at the age of 16. A well-built youth, he had no difficulty in convincing the recruiters that he was 18.

Frustrated by the slowness of repatriation, Guthrie took his trademark direct action, resigned from the RAF and bought his own ticket home in January 1919. He may have recouped his investment, so the story is told, in exercising his lifelong love of cards while crossing the Atlantic Ocean and winning $2,000 at craps and poker. However, the 18-year-old veteran returned to a family trauma. His father, the Rev. Donald Guthrie, had been a padre during World War I and was greatly affected by his experiences in France to the extent that he left his family, went to New York State and joined the Roman Catholic Church. The eldest brother, another Donald, went to California, leaving Ken to support the family.

First trying his hand at several jobs, including selling cars, Guthrie eventually obtained a commission as lieutenant quartermaster in the Royal Canadian Army Medical Corps at Cobourg, Ontario. Thanks to his flying experience, he was able to transfer to the CAB with the rank of pilot officer on 1 January 1921. Perhaps his agreement to serve, initially, in the Technical Branch (Stores) without pay influenced the decision. He soon moved to flying duties at Camp Borden, Roberval and Winnipeg, flying HS-2L and Viking flying boats on forestry patrols, photo surveys and general transport duties in Manitoba, Saskatchewan, Ontario, Quebec and the Northwest Territories. In July 1923 he had the only accident in his flying career, hitting a submerged deadhead with a Viking while taking off from the Ottawa River at Rockcliffe. He was unhurt, but had to make the journey to Winnipeg by train while the aircraft was being repaired.

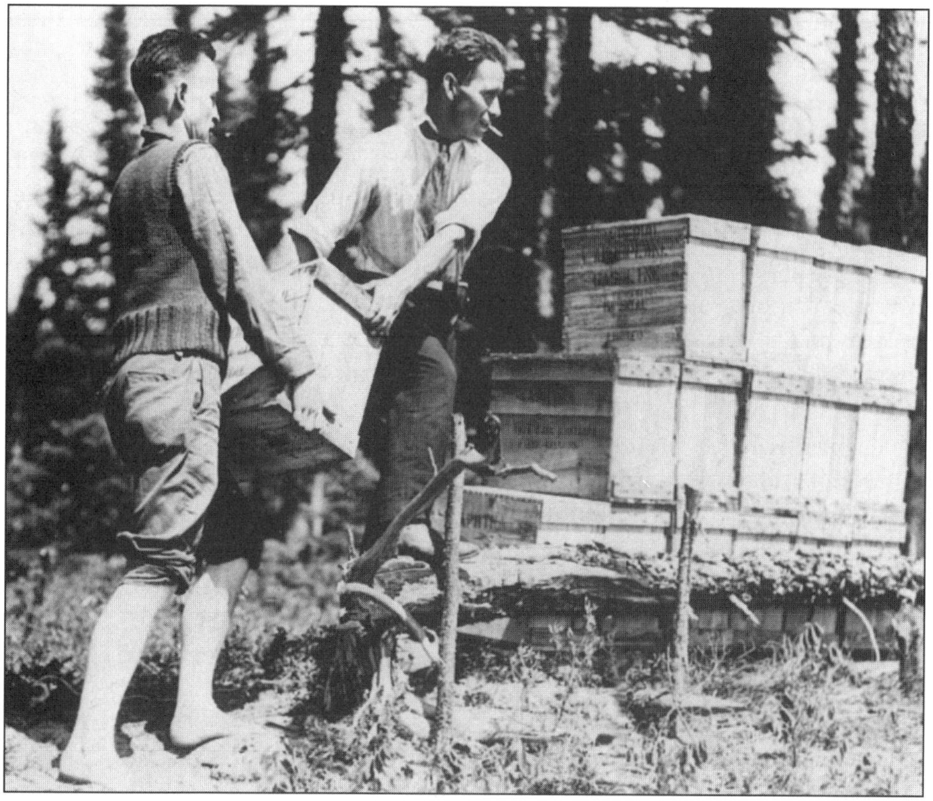

In a time when personnel had to be versatile, Guthrie, a commissioned officer, was identified as "storekeeper" when photographed with "air pilot navigator" T.A. Lawrence while stacking cans of gasoline at Stacker Lake, Quebec in 1921. Guthrie, on left, is smoking a pipe, Lawrence a cigarette. Both men survived the fire hazard and went on to become air vice-marshals.

Lawrence had joined the Canadian infantry in 1915, hoping to transfer to the RFC, but served in the trenches for two years until he was assigned to pilot training. He had to enlist in the CAB as a trainee mechanic after the war, but was soon promoted pilot. *NAC PA123018*

Guthrie was promoted flying officer on 1 January 1924 and became one of "the originals" in the RCAF on its formation 1 April 1924. He spent another year at Winnipeg on photo surveys and forest patrols before his appointment as senior personnel officer at AFHQ in Ottawa. Rescued from the gypsy-like life of a photo pilot, Guthrie could now pay attention to more personal matters. He married Catherine Mary Fidler at St. Clements Church, Toronto, in October 1926, and their only child, Ann Elizabeth, was born in 1928. Ann remembers her father teaching her to play his favourite games of tennis and badminton, at which he excelled. He made her a magnificent Victorian dollhouse complete with furnishings: "He was simply a good father." Ann married Fred Pollak, and their daughter Susan Schultz delivered a touching eulogy, from which much of this information is drawn, after her grandfather died.

Still at Ottawa, Guthrie was promoted flight lieutenant in December 1929. He went to RAF Staff College at Andover before being posted to Winnipeg in 1933 as air staff officer at the army's Prairie Command. In 1934 he returned to Ottawa as a squadron leader and the post of assistant director, Military and Air Force Intelligence, Army General Staff—a posting he welcomed, as it gave him an ideal opportunity to promote army–air force co-operation and to pursue his career-long battle against parochial jealousy within the services. In 1938 he returned to flying duties commanding RCAF Station Rockcliffe with the rank of wing commander.

At the outbreak of war, Guthrie was sent from Rockcliffe on "temporary duty" as senior air staff officer at Eastern Air Command, Halifax. He was on the east coast for three years, playing a key role in the Battle of the Atlantic. He was CO at Gander, Newfoundland, during most of 1941, overseeing the massive buildup for transatlantic ferry flights. In April 1942, a time when there was great apprehension about Japanese intentions, he was promoted group captain and sent to Western Command in Victoria as chief staff officer. Later that year he returned to Ottawa as an air commodore to be deputy member Air Staff (Plans). He was part of the Canadian team with Prime Minister Mackenzie King at the Quebec Conference in 1943 where he met Winston Churchill and the senior members of the Allied air forces.

The staff positions held by Guthrie over many years may have left the impression that he was a pretty dry character. Far from the truth. A/M C.R. Dunlap recalled meeting him for the first time at Lac du Bonnet in 1929: "That effervescent character Ken Guthrie was in charge at the time of my visit and I might say that the place was jumping both by day and by night."[20] Always

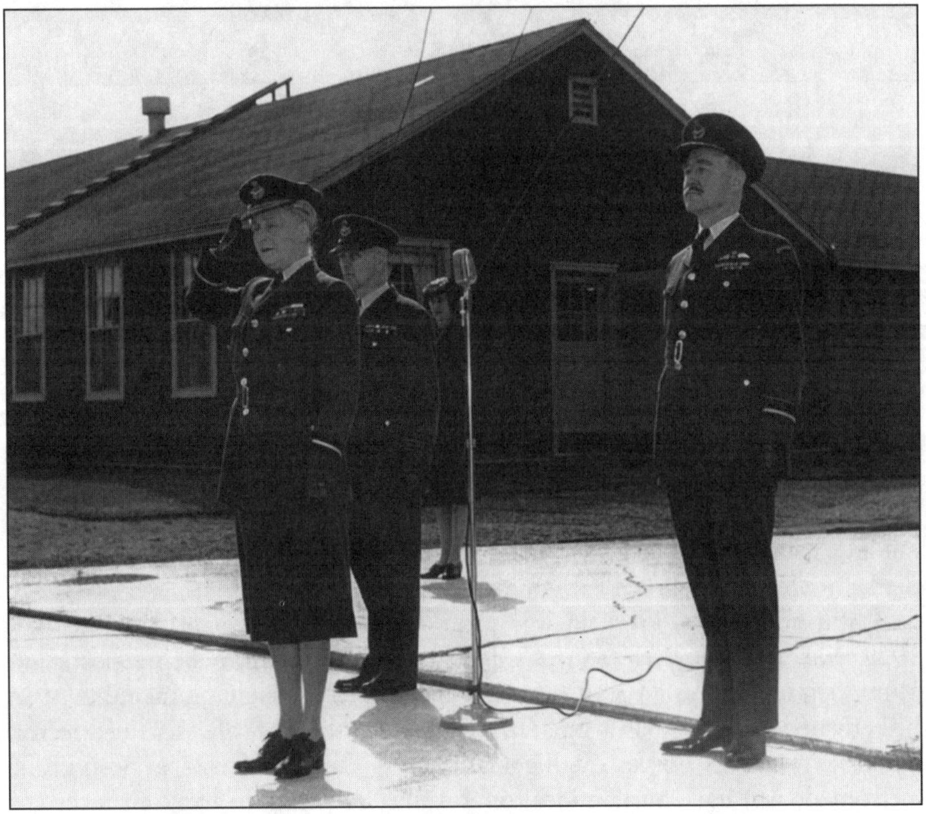

Continuing his close association with the British Commonwealth Air Training Plan, A/V/M Guthrie took command of No.2 Training Command in Edmonton in May 1944. Here, he is on the right with H.R.H. Princess Alice and station CO W/C T.F. Cooper at No.8 Repair Depot, Winnipeg, in 1944.
IRENE COOPER

20 C.R. Dunlap, op. cit.

on the move, he was known as the Green Hornet. His granddaughter Susan Schultz noted that his zodiac sign was Leo, which could explain why he was sometimes accused of roaring his way through life. She quoted him: "After all, who can hear a mouse on a parade ground." His daughter Ann remembers his love of evenings in the mess with "the boys," often not returning home until the small hours, once with a dislocated shoulder resulting from energetic horseplay on the staircase at Victoria, BC.[21]

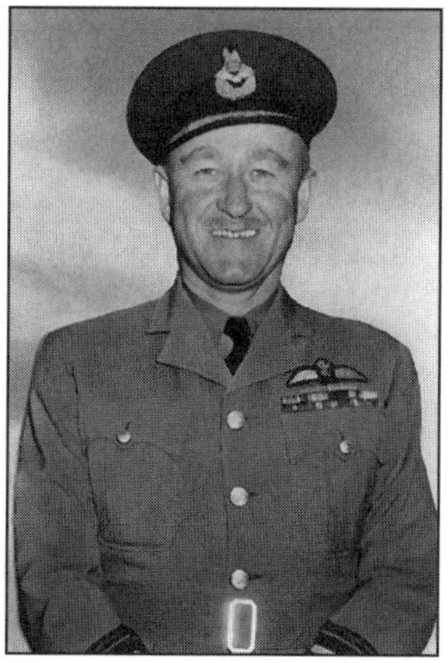

With reorganization of the RCAF after hostilities ceased in 1945, A/V/M Guthrie was appointed air officer commanding Northwest Air Command, responsible for operations from the Lake Head to the Pacific Ocean, until his retirement in November 1949.
ANN POLLAK

Guthrie Lake in central Manitoba was named for Ken Guthrie in recognition of his photo pilot days. He received the CBE in 1942, the US Legion of Merit in 1943 and the CB in 1944. He was an honorary Texan, honorary wing commander of No.418 (City of Edmonton) Auxiliary Squadron, and a keen supporter of the air cadets and reserve air force. His many testimonials were summed up by his CBE award:

> Air Commodore Guthrie has rendered outstanding and devoted service since the commencement of present hostilities. While employed as senior air staff officer at Eastern Air Command by his high endeavour and devotion to duty he contributed greatly to the efficient organization of the Command and to the success of East Coast operations. Since his appointment as deputy air member for Air Staff (Plans) the wealth of knowledge and experience gained by him during his extensive service have been of inestimable value to the preparation of plans for all Air Defence of Canada for which he is responsible. Air Commodore Guthrie is the chairman of the Combined Staff Planners dealing with major problems of Canadian defence.

In a complete change of careers after retirement from the RCAF, Guthrie spent several years as secretary/treasurer of a small oil and gas exploration company before retiring again to concentrate on his military interests and service on hospital, art, museum and symphony boards. He was first vice-president of the World Veteran's Federation, president of the RCAF Association, and an active member of the World War I Flyers. He was a stalwart supporter of the National Aviation Museum and was seen at most events proudly sporting his RFC tie. He died on 14 March 1993, the last surviving officer of "the originals" who signed on with the RCAF on 1 April 1924.

21 Ann Pollack, personal communication.

Alfred Alexander Leitch

Early days in Winnipeg. Outside the store of Michael Ert Limited, "Everything For Motorists." Archie Leitch is second from the right, c.1917.
SUZANNE M. HIGGINS

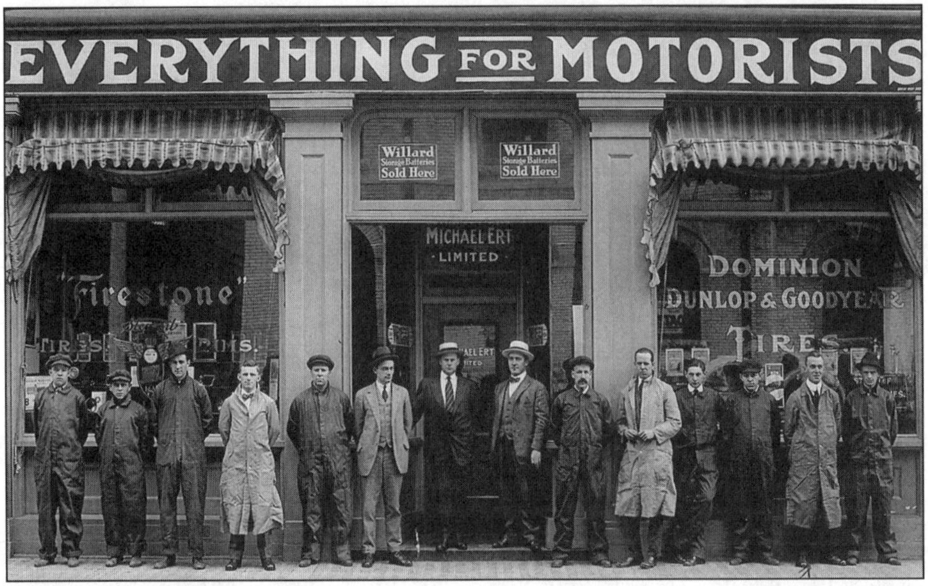

SON OF MILLWRIGHT Alfred Rawson Leitch, Alfred Alexander Leitch was born in Killarney, Manitoba, on 5 February 1894. Work opportunities took his family to Winnipeg where he attended the University of Manitoba. In June 1916, his brother Roydon Charles, serving with Princess Patricia's Canadian Light Infantry in France, was reported missing in action. No further news was heard, which prompted Alexander to volunteer for the RFC on 15 May 1917 in the naive hope that he could get to Europe and find his brother. After pilot training at Camp Borden he was posted to No.65 Squadron in France. Cryptic squadron reports give an indication of Camel pilot Leitch's precarious existence at "the Front," for which he received 14 shillings per day:

5.2.18 Attacked Albatross scout which he shot down. He was then attacked by other e.a. and was forced to cross the lines at about 1,200 ft.

16.5.18 Attacked one e.a. triplane which is confirmed by an A.A. Battery to have crashed.

25.5.18 Brought down an e.a.

29.5.18 MC to 2/Lt Leitch.

26.6.18 e.a. activity slight. Capt. A.A. Leitch, No 55 [sic] brought down one e.a.

1.9.18 DFC to Capt. A.A. Leitch, MC.

What is hidden behind these bland reports defies imagination today, but No.65 Squadron diary gives a little colour to the bare statement above regarding events on 16 May 1918:

16.5.18 Camel C.8280. 8.10 pm S.E. of Albert. O.P. 16,000'. Dest 1. Leader of O.P Attacked two formations of 6 triplanes at 13,000' and drove them E. Returned to line and climbed to 16,000', again attacked E.A. S.E. of Albert; fired on nearest which dived and disappeared. Rest retired E. At 8.30 attacked same formation again; they split up and dived S.E. A.A. Battery saw triplane crash in flames at 8.10 near 62D F2.

A.A. Leitch carried a lasting memory of his wartime experiences in his nickname of "Ack Ack," derived from the coincidence of his initials and the common term "ack ack" for anti-aircraft fire. Originally called "Alfie" by his family, for the rest of his life he was known as "Ack."

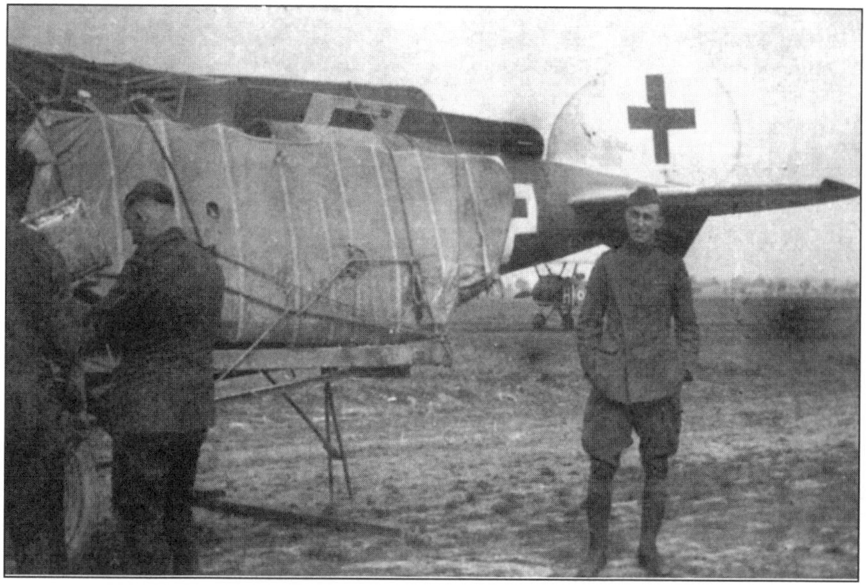

A.A. Leitch stands beside a German aircraft he shot down, France, 1918. *SUZANNE M. HIGGINS*

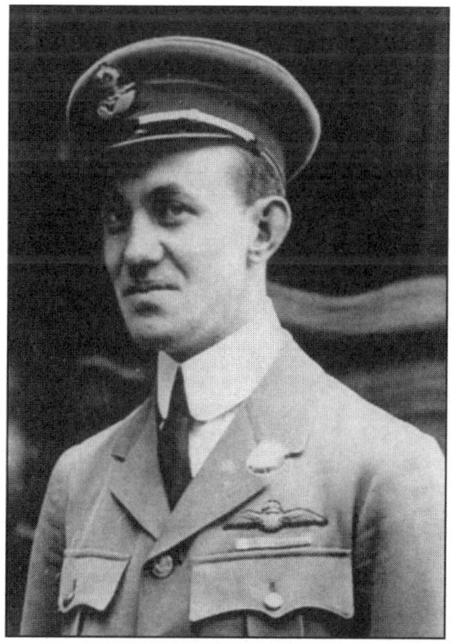

A.A. Leitch at Buckingham Palace investiture to receive his MC and DFC, 1918. *SUZANNE M. HIGGINS*

A.A. Leitch was promoted temporary squadron leader on 1 April 1929 when sent to command RCAF Station High River and was confirmed in the rank the following year. He had been courting Evelyn Rachel Robertson since his early days there, although his visits were somewhat irregular due to prolonged absences on duty. He usually arrived after a northern expedition, resulting in him being known as "Eve's Eskimo." Eve grew up on the Robertson family ranch on the eastern fringe of High River, and Ack's courtship must have generated mixed feelings as her mother, Mrs. T.W. Robertson, was still seething over the transformation of a large slice of her peaceful and successful ranch into the airfield. Nevertheless, Eve and Ack were married at High River in June 1929. They had three daughters, Evelyn Margaret, Suzanne Grace and Diane Mary. *DNDPU PL11750*

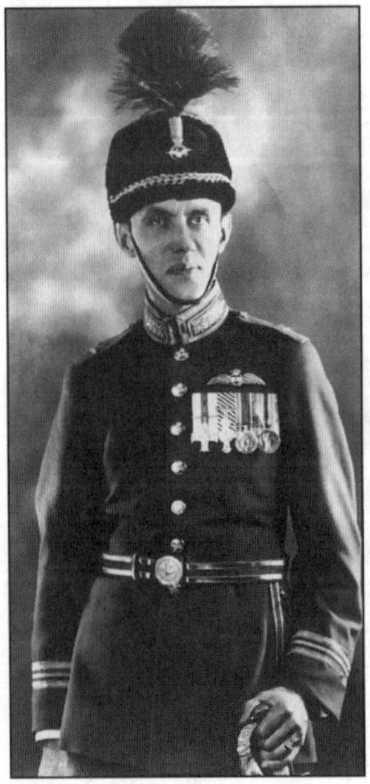

Leitch participated in one of the futile expeditions sent by France and Britain to help the White Russians against the Communists in 1919. In the process, he was one of the pioneers of flying aircraft from a ship—the foundation of international power today. His diaries record taking off from HMS *Argus* on 5 June 1919 with six other aircraft and he was one of the three to return successfully to the deck. On 24 July, HMS *Argus* set sail from the Firth of Tay in Scotland for Archangel in Russia with aircraft and crews aboard—a forerunner of relief convoys during World War II. There, he had a serious accident when a broken aileron caused his Short Seaplane to spin into the harbour. His observer, Lieut. Patterson, was cut about the face by his Lewis gun as he was thrown out of his cockpit, but Ack was more seriously injured, with a fractured skull and other injuries. They were picked up by a launch from the *Argus* and Capt. Leitch was kept in the sick bay until transferred to South Queensferry Hospital when the ship returned to England in August. He was released from the RAF in September 1919 and sent home for three months' leave. No records were apparently kept of his injuries, an omission that caused Leitch a great deal of trouble in later years.

Back in Canada, Leitch exercised his mechanical skills in the repair, maintenance and sale of gasoline engines for several companies in Winnipeg before returning to his love of aviation as manager of Winnipeg AirCo Ltd's Air Harbour at River Park. He and buddy Mel Dover, returned from service with the RNAS, had a brief career as "barnstormers." Service life called,

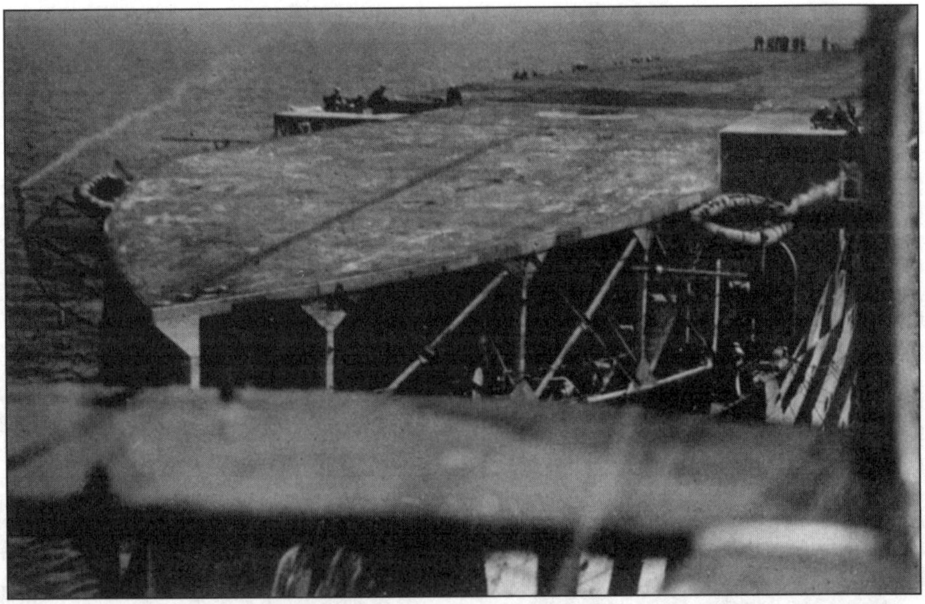

Taken by A.A. Leitch from a D.H.9A after takeoff from HMS *Argus*, Firth of Tay, Scotland, 1919 *SUZANNE M. HIGGINS*

HMS *Argus* was the world's first fully flat-deck carrier, constructed on the hull of Italian merchant ship *Conte Rossi* during World War I and continued in active service to 1946.

however, and his performance in France made him a welcome recruit to the CAB in November 1920. Promoted flying officer in 1922, he flew at various bases, including Jericho Beach, where he took the seaplane course in 1924. He operated mainly from High River, Alberta, although he unsuccessfully requested a posting to Manitoba in June 1921 to help care for his widowed mother. He was one of the pilots charged with dropping mail for the Prince of Wales at his E.P. Ranch, near Pekisko 17 miles southwest of High River, when the roads were impassable. In June 1924 he took George Young, the mayor of High River, on a "courtesy flight" as a public relations gesture. In 1925 he was appointed a civil aviation examiner, giving him the authority to approve pilot licences.

Moving through the successive reorganizations of the CAB, CAF and RCAF, Leitch established a record as an exceptional pilot and was assigned to one of the six Fokker Universals of the Hudson Strait Expedition of 1927–28. Spending the winter at bases at the entrance to Hudson Bay to photograph ice movement from single-engine aircraft to establish the period of access to the new port of Churchill was hazardous duty for the aircrews. Their very survival was a tribute to their skill and the abilities of the mechanics. In a typical event, F/L Leitch, caught in a snowstorm, had to land on an ice floe and wait overnight for clearer weather, making it back to Wakeham Bay with only about one quart of fuel remaining and his crew suffering from frostbite. Commander of the RCAF contingent, S/L T.A. Lawrence, reported on Leitch: "He showed a high standard of efficiency in charge of an operational unit. He has a very fine flying sense and tends to impart it to others to good use."

For many years, performance reports regularly iterated Leitch's keenness, efficiency and flying ability. But his lack of sympathy for the reports and procedures that are an inescapable part of service life were hinted at with tongue-in-cheek comments such as, "He has made most satisfactory progress in administrative knowledge."[22] He went to the RAF armaments course in 1925 and the Royal Navy Staff College in 1931.

At 102 years of age (in 1998), Ack's widow Eve had clear memories of his eccentric inventive genius over half a century previously. He converted a Mix-Master motor into a lawn edge trimmer, designed a garden hose reel, and wired their whole house for sound from a single radio. Their family home still has the "disappearing staircase" mounted on old roller skates that he built to give access to two bedrooms in the attic. A motorcycle enthusiast with a limited budget, Ack installed a small gasoline engine on his bicycle and towed trailers built from abandoned baby carriages. Eve remembered summers spent at Eden Valley Ranch, on the Highwood River southeast of Calgary, where she kept a horse. Ack did not share her love of horses and preferred to fish if he couldn't fly. His daughter, well known Calgary alderman Sue Higgins, still laughs about the commotion when Ack was caught trying to keep his distance and saddle a horse from the adjacent stall.

22 NAC Personnel File (and subsequent quotations not otherwise cited).

In 1935, Leitch was in charge of the bombing school at Trenton when critical remarks started to appear in his assessments noting occasional irrational and erratic acts and attributing them to his World War I wounds.

This was the beginning of an unfortunate period for Leitch. He had been an exemplary pilot in the RFC and RAF, and he flew almost every aircraft type in the RCAF inventory between the wars. From 1918 to 1924 alone his log shows an accumulation of 1,240 hours on the Curtiss JN-4, JN4A, Avro 504K, Sopwith Two-Seater, Bristol Scout, Camel, Salamander, Deck Camel, Short Seaplane, D.H.9A, D.H.4, Bristol Fighter, Avro Lynx, Avro Viper, and Siddeley Siskin. His "erratic" performance now compounded by a drinking problem, he was taken off the flying list and put in charge of the Trenton workshop.

It is not clear that Leitch had ever received any medical attention for his injuries, likely exacerbated by his Hudson Strait hardships, but he did enquire about a disability pension in May 1938. The official response admitted that there was an "impression" that he had received a skull fracture and other minor injuries, but "a survey of RAF and RCAF records has failed to produce any evidence to support the impression." In fact, there is remarkably little in his files about his World War I experiences, not even an accounting of enemy aircraft shot down, although squadron diaries indicate that he had at least five victories. Leitch was asked to document the accident.

This he did: "The injuries consisted of a fractured base of skull, broken between left ear and eye and between both eyes at the top of the nose." The "minor injuries" he listed as, "lacerations to face, two broken bones in left hand, injuries to left leg and feet and internal injuries." It appears that he had hitherto played down his injuries, although there were consistent medical reports of one pupil being noticeably larger than the other, and he was discouraged from attending parades because his "badly deformed" foot caused him to limp. Although not officially recorded, the nature of Leitch's injuries appears to have been well known throughout the service. For example, Leitch flew PP/O Ralph McBurney for his first parachute jump and McBurney vividly remembered his "evil grin over the side of the cockpit due to having one side of his face pushed in at Archangel in a bad crash."[23]

Leitch's daughter Sue remembered that he "only smiled with half his face"[24] and believed that the constant pain he suffered was a critical factor leading to his over-indulgence in alcohol. He was released from the RCAF in September 1938 with a pension from which $140 per year was deducted as he had missed premium payments for nine years.

Throughout World War II and until 1946, Leitch commuted from his home at High River to Edmonton where he was a civilian safety officer at No.16 Technical Detachment.

Ack's final gesture was typically eccentric: he died in his sleep, and his family was never sure if he left them on 31 December 1954 or 1 January 1955. S/L A.A. Leitch, MC, DFC, was buried with full military honours at High River.

23 Ralph McBurney, personal communication.

24 Sue Higgins, personal communication.

Colin Stone Macdonald

GREAT RELIANCE was placed by the air force in its early years on a select group of civilian surveyors. They undertook most of the navigation on long flights over unknown territory, often doubling as photographer—in addition to sharing the refuelling, loading, unloading, cooking and maintenance chores required of all aircrew members.

One of this intrepid group of dominion land surveyors was C.S. Macdonald, born in Ottawa on 26 May 1887. He had only completed grade VII when his father died and he had to leave school to help support the family. There was little job opportunity for an 11-year-old, so he worked as a "gofer" in a newspaper office and a dry goods clerk in Charles Ogilvy's general store, Mr. Ogilvy remaining his friend for life. Learning Morse code earned him a post as a telegraph operator. Always attracted to the outdoor life, he worked as a labourer on the National Transcontinental Railway north of Lake Superior. By 1904 he was a lumber camp clerk for the Fraser Lumber Company, spending the winter in the forest north of Mattawa.

Evidently he made early connections with the surveying fraternity as he spent the summer of 1905 as a chainman on the dominion lands survey in Manitoba. Finding his vocation, he expanded his knowledge by operating a transit on surveys while studying for the dominion land surveyor examinations. In 1906 he supplemented his income by joining the militia and became a sergeant. Still in Manitoba, he passed the preliminary surveyors' exams in 1910 and was sent as assistant on surveys in northern Manitoba and Saskatchewan. In 1914, he achieved his chief ambition and received his DLS commission, continuing much the same work for the next three years. However, he turned down an invitation from the Department of the Interior in 1918 to lead a survey of school lands in Manitoba in favour of accepting a commission in the Royal Canadian Engineers.

Macdonald spent most of the 1920s surveying in the northern reaches of central Canada for the Topographical Survey Branch of the Interior department. His first introduction to aerial photography was at Brochet,

Sent overseas on the S.S. *Saturnia*, Lieut. Macdonald was fortunate to make friends with fellow passenger Nursing Sister Margaret McCullough, RCAMC. She, however, developed Spanish 'flu and was taken off the ship unconscious in England. She did recover and, after an anxious search, Macdonald finally located her at a hospital near Buxton, Derbyshire, nursing repatriates from France. They were engaged shortly afterwards and married on their return to Canada in 1919. They had four sons and one daughter.
RODERICK W. MACDONALD

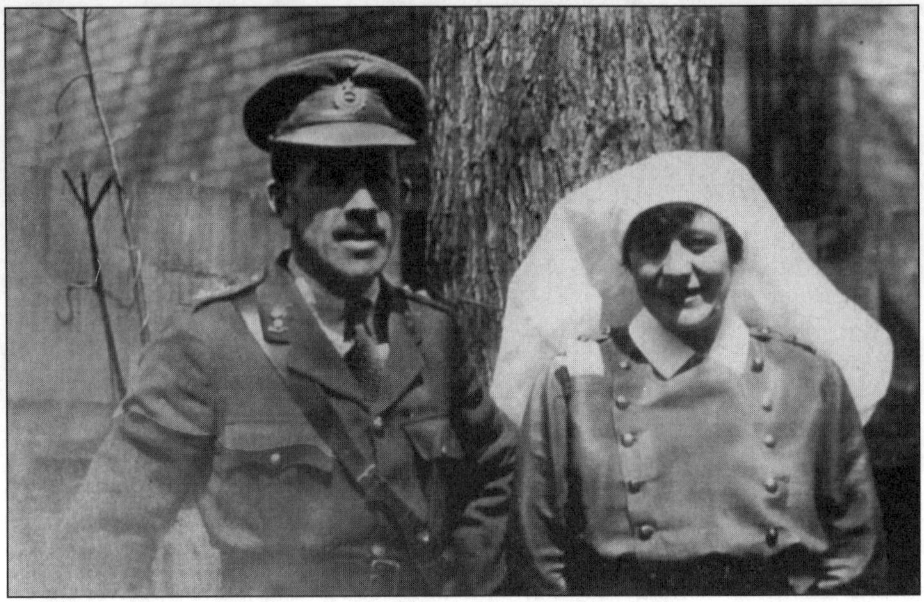

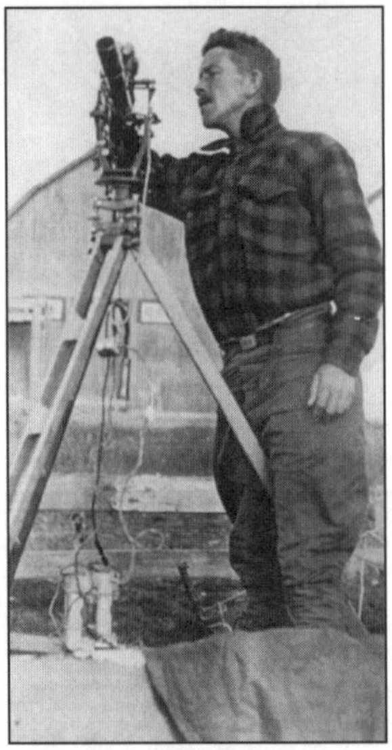

Colin S. Macdonald, DLS, preparing to establish his position in the 1920s.
RODERICK W. MACDONALD

Manitoba, in July 1924, while laying down a ground traverse as the basis for maps of the Reindeer Lake area to be plotted from aerial photographs. When the Vickers Viking piloted by S/L Basil D. Hobbs arrived to take the photographs, he spent many hours comparing notes with his colleague Robert D. Davidson, DLS, navigator of the aircraft. Each surveyor revised his plans in the light of information provided by the other. This event certainly convinced Macdonald that aerial photography was of significant assistance in map-making and probably influenced his agreement to be the navigator of a more ambitious undertaking to the Northwest Territories in 1930.

While on summer camping trips along the Rideau River, Macdonald's young sons realized that their father was not troubled, as they were, by the ever-present pesky blackflies. He had gained this immunity, or tolerance, through years of exposure on northern expeditions, which made him the envy of his tormented colleagues. There was a downside, however, to these idyllic years spent either with his family or on surveys in the north, both occupations that he loved. In 1928, and again in 1929, Macdonald hurt his back in a canoe accident while on survey, injuries that came back to haunt him in later years.

Macdonald's extensive northern experience resulted in recruitment by the RCAF in 1930 as navigator on an epic 13,000-mile mission reaching north down the Mackenzie River to the Beaufort Sea and east to Repulse Bay in Hudson Bay. The expedition was commanded by one of the leading "characters" of the RCAF, F/L F.J. "Mawdie" Mawdesley, flying a Fairchild 71, accompanied by F/Sgt. H.J. Winny in a Vickers Vedette. Mechanic-photographer A/Sgt S.C. Dearaway completed the team. Their objectives were to explore and photograph the main water and air routes in the

Northwest Territories and to inspect the RCAF's 32 fuel caches laid down by ground and water transport, usually at small settlements, throughout that vast territory. The latter task involved first locating the caches, some of which they were not even sure existed, as confirmation had not been received from the contractors charged with their delivery.

Macdonald's journal chronicles his attempt to resign from the expedition at Aklavik after he believed Mawdesley had endangered his life by taxiing his Fairchild at high speed with the surveyor desperately hanging onto the float struts. Macdonald failed to appreciate Mawdie's "joke" and simmered until they reached Aklavik where he tried to resign by telegraph. Unable to get an acknowledgement from Ottawa, he decided, "against my better judgement, I guess I will go."[25] He did, however, insist in flying with Winny in the open-cockpit Vedette instead of with Mawdesley in the relative comfort of the Fairchild cabin.

The three-month-long expedition returned to Rockcliffe on 1 October, having flown about 13,000 miles over largely unknown territory and taken 3,100 photographs. It was a major achievement and established northern flying and aerial photographic procedures for the next decade. Macdonald returned to his post with the Topographical and Air Survey Branch, Department of the Interior. He published a detailed account of his journey which, typically, makes no reference to his difficulties with Mawdie and concludes by crediting the expedition's success to the foresight of the officers concerned and "the excellence of the aircraft used and the skill of the pilots."[26]

In 1931, he took part in a 2,800-mile aerial photographic survey in the Northwest Territories, part of a concerted effort to map the Arctic coastline and islands. His movements, however, were increasingly constrained by back pain and weakness of his legs, which he attributed to his injuries on surveys a few years previously. In 1932 his difficulties were severe enough for the proud and independent man to enquire about thc possibility of a disability pension. Although his injuries had been documented at the time by witnesses and by doctors who examined him at The Pas and later in Ottawa, assistance was denied because, "no claim would be maintained unless commenced within six months of the date of the accident"—and in the province in which it occurred.[27]

Of financial necessity, Macdonald continued to work, but laboured under great difficulty. His professional accomplishments were recognised by his peers in 1935 when he was elected president of the Canadian Institute of Surveying. While he never flew again with Mawdesley, his interest in new methods of surveying continued and his presidential address in 1936 concentrated on the revolution in surveying methods brought about by aircraft and wireless. At the same meeting, he introduced the keynote speaker, Lord Tweedsmuir, Governor General of Canada.

Regressing inexorably from cane to crutches, Macdonald was forced to resign from the interior department in 1937 because he could no longer walk.

25 Transcript of C.S. MacDonald's daily journal by Jean Cockburn (unpublished).

26 C.S. MacDonald, "Thirteen Thousand Miles by Air," *The Canadian Surveyor* V, no. 1 (1934): 4–8.

27 Jean Cockburn, private records.

He was only 50 years old, unable to work and with only a small pension. This left his family in poor financial shape. By 1940 he was immobile, barely able to move his head. Dr. Mount, an Ottawa neurologist, referred him to the Montreal Neurological Hospital where Dr. Wilder Penfield did extensive spinal surgery, but with no positive result. His nurse wife cared for him at home for 15 years, alleviating her financial difficulties somewhat by taking in elderly boarders and providing them, too, with nursing care. Her anxieties were compounded during the war when one son was killed by a land mine in Britain while a second was navigating de Havilland Mosquitoes over Europe, and the third contracted multiple sclerosis when training to be a pilot.

The radio was a lifeline for the invalid, but it had to be tuned for him—and he was stuck with whatever station selected while his wife and children were busy with other duties. Daughter Jean recalls with conflicting feeling the peremptory whistle, his only way to ask for relief from discomfort or loneliness, that invariably welcomed the children's return from school. One of the few things he was able to enjoy was his pipe, carefully cleaned, recharged with Old Virginia tobacco and lit by one of his children. Visits by surveyor colleagues were regular events that gave him something to anticipate. Making the best of his sad situation, Macdonald managed to read books with the aid of an "automatic page turner" that required his family to fasten clips to each page so he could turn them through a linkage system by moving his chin. The family encouraged his efforts to market "Prisma Glasses," a device for people who could not hold a book, as it gave him an interest, but minimal financial return.[28]

The Macdonalds' financial situation did eventually improve in January 1953 after the government belatedly took another look at his claim and approved the award of a disability pension. Most of his accumulated medical expenses were covered and he was admitted to the Department of Veterans Affairs Hospital in Ottawa, where he died on 3 July 1954.

28 Ibid.

Frederick Joseph Mawdesley

"MAWDIE," as he was known to everyone, was probably the most colourful character in the pre-World War II RCAF. He was born on 11 March 1891 at Parkanaur Castle, Caulfield, in County Tyrone, Ireland, where his father Robert, a civil engineer, sometimes took his son with him on contracts in South America, the British West Indies and the Orient. Educated in England at Hill's House Preparatory School, Windsor, and Beaconsfield College, Plymouth, Mawdie obtained his senior matriculation before working as a ledger clerk at the London Joint City and Midland Bank for three years. He emigrated to Canada in 1914 and joined the Grand Trunk Railway as an "accountant." Rumours, however, persisted that he was a "remittance man," paid by his family to stay away from Britain. Whatever the truth of this matter, Mawdie enlisted in the Canadian Expeditionary Force at Niagara Falls in 1915 when he, strangely, listed his previous employment as "caterer." He returned to England in March 1916 on his way to France in July as a gunner with No.2 Section, Third Division Ammunition Column.

Mawdie's experience during World War I was typical of many Canadians who enlisted as private soldiers and ended up in the RFC or the RAF. He was promoted corporal and then acting sergeant before posting to the Canadian Artillery Depot in England for officer training in May 1917. Temporary Lt. Mawdesley went back to France with the 123rd Pioneer Battalion in November 1917. He was seconded to the RAF and returned to England in May 1918 for training at Reading as an observer on probation. Posted to 49 Squadron in France in August, he was invalided back with an injured ankle within two weeks to the British Officers' Hospital in Eaton Square, London, and then to Edinburgh Hospital.

After completing courses at No.1 School of Aeronautics and the Wireless School, Mawdie was released from the RAF in May 1919. He had a total of 280 flying hours in his logbook as observer in Bristol Fighters, D.H.9s and Handley-Page aircraft. He sailed for Canada in December 1919, probably after visiting his father, who had moved to Wallasey, Lancashire.

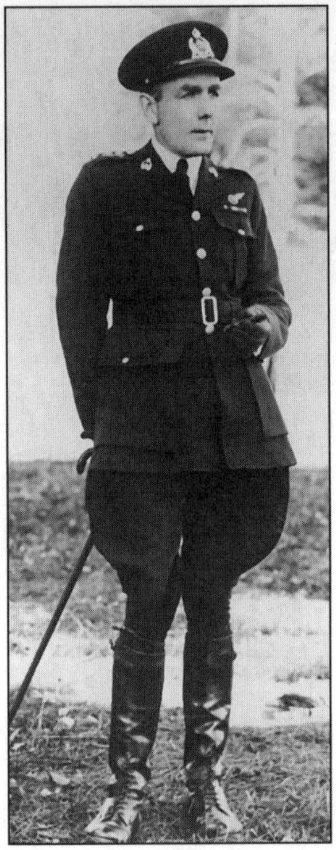

Mawdie rejoined the Grand Trunk Railway, but accounting was, apparently, too tame for his tastes and he signed up with the CAF as a lieutenant on 3 January 1921. He is wearing the dark blue and silver uniform of the CAF with his observer wing, 1921. *DNDPU RE12809*

Mawdie was the adjutant at Camp Borden until promoted flight lieutenant in September 1921 and posted to Ottawa as assistant to the staff officer, personnel. Still haunted by his accounting experience, Mawdie was assigned special duties relating to the drafting of Pay and Allowances Regulations and King's Regulations and Orders for the planned RCAF.

When the RCAF was formally established on 1 April 1924, Mawdie was demoted to flying officer and told in no uncertain terms that, in accordance with new policies, he would have to learn to fly within one year in order to retrieve his flight lieutenant rank. Within four days he was at Camp Borden to start flying training. His progress was slow. In October, his instructor, F/L Brocker, reported on a total of 34.20 hours dual and 43.15 hours solo, noting that his approaches and landings were erratic and often dangerous. After a check flight by S/L Anderson, Mawdie's flying was discontinued, "pending further orders from AHQ." Mawdie was keen to fly and had not crashed or damaged an aircraft, but appeared to like flying "on the edge" and was unable to qualify as a pilot. This was somewhat of an embarrassment to the school as he was also on the instructor staff teaching army organization, air force law and administration to second-term pupils. G/C J.S. Scott stepped in and authorized three more hours' dual instruction for Mawdie and, perhaps indicating that a personality conflict existed, added that he could choose his own instructor.[29]

Three weeks' leave over Christmas 1924 gave Mawdie time to reflect and he returned to fly 3.55 hours dual with F/L Roy Grandy and another 1.05 hours solo. Grandy reported that he had passed some of the tests satisfactorily, but others remained to be completed. Grandy considered that Mawdie showed outstanding keenness, but still lacked judgement in distance, was overconfident and had little natural flying ability. Headquarters concurred with Grandy's recommendation for a further two hours dual and two hours solo. On 25 March 1925, Mawdie was at last authorized to replace his observer wing with those of a pilot. After his 280 hours as an observer, he had flown 43.55 hours dual instruction and 48.30 hours solo over almost one year. It was rumoured that the popular Mawdie also had the benefit of some unofficial tutelage from his friends. Probably a record for the longest time for a pilot to win his wings. Col. Logan commented, tongue in cheek, "When the time came that observers had to learn to fly, Mawdesley had a very hard time. Grandy proved to be the only one who could teach him to fly. Eventually, to everyone's relief and amazement, Mawdesley received his pilot's wings. Grandy received the OBE."[30]

While Mawdie's performance as a trainee pilot is remembered as something of a joke, there was likely a serious underlying cause. He had threatened to resign over what he considered unjustified demotion that lost him both pay and almost two years' seniority. He was particularly incensed over the fact that "certain officers"[31] were placed senior to him who were pilot officers or flying officers

29 F.J. Mawdesley, Biographical File, Directorate of History, Dept. of National Defense.
30 Ibid.
31 NAC Personnel File.

when he was a flight lieutenant. This probably affected his performance in 1924, perhaps deliberately, because his piloting skills afterwards were consistently rated highly. He undoubtedly carried this grudge throughout his career, climaxing in a stormy session with W/C G.O. Johnson in January 1931 over his non-promotion to squadron leader. Even in 1942, he was complaining that he had been passed over for group captain, citing his 1924 demotion from flight lieutenant, and generating comments in annual assessments such as, "He is something of a character who looks older than he is."

Shaking off the dust of Camp Borden in July 1925, Mawdie went to Jericho Beach for seaplane training and subsequent patrol operations on the west coast, and then to Winnipeg in May 1926.

Mawdie renewed his acquaintance with the RAF over the years. Early in 1927, he accumulated 107 hours, mostly on Southampton flying boats, during a four-month period on the Air Navigation Course at Calshot. From 1933 to 1935, he was on exchange posting with RAF No.210 Squadron at Pembroke Dock. During this posting, he delivered a Singapore III flying boat to Singapore, necessitating 18 stops at such exotic places as Athens, Basra, Karachi, Allahabad, Rangoon and Penang—a total of 8,377 nautical miles in 105 flying hours from 12 January to 2 April 1935.

Mawdie's posting to the RAF concluded with the delivery of the flying boat to Singapore and he returned to Canada via Hong Kong on the S.S. *Empress of Canada* to take up his appointment with No.4 Squadron at Jericho Beach, Vancouver. This round-the-world trip was a feather in Mawdie's cap, but hindered his promotional prospects. He missed annual Canadian promotional examinations held in March and, although special arrangements were made, his schedule did not permit him to sit them in India.

Despite his promotional problems, Mawdie was generally very lucky, escaping unharmed from many tight corners. Historian F.H. Hitchins noted that he was reputed to have the homing instincts of a carrier pigeon and recounted the story of his becoming lost over unknown territory, with only sketchy and inaccurate charts and a wayward compass to guide him:[32]

> Hopefully, Mawdie followed a river for some time until a dwindling fuel supply made it imperative that he should be sure he was on the right course. Landing on the river, Mawdie went ashore, climbed to the top of a hill and surveyed the landscape, and then carefully studied his big gold watch to determine the right direction to fly. Returning to the Fairchild, Mawdie took off again, but to his crewmen's surprise, headed in the opposite direction along the river. Whether it was his sixth sense, sound judgement, or sheer luck, Mawdie finally reached his destination with scarcely enough fuel left to fill an eyedropper.

During 1929, Mawdie was a busy pilot. He put in 533 flying hours involving 566 landings. On a mission to the Oxford Lake area, he discovered

32 F.H. Hitchins, op. cit., p. 243.

a serious epidemic was raging in the native community and promptly flew in the Indian agent, doctor and RCMP constable. In May, he took off on another mercy flight from Cormorant Lake for York Factory on Hudson Bay with a doctor and two nurses to attend to Archdeacon Faries, who was seriously ill with typhoid fever, and to look after his wife who was breaking down under the strain. Caught there by a storm, Mawdie made a circuitous takeoff between snowdrifts as high as the wings of his Fairchild for the 5-hour return flight. The bishop of Keewatin expressed his appreciation for the "fine service . . . which evidently was the means of saving the life of a valuable missionary."[33] Nothing was reported about the medical attention given to his wife.

Mawdie was one of the very few pilots who liked the lumbering Varuna flying boat. One of his memorable performances was to lash a four-bladed propeller to the hull centre section of his Varuna and fly it 200 miles from Berens River to Cormorant Lake to the rescue of a survey party heading for Pukatawagan in another aircraft. Instead of waiting at least a week for ground transportation, the party was on its way north within a few hours.

S/L N.R. Anderson, CO Winnipeg, detailed Mawdie's "specially meritorious service" during 1929 and recommended him for the Cross-Canada Trophy in 1930. He was not selected, but his leadership abilities and reputation for success against the odds probably won him command of the most ambitious undertaking of the time. In July 1930, he set out from Rockcliffe in a Fairchild 71, accompanied by Sgt. H.J. Winny in a Vedette, to survey the main water and air routes in the Northwest Territories and inspect fuel caches laid down through the vast territory. Reaching as far northwest as Aklavik and then east to Repulse Bay, the detachment flew approximately 13,000 miles without serious problems.

While Mawdie's idiosyncrasies entertained most of his air force colleagues, a civilian surveyor, Colin S. Macdonald, navigator for the trip, was not amused by his bad language or propensity for practical jokes. Willingly sharing in the myriad of tasks necessary to keep the four-man expedition operating, Macdonald noted in his journal that he was balanced on the float of Mawdie's Fairchild, hauling in a tow rope, when Mawdie suddenly opened the throttle. Almost thrown into the water, he could not open the cabin door and was about to abandon ship when Mawdie throttled back enough for him to clamber into the cabin only to be greeted with verbal abuse. Incensed, Macdonald tried to resign, but was could not obtain confirmation from Ottawa, so he reluctantly completed the three-month expedition.

Mawdie was posted to Ottawa in 1931 as superintendent of personnel where he incurred the wrath of W/C G.O. Johnson, A/Director of Civil Government Air Operations, over his insistence on promotion to temporary squadron leader. Further annoying Johnson, he went over his head to senior levels without approval. Johnson reported that Mawdie "displayed considerable temper . . . and wilful and deliberate disobedience of

33 NAC Personnel File memo, 18 January 1930.

instructions."[34] Mawdie's pending appointment to superintendent of civil flying operations was cancelled, W/C Breadner declined his services in the RCAF Directorate, and Mawdie remained in his job as a flight lieutenant. It was September 1936 before he was promoted squadron leader.

In 1937 Mawdie was given the prestigious job of flying Governor General Lord Tweedsmuir on his tour of British Columbia. In the same year, he made the headlines with a mercy flight to evacuate an injured employee of Consolidated Mining & Smelting from Anyox about 80 miles to hospital in Prince Rupert.

Subsequent postings took him to command of the air navigation and seaplane school at Trenton and as an instructor on Supermarine Stranraer flying boats at Dartmouth. In April 1939 he returned to his old job at Trenton, and in July, W/C Mawdesley took command of No.4 (BR) Squadron at Vancouver. He went back to Trenton in 1940 as CO of the School of Administration, responsible for training officers to administer the burgeoning British Commonwealth Air Training Program, and then to various senior positions in Western Air Command.

Ann Pollak, daughter of A/V/M K.M. Guthrie, remembers her father telling, many times, the story of his inspection visit to Mawdesley's station at Rivers, Manitoba, in 1945. Anxious to make a good impression, Mawdie imported, at his own expense of about one thousand dollars, fresh flowers from Winnipeg, and planted them in the remnants of the winter snow bordering the parade ground. Guthrie managed to keep a straight face during his inspection, but could not restrain his amusement on subsequent recounting of his springtime experience at Rivers.

Tales of Mawdie's eccentricities abound, many of them chronicled by Jack Hunter in the CAHS *Journal*, Summer 1995. One memorable event occurred at Jericho Beach: Sergeant Major Barton had assembled all the personnel for inspection by tardy duty officer Mawdesley, who eventually arrived by taxi in which he had changed into uniform after a night on the town. Resplendent in his air force uniform—but wearing a bowler hat—Mawdie marched off to receive Barton's salute. Returning it, his hand brushed the brim of the bowler. He did a smart about-turn, climbed back in the taxi and was not seen for the rest of the day. Also at Jericho Beach, Mawdie obtained permission to study at home for his promotional exams and checked in by telephone each day, giving the assurance that if needed he could be on duty in half an hour. Earl McLeod, his commanding officer, nearly fell off his seat while watching the Pathé News at the cinema in Vancouver. There was Mawdie, large as life, in the front row of the opening ceremonies at the new Hialeah Race Track in Los Angeles.

Hunter also indicated that the tales of Mawdie's piloting inadequacies may have been exaggerated and tells the story of his having to make an emergency landing with a dead engine in a tiny lake. The engine was soon repaired but "the pothole wasn't big enough for a straight takeoff run."[35] Mawdie solved

34 NAC Personnel File memo, 29 January 1931.

35 J.D. Hunter, "Maudie," *CAHS Journal* 33, no. 2 (1995): 65.

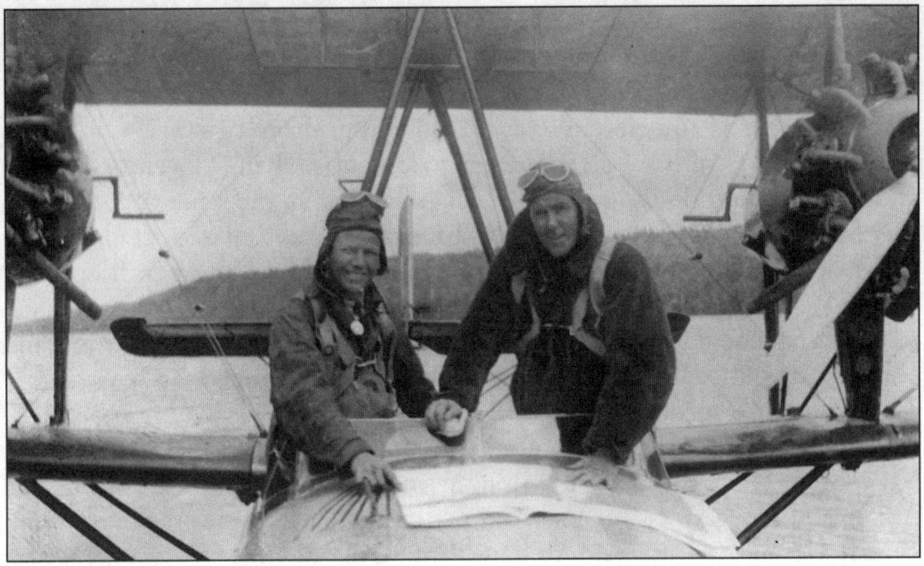

Notoriously camera-shy, Mawdie (right) is shown here with Sgt. Harry Bryant in a Canadian Vickers Vancouver. In his *Sixty Years*, Larry Milberry quotes Bryant as not remembering why they wore parachutes, normally kept stowed in the cockpit, because they would only be required if caught in a fog when the crew would have plenty of time to put them on. Bryant was surprised that, while he was wearing his Perrin's life belt, Mawdie, who could not swim a stroke, was not. Bryant noted the watch worn around his neck where it could be readily seen; the lines under his right hand used as an aid in maintaining a photo line; the ever-present cloth in Mawdie's hand used for wiping rain and fog condensation from the windshield; and the hand-powered cranks inboard of both engines—"We had no power starters."
DNDPU RE 19227

the problem by running the Vancouver round and round the perimeter of the small lake until he judged it had adequate speed to make a quick dash across the centre, get on the step and fly. It worked.

W/C R.I. Thomas told the story of Mawdie making a series of dubious decisions and atrocious landings during a west coast tour to the extent that he had his crew in fear of losing their lives. Pressing on to the Queen Charlotte Islands though rain that peeled off the paint from the fuselage, Mawdie opened up a hinged part of the windshield as he attempted a landing. Rather than improving his visibility, the rain struck him in the face and covered his spectacles. Disconcerted, he almost stalled the aircraft about 100 feet above the water. The crew really believed they were done for this time, but Mawdie recovered, pushed up his spectacles and made a perfect landing without a bounce, in spite of the very rough water. "Of course, everyone in the aircraft hated his guts for getting away with it."[36]

Some completely unfounded stories of Mawdie circulated, such as his being navigating officer of a Royal Navy submarine during World War I. No doubt improved over the years, but probably with a large grain of truth comes the undocumented story of Mawdie's initiative at Jericho Beach when the air force doctor arrived to check out the pilots. Mawdie feared that his eyesight was failing so he sent an airman in uniform, but wearing a red tie, to fish from a rowboat way out in the bay. He then invited the doctor to a tour of the base and just happened to notice the fisherman and complained loudly about him wearing a red tie while in uniform. The doctor could barely see the boat, so he was astounded to see when the airman, on cue, rowed back to the beach that he was indeed wearing a red tie. So impressed was he that he saw no need to test Mawdie's eyes.

36 R.I. Thomas, *Stories for a Grandson*, unpublished (ca. 1980).

W/C William Skelding spoke to the Toronto chapter of the CAHS in 1984 of his years as a mechanic and photographer and, of course, had a Mawdie story to tell. This one about the "lung tester," a shop-made gadget about the size of a blowtorch with a spout mouthpiece on the side and a free-turning fan that was supposed to indicate by its speed the lung strength of the person blowing into it. Mawdie had been caught by a favourite trick of filling the reservoir with lampblack, which a good hard blow would distribute all over the victim through small vent holes. He was determined to obtain his revenge and selected the station CO, rumoured to be the most humourless man in the air force, as his victim. Mawdie had figured out how to hold his fingers over the vent holes and demonstrated how to use the device with a good hard blow. Unfortunately for him, someone had drilled a few extra holes and he was again covered in lampblack. He did get some satisfaction, however, as it was said to be the only time the CO had been known to laugh.

Mawdie's generosity was legend. In his early days when CO at Lac du Bonnet, all the personnel were refused pay and Christmas leave because, despite best efforts, all the necessary work for winter storage of their aircraft for the winter had not been completed. Mawdie wrote all his men personal cheques and allowed them local leave. A personal recollection of Jack Hunter concerned an accidental meeting with Mawdie in Southampton, England. Discerning that Hunter had little money, Mawdie signed a cheque: "I'll leave it open. You can fill it in for whatever you want and when we get back to

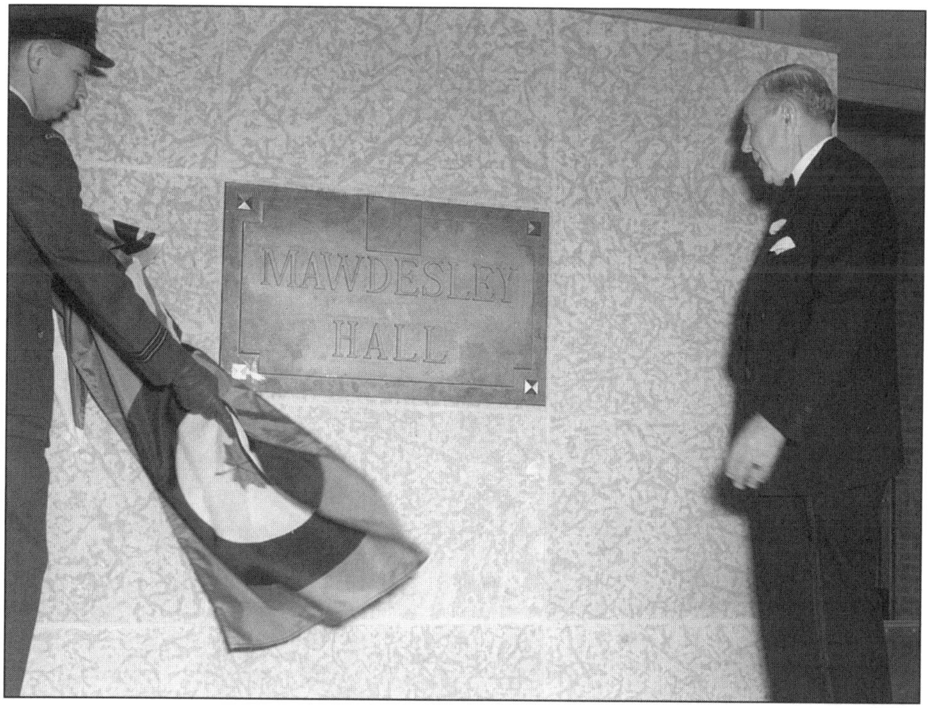

G/C F.J. Mawdesley unveiling the plaque dedicating Mawdesley Hall School of Navigation at RCAF Station Winnipeg in his honour, 7 October 1958. Assisting is F/L E.M. Garret, an instructor at the school. Mawdesley Lake, northwest of Lake Winnipeg, also carries his name.
DNDPU PL 93944

Ottawa, you can pay me back sometime." Afraid he would lose it, Hunter tore up the cheque, but the gesture left a lasting impression.[37]

Mawdie was promoted group captain in 1943, and awarded the Air Force Cross in 1945 when he was CO of No.5 B&GS. His citation read:

> This officer as commanding officer of a Bombing and Gunnery School, by his energy and good leadership, has brought the functioning of this station to a high level of efficiency. Throughout a long flying career he has made an excellent record as a pilot and instructor and has at all times taken a keen and active interest in the progress and welfare of those undergoing flying training. By his personal example of leadership, ability and devotion to duty he has set a high standard for all officers to follow.

After release from the RCAF on 6 July 1945, Mawdie served with the International Civil Aviation Organization in New York.

It was probably the combination of his generosity and gambling habits that led to Mawdie's eventual downfall, perhaps with a little assistance from alcohol. Also, despite his reputation, he was a shy man who would never burden anyone else with his problems and he never called on his many friends in his hour of need. Although for many years his brother Charles in London, England, was listed as next of kin, one of Mawdie's last messages to DND had a sad note: "I am a single man; I have no next of kin, they are all deceased."[38] He died on 12 May 1968 in the Orange County Memorial Hospital, New Jersey, intestate and destitute. His burial expenses were paid by the St. George Society of New York, which was partially reimbursed by payment of his outstanding pension.

37 J.D. Hunter, op. cit., p. 65.
38 NAC Personnel File.

Alexander John Milne

ALEX MILNE was born in Scotland in 1900 and came with his parents to Canada in 1904. He returned across the Atlantic Ocean in 1916 with the 5th Canadian Infantry Regiment to the trenches in France. While he was recovering from a wounded hand, the big push at Vimy started and he was among the walking wounded rushed forward to reinforce the attack. He was again wounded, more severely this time with shrapnel in his left leg, and spent eight months in hospital before returning to Winnipeg. There, he was discharged from the army in July 1919, but he continued receiving treatment for several years.

In 1920, Milne married Elin Stenvor Olsen, who was born in Iceland. They had one son, Alexander, starting a tradition that continued with a grandson and great-grandson with the same name. Milne worked at various engineering jobs in Winnipeg until he joined the CAB in June 1923 as a fitter. His first job was to help roll up and put into storage two "blimps" that came from Britain with the Imperial Gift, but were never used. He was then sent to join the group operating ex-US Navy Curtiss HS-2L flying boats from Victoria Beach on the eastern shore of Lake Winnipeg where conditions were somewhat primitive. In his 1977 address to the CAHS, he recalled that they did have a workshop lorry equipped with a Coventry gasoline engine to generate electricity and drive a lathe—but that was about all. They had to provide their own hand tools and the wrenches fitted only British nuts, but the HS-2Ls had US hardware: "We collected half our tools and took them to the technical school in Winnipeg, where we had them partly weld up some of the openings . . . and got busy with vises and files, making them fit American nuts and bolts."

Milne accompanied the first HS-2L sent in 1923 to open a sub-base at Norway House, some 385 miles north of Winnipeg, to fly forestry patrols on behalf of the Manitoba government. He received a monthly salary of $120, "two dollars a day for grub" and 75 cents extra for every flying day. Good pay for the times, but not enough to buy food from the Hudson's Bay Company store. So they ordered groceries from "old man Finklestein" in Selkirk who shipped supplies weekly on the steamer *Kenora*. They took turns cooking, adding flavour from their oil- and grease-impregnated hands.

Cpl. Milne, and the rest of the Viking crew at Norway House, were given the unusual appointments of Indian agents in 1923. Assigned to transport Dr. Stone of the Department of Indian Affairs to make treaty payments to Indian bands in Northern Manitoba, the airmen were put to work. Milne was assigned to check off on a list each Indian who received $5 from Dr. Stone. He commented that the factor from the local HBC post was invariably lurking nearby to relieve the Indian of his "Queen's Money" in payment for accumulated debt. He was amused to add that the Hudson's Bay Company was better known at Norway House as "Here Before Christ."[39]

The obvious disappointment of many Indians in this modern method of payment generated some sympathy from Milne, who enjoyed a ceremony. Arriving without warning out of the blue, the Viking ruined a long-standing tradition of greeting the slow-moving, canoe-borne treaty payment party in full ceremonial regalia and with drums beating. Some of the more distant Indians did not even get to the village before the Viking roared off to make the next delivery.

Under the command of S/L Basil D. Hobbs, Milne served as the mechanic on the pioneering flight of Vickers Viking G-CYET to Reindeer Lake and Churchill River Districts in 1924. Photographer P/O Jimmy Cairns and navigator Bob Davidson made up the crew. He kept the aircraft serviceable in very primitive conditions despite some serious problems. Even starting the engine by hand, he recalled, was difficult: "You took that crank in one hand and hung onto a strut with the other; hung on and wound like hell until the engine got going." But, one day, no amount of effort could turn the crank. Thinking he may have to take off the front of the Eagle engine to free the crank, but lacking the necessary tools, Milne resorted to "the old hammer tuning" and vibration finally solved the problem. "Boy, did I squirt Castrol in! It had the same effect on that engine as it had on me when I swallowed it."[40]

Milne saved the day again at Stanley Mission when 'ET could not take off because of the drag imposed by torn fabric under a wing root. Dipping into his tool kit, he found enough fabric for a patch, and a tube of Ambroid glue in which he had great faith: "You could do anything with Ambroid—stick anything—under any conditions! You could even stick yourself to the aeroplane, if you didn't watch your hands." With the other three crew members perched on the opposite wing to tilt the aircraft and give him working room, he stripped to his shorts and spent three hours in the frigid water sewing on a blanket patch coated with Ambroid, "which tightened up the fabric beautifully."[41]

Milne's lifelong fascination with the native people was fuelled on the Reindeer Lake foray. The Viking was the first aircraft to land at most of the

39 Alex J. Milne, "Pioneer Survey Flights in Northern Manitoba and Saskatchewan," *CAHS Journal* 21, no. 4 (1983): 102.
40 Alex J. Milne, "Pioneer Survey Flights in Northern Manitoba and Saskatchewan," *CAHS Journal*, 22, no. 1 (1984): 5.
41 Ibid., p. 9.

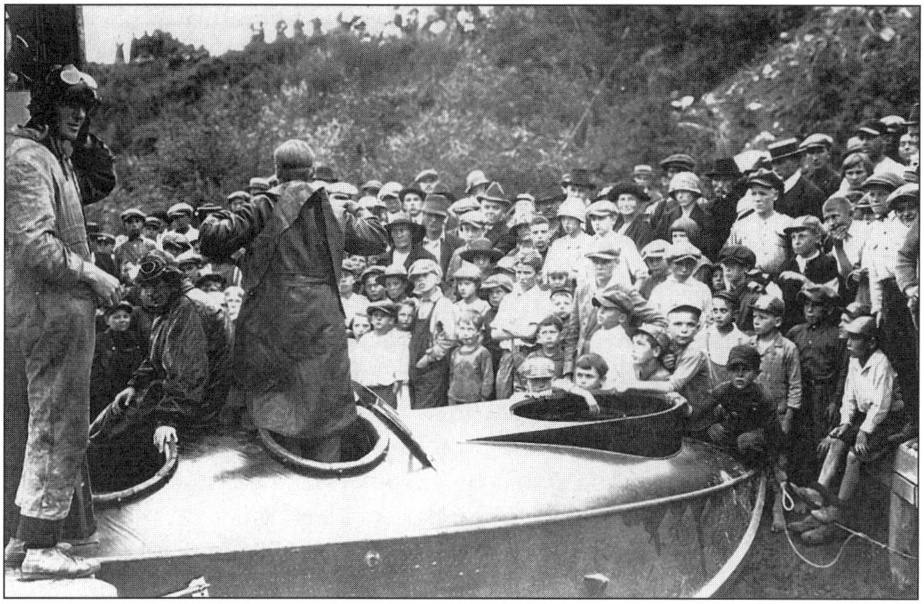

The arrival of an aircraft at Prince Albert, Saskatchewan, in 1924 was a momentous event. Alex Milne is standing on 'ET at left and Bob Davidson is seated while Basil Hobbs addresses the welcoming committee. In the centre of the crowd wearing a fedora is a young local lawyer, John G. Diefenbaker, who became prime minister of Canada in 1957.
ALEX MILNE

The *Edmonton Daily Herald Tribune* waxed lyrical: "Soaring through the skies and swooping downwards from the heights above the pine-crested hills miles north of Prince Albert four intrepid aviators descended on the city at seven o'clock last evening in their twelve-cylinder Vickers Viking."

isolated refuelling bases. "When we arrived they would run into their huts or wigwams and come out gaily decorated, dressed in their most brilliant colours." Drum dances would be part of the entertainment, and they would sit watching the aircraft for hours: "When 'ET moved with the wind and swung closer to them, they would all get up from where they were sitting and move back, gradually creeping forward until it moved again. . . . It was fascinating and I wish I had had my movie camera with me."[42]

The Reindeer Lake flight was a notable achievement and did much to establish the ground rules for subsequent photo flights. S/L Hobbs gave full credit to his crew, who took a very fragile piece of machinery into the unknown. He reported on Cpl. Milne: "Our mechanic was always ready for any emergency. The condition of the machine and engine on our return from this flight is a testimony to the care with which he did his work. Refuelling, mooring and the necessary maintenance were often carried out under difficult circumstances, which threw an additional burden on him. His conduct was at all times a credit to the service."[43]

The remainder of Milne's first stint of air force service was spent keeping equipment working at a variety of locations, including route surveys for the new railway from Winnipeg to Hudson Bay. He left the air force in February 1929 to join Dominion Explorers Ltd. as chief mechanic.

On one of his first jobs in "civvy street" he accompanied the ill-fated 1929 MacAlpine Expedition. A Fairchild FC-2W2 and a Fokker Universal set out from Winnipeg on a mineral survey expected to take them up Chesterfield Inlet to Baker Lake, Pelly Lake, Bathurst Inlet, Coppermine, Great Bear Lake, Fort Norman and north to Aklavik, returning by Fort

42 Alex J. Milne, op. cit., p. 8.
43 B.D. Hobbs, "Report on Reindeer Lake and Churchill River Photographic Operations 1924," NAC Accession No. 912040 (Winnipeg: n.p., 1924).

Simpson and The Pas to Winnipeg. When they did not return as planned, an aerial search was initiated, but with some difficulty as the searchers had no idea where along this long route the expedition had run into trouble. MacAlpine's men were, in fact, relatively safe, but unable to communicate with the "outside" world. They were stranded without fuel alongside an Eskimo camp at Dease Point in Queen Maud Gulf. Survival was the order of the day. With the willing help of the Eskimos, they were able to reach Cambridge Bay on Victoria Island, without being spotted by the searching aircraft, 58 days after being "lost."

One of Milne's many stories of his Dominion Explorers days involves having to put down in white-out conditions "up north." The weather was so bad the crew decided to overnight in a trapper's cabin which, of course, involved draining the Fairchild FC-2's engine oil, keeping it warm and replacing it the following morning. In the process, Milne noted that the oil level was dangerously low. Bucking the expected headwinds was not going to improve matters so he advised the pilot to head for a Bellanca that he knew had been abandoned nearby. Fortunately, they were able to locate it. Building an igloo over the engine, he was able to generate enough heat with a small oil burner for the oil to flow into a can and quickly transferred it to the Fairchild. With a full supply of oil, the journey was completed without problem. During the salvage of another Fairchild, Milne exhausted his supply of welding rod while repairing its undercarriage. Nothing daunted, he confiscated all the change carried by the recovery crew and finished the job with dimes and quarters.[44]

During the 1930s, Milne worked as chief mechanic at the Winnipeg Flying Club. He had a spell as a businessman, running his own automobile service station, but aviation called and in 1938 he moved to Fort Erie, Ontario, in charge of the final assembly department at Fleet Aircraft Limited.

44 Alex Milne, personal communication.

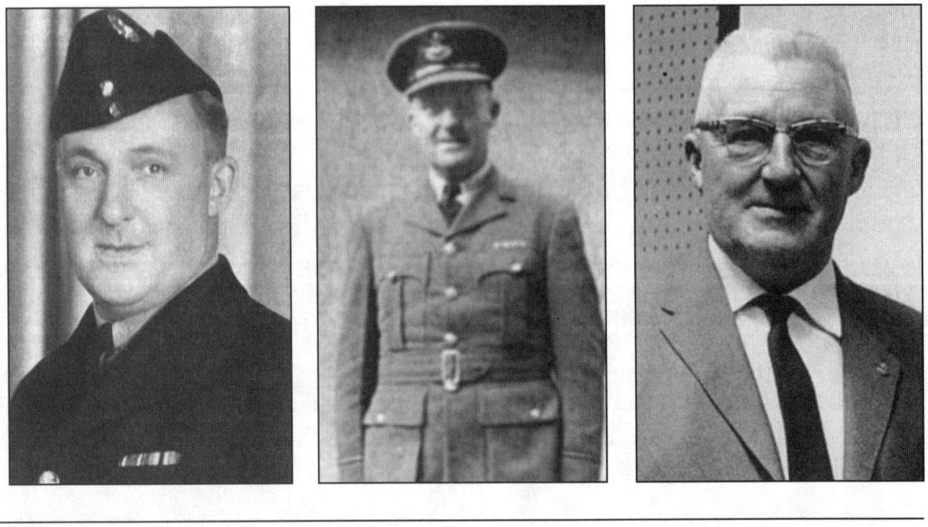

From the Milne family album.
ALEX MILNE

In January 1940, Milne re-enlisted in the RCAF as a sergeant fitter and was commissioned as an engineering officer in 1943. He spent most of the war based at Toronto where his reputation for getting stubborn machinery and people working efficiently resulted in him having a roving commission as a "troubleshooter." With reorganization of the RCAF in 1945, he reverted from flight lieutenant to flying officer and did a six-month stint in Washington, DC as an inspector, helping prepare aircraft for the Berlin Airlift. Back in Toronto, he served as technical officer at No.1 Supply Depot until his retirement in 1949.

Milne then went to Orenda Engines at Malton, again as an inspector, until the demise of the Avro Arrow in 1959. Not one to be idle, Milne transferred his energies to the Toronto chapter of the CAHS where he was a leading figure and treasurer for many years.

Alexander John Milne, British War Medal, Allied Victory Medal, CVSM, 1939–45 War Medal, CD, died on 23 May 1991. He was the last survivor of the epic Reindeer Lake Expedition. He is fondly remembered as personifying the combination of enterprise, get-on-with-the-job ethic, skill, versatility, humour and team spirit that were the characteristics of the early air force.

Arthur Lawrence Morfee

"LAURIE" MORFEE, born in 1897 in London, England, was an aviation enthusiast from a very early age. He recalled cycling many times as a young boy to the motor racing track at Hendon to watch the RFC experimenting with early model de Havillands, Farmans and Royal Aircraft Factory FEs.

He emigrated with his family to London, Ontario, in 1913 where he joined the Canadian Expeditionary Force in 1915, serving with the 135th, 134th, 19th and 3rd Battalions. In France, he was soon introduced to the horrors of the "rolling barrage." In theory, the shells fell in front of the advancing infantry, but Morfee remembered, "Sometimes the shells would land behind us . . . perhaps this was done to egg us on!"[45] As a scout, he was assigned to map enemy trenches. He was promoted sergeant before being wounded at Hill 70. After recuperating in England, his demonstrated aptitude for air photo interpretation assisted in obtaining a commission in the RFC in 1917. He trained as an observer, and was taking a photographer course when the Armistice was signed.

Demobilized and back in Canada, Morfee worked in commercial air photography in Montreal. His expertise led to an appointment at Camp Borden in 1921 as an air photographer instructor. He accepted the offer of a commission in the RCAF on its formation in 1924, largely because he was promised training as a pilot. When he received his wings, he flew as a photo pilot from most of the air force stations across the Dominion during the next six years.

In the days before airborne radio, a couple of pigeons were supplied to the west coast HS-2Ls for emergency communication with the base at Jericho Beach. Experiencing engine trouble off Texada Island in the Strait of Georgia, Morfee landed, scribbled his position and slipped the messages into the special containers fastened to the pigeons' legs, and released them. As he remembers, "The pigeons would not leave the aircraft! They were running around on the wing of the flying boat with my aero-engine mechanic chasing them; by the time they left, we had already tapped into a telegraph line on the

45 Frank Honey, personal communication.

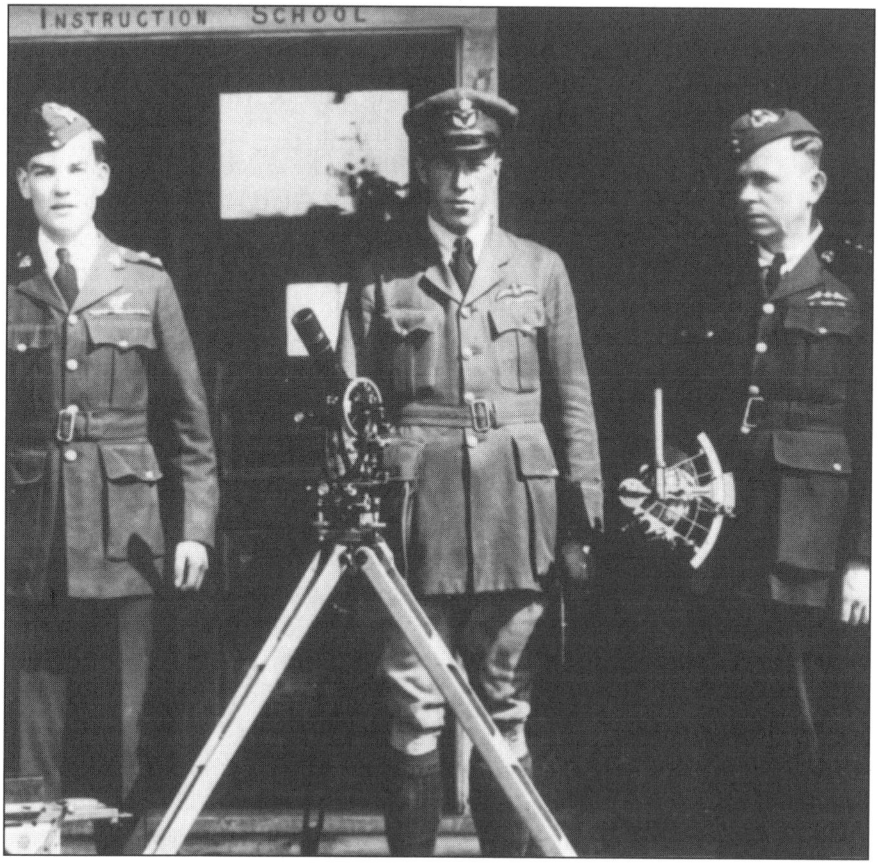

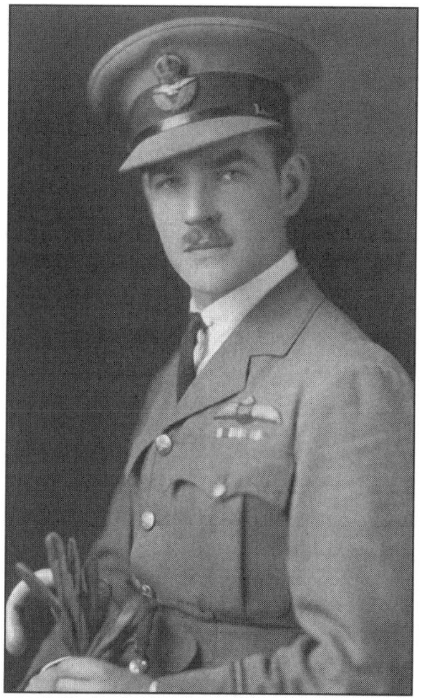

Left: On the left, Morfee is wearing his observer wing. Centre is A. Tapping. R.A. Logan is on the right.
DNDPU RE-17684

Above: Morfee was promoted flight lieutenant in 1928 and squadron leader in 1935. Photo c.1926.
FRANK HONEY

island, and the other aircraft came and picked us up. We were home before the pigeons."

What was possibly the first attempt to photograph an eclipse from an aircraft was undertaken from Camp Borden on 23 January 1925. F/L G.E. Brookes and F/O A.L. Morfee in one Avro 504, and F/L R.S. Grandy with *Toronto Star* reporter F.G. Griffin in a second, met with only limited success. Grandy's aircraft had to land with engine trouble near Newmarket. Brookes and Morfee met solid cloud and climbed to 9,400 feet where, through gaps in the cloud, Morfee was able to take a few exposures of the sun emerging from behind the moon.

The Canadian Vickers Vikings had a series of accidents that resulted in strengthening of the fuselage. The first indication of trouble was experienced, in July 1926, by Morfee and his crew. The hull of G-CYEY broke in two after a wing tip float dug into a wave when landing at Kashabowie Lake, Ontario. They were surprised, but uninjured.

Early experiments with skis introduced Morfee to Billy Mitchell, the famous US Army proponent of air power, who arrived at Camp Borden to see for himself what was happening. Two shiny, new D.H.4s on wheels ignored the

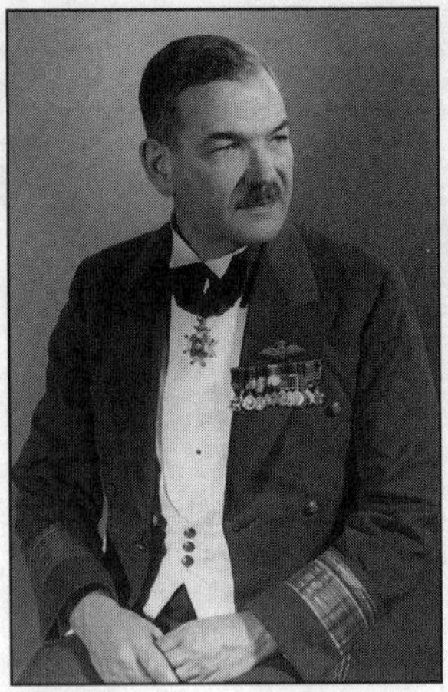

A/V/M A.L. Morfee. Five geographical features from British Columbia to Manitoba were named in honour of his days as a photographic pilot during the 1930s. The public school at Greenwood, Nova Scotia, also carries his name. *NAC RE4405*

marked landing strip on packed snow, preferring to land where grass peeked through at the far side of the field. Unfortunately, the grass was three feet tall and both aircraft ploughed in, breaking their propellers. Mitchell was reported to be undismayed: "We have lots more ships like these."[46]

Morfee is probably entitled to a *World Book of Records* entry as pilot of the first aircraft in which a baby was born. On 29 March 1932, he and his crewman, LAC Link Bailey, were assigned to fly an expectant Indian lady in a Fairchild FC-2 from Mile 214 on the Hudson Bay Railway to hospital at The Pas, Manitoba. En route, the baby decided to make an unscheduled arrival. Unable to leave the controls or to land the aircraft on the rocky ground, Morfee left the problem in Bailey's apparently capable hands. The aircraft and its "airborn" baby arrived safely at The Pas. Morfee's logbook contains the laconic statement: "Sick Indian woman to Pas. Baby born en route." The mother marked the occasion by naming her son Lindbergh Wright Cook.[47]

Morfee spent 1933 at the RAF Staff College in Andover, returning to various staff positions at AFHQ in Ottawa until 1938 when he was promoted wing commander and appointed senior personnel staff officer at No.1 Training Command in Toronto. At the outbreak of World War II, he took command of RCAF Station Dartmouth as a group captain, moving to Eastern Air Command at Halifax in 1942. He went to St. John's, Newfoundland to command No.1 Group in 1944, and became air officer commanding Eastern Air Command in 1945 with the rank of air vice-marshal.

Awards included the OBE in 1943 and CB in 1948. He also received the CD and Clasp, and was made a Commander Legion of Merit (US) in 1950 for "exceptionally meritorious conduct" during the war years and his post-war work as a member of the Permanent Joint Board of Defence. His CB citation summed up his wartime service:

> A/V/M Morfee held a number of positions of responsibility throughout the recent war. Since 1941 he served on the East Coast, first as Chief Staff Officer, Eastern Air Command, later as Air Officer Commanding-in-Chief, Eastern Air Command. His administrative ability based on many years of efficient and valuable service contributed in a large degree to the successful development of a command which played an important role in the Battle of the Atlantic. By his sound judgement, energy and unceasing vigilance he kept his command at a creditable state of efficiency in spite of rapidly changing commitments during the latter months of the war. His confident leadership has been an inspiration to all those under his command.

When the RCAF's five commands were reduced to two in 1947, A/V/M Morfee was posted to Ottawa, responsible for service and operational requirements, organization and establishments, and intelligence. He was a key

46 Frank Honey, personal records.
47 Ibid.

player in the selection of the F-86 as Canada's first jet fighter and its transformation into the Canadair-built, Orenda-powered CF-86 that made the RCAF king of the European skies during the 1950s and '60s. He was also a member of the Inter-Departmental Committee on Civil Aviation.

Upon retirement in 1948, A/V/M Morfee settled near Annapolis Royal where he was emergency measures co-ordinator and honorary superintendent of historic Fort Saint Anne National Park. He also served on the Annapolis Royal town council. He was the first national president of the RCAF Association, a director of the Air Cadet League of Canada, and vice-chairman of the Nova Scotia Corps of Commissionaires. After an unsuccessful attempt to obtain a parliamentary seat for the Liberals in 1957, Morfee started an antique business in Annapolis Royal in partnership with his wife, Cecile (Hurd). He retired for the last time in the mid-'60s, and led a quiet life, "sitting in his favourite chair and smoking his pipe," until his death on 19 March 1988.

Edward Rosser Owen

BORN ON 30 August 1894 in London, England, where his father had a newsagent's and tobacconist's shop at Finchley, Owen attended Trinity College and Haverstock Schools. From 1908 to 1910, he studied art and photography at the London School of Photo Engraving and Lithography. He then served a two-year "improvership" in the experimental department of photo supplies company A.W. Penrose, assisting with the development and testing of new photographic equipment and processes.

Recognizing an opportunity to capitalize on his state-of-the-art knowledge, Owen emigrated to Canada in 1913, signing on as a photographer with the Topographical Survey Branch of the Interior Department on 5 September at a salary of $800 per annum. His duties included lithography, photo engraving, photo processing and map making. He also obtained his St. John's Ambulance first aid certificate and joined the Canadian Army Medical Corps branch of the militia. His civil service job lasted only one year, as he enlisted in the Canadian Expeditionary Force in September 1914 and returned to England in 1915 as a hospital medical attendant.

Promoted corporal in February 1916, he was posted to France with the Canadian Cavalry Corps in July on "sanitation and medical" duties. He soon suffered the first of a series of injuries requiring him to change roles in hospital and become a patient. In October his right leg was injured when he was thrown from his horse, in February 1917 he had a suspected fractured left tibia resulting from an energetic rugby tackle, and in March did fracture his right leg in another sports accident.

Nova Scotian Stuart Graham's record of service at Cattewater was published in the *CAHS Journal*, Summer 1999. He described a patrol on 17 March 1918 in a Short 184 Seaplane with Owen in the observer's cockpit. Launched by crane from the Plymouth breakwater and loaded with two 230-pound bombs, they had an uneventful patrol until turning for home, some 40 miles out at sea, at which point the single Sunbeam-Coatelen engine lost power. They were unable to maintain altitude and were too low to risk

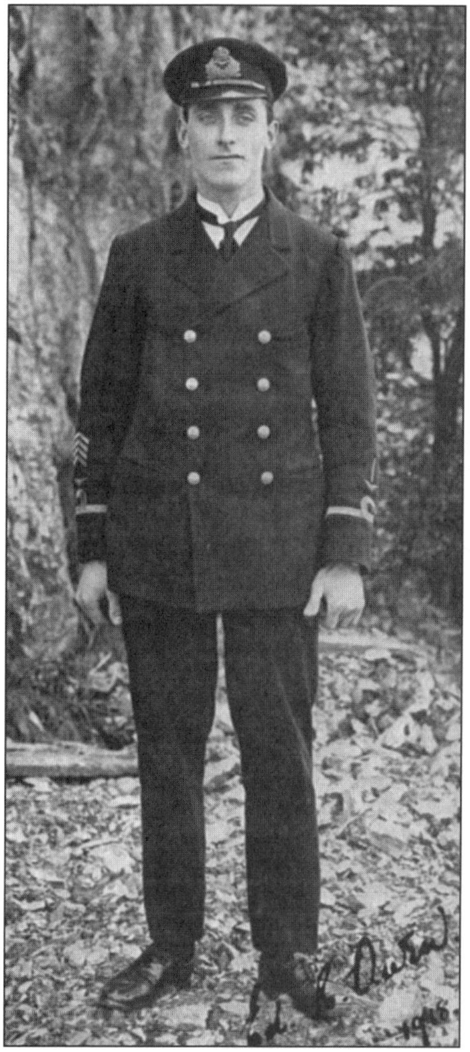

In September 1917 Cpl. Owen's photographic knowledge enabled him to obtain a commission as temporary flight sub-lieutenant in the RNAS where he was trained as an observer. He was transferred to the RAF in April 1918. *DAVID L. OWEN*

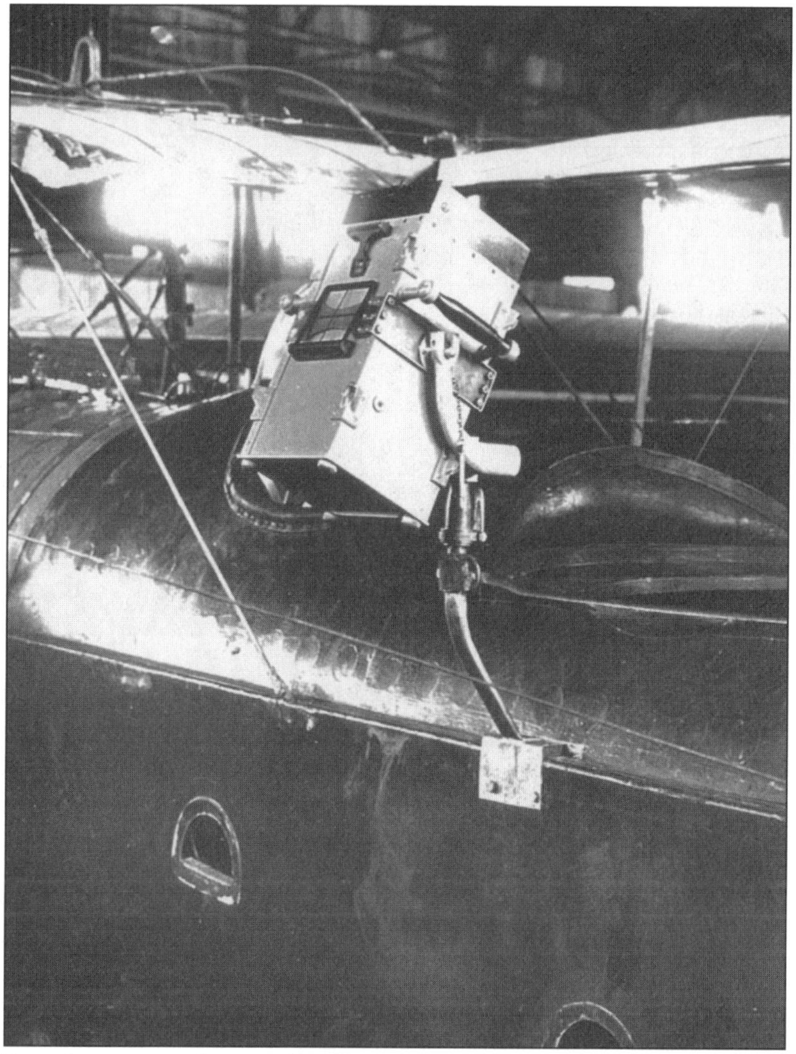

Lieut. Owen was a photographic officer at RNAS (RAF in April 1918) Station Cattewater, Plymouth, Devon when Lt.-Col. Philip Shepherd recommended him for the DFC: "He is an officer of much ability and gallantry and was always a fine example to young and inexperienced officers."[48] He did not receive the DFC, but in January 1919 was Mentioned in Dispatches for "distinguished conduct." This is Owen's camera installation in a Short Seaplane. *DAVID L. OWEN*

jettisoning their bombs, so they prepared to land while Owen attempted to send off a distress radio signal. Not sure if the message had been received, they also released their pigeons with messages attached. Night fell and it was clear that no help was on the way. Fortunately, the sea was reasonably calm. The engine continued to fire, albeit weakly, so they taxied slowly towards land. A Royal Navy sub-chaser spotted their flaming exhaust, came over to investigate and led them back to harbour. Graham reported that their engine quickly went on the scrap pile.

48 David Owen, private papers.

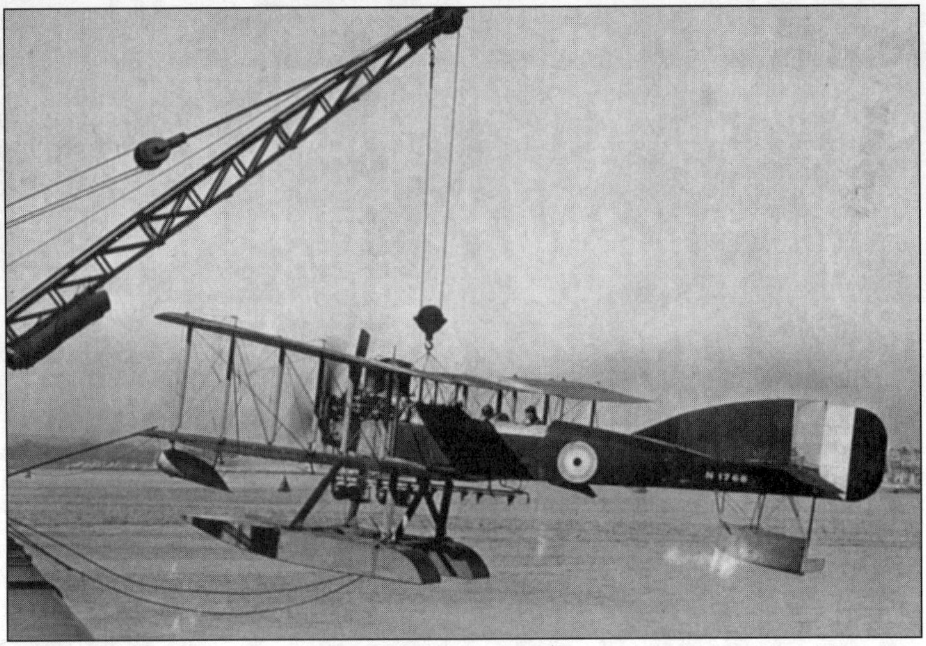

Pilot Lieut. Graham and Observer Lieut. Owen being launched from the breakwater at Plymouth for anti-submarine patrol off the English coast in 1918. They were both involved in several attacks on suspected U-boats. Note the two bombs slung between the floats.
DAVID L. OWEN

It was a coincidence that Owen and Graham served together in the RNAS at Plymouth, as both subsequently made important contributions to Canadian aviation. Owen was a key player in the development of aerial photography. Stuart Graham is recognised as one of Canada's premier bush pilots and held many influential positions in both the RCAF and civil aviation.

After returning to Canada in the troopship *Toloa*, Owen was demobilized in March 1919 and resumed his job with Topographical Survey in Ottawa. He was clearly intrigued by the possibilities offered by aerial photography and successfully applied in April 1920 for transfer to the CAB as air photographic inspector. Dr. E. Deville, surveyor general, who recommended his appointment to Lt.-Col. Leckie, superintendent of flying operations, commented: "I think the Air Board is to be congratulated for securing the services of Mr. Owen. He has not only the technical qualifications for the position, but he has had wide experience in Air Photography and his personal characteristics are everything that can be desired."[49] He reported to Camp Borden to organize and command the photographic section.

Owen soon had his first of many battles with "The System." Correspondence indicated that he wore a pilot's wing on his uniform because no observer wings were available. Chastised for being improperly dressed, he submitted his resignation, but calmer heads prevailed and he stayed in uniform—with an observer wing. A second brush with authority came in July 1921 while he was at Dartmouth, Nova Scotia, installing a camera in an HS-2L flying boat. A terse telegram from Lt.-Col. Robert Leckie, director of flying operations, "No reports received from you since leaving Ottawa. Why?"

49 NAC Personnel File.

generated a lengthy response. Listed in some detail were the unavailability of workers, a strike of moulders and fitters, and the necessity for him to put in 14-hour days. Owen concluded with a query as to whether he should start modifications of a second aircraft, generating an even more brief response from. Leckie: "No."[50]

The remainder of 1921 and the first few months of 1922 were spent with Capt. H.L. Holland, superintendent of the Rockcliffe station, photographing locations in Ontario and Quebec. Notable flights were made along the St. Lawrence River, providing photographs for the planning of the St. Lawrence Seaway.

Disaster struck on 23 April 1922. Holland and Owen were en route from Camp Borden, to complete photographic coverage of London started the previous year, when their D.H.4B crashed near Brantford, killing Holland and seriously injuring Owen. His injuries included cracked ribs, facial injuries, contusions of one leg and burns to his upper body. He was in London Hospital, suffering also from shock, for several weeks before returning to duty. A medical board decision that he was permanently unfit to fly posed a serious problem because Owen, as CO of the Photo Unit, was on the General List, meaning he could be called upon to fly at any time.

In March 1923, F/L G.O. Johnson confirmed previous discussions with Owen, suggesting that his flying duties would be kept to a minimum because of his disability and, strangely, offering to arrange pilot training for him. Internal correspondence indicates a much harder position by the Air Board. Secretary J.A. Wilson went on record stating that Owen was not medically fit to be commissioned into the RCAF, but his qualifications were such that it would be difficult to replace him. As his value to the organization was very great, Wilson recommended that he be temporarily retained until a suitable replacement was found. Owen believed he was being asked to resign and enquired about compensation for his injuries. The outcome was agreement by a departmental committee to offer Owen a commission in the RCAF "without award for disability."[51] Owen was evidently content with the trade-off and commanded the photo school until 1936, being promoted flight lieutenant in 1925 and squadron leader in 1930.

A footnote to this story concerns Owen's kit, destroyed in the 1922 crash. He put in a claim for $201.50, the depreciated value of his personal equipment that had originally cost over twice this amount, but had to settle for $149.35.

On 12 November 1924, he married Bertha Lyall Smeeton at Ottawa. Two children died in infancy and they adopted a son, David.

Owen attended the Staff College Course at the Royal Military College, Kingston, during the winter of 1931–32. He served on many boards and committees while stationed at Rockcliffe, including the Committee on Flying Clothing and the Committee on Aerial Photography for Mapping Purposes. He was also a trustee of the RCAF Benevolent Fund. After leaving the photo

50 Ibid.
51 Ibid.

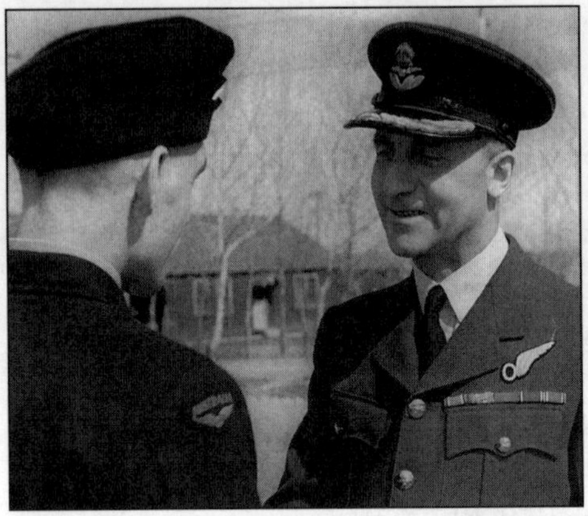

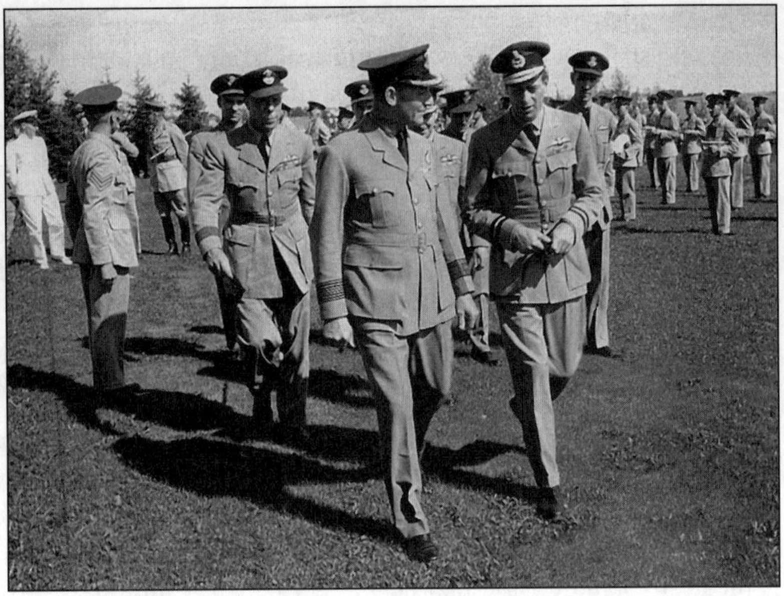

Above: G/C E.R. Owen, Commanding No. 2 WS, congratulates a graduating student. Calgary, 1943. *DAVID L. OWEN*

Right: While in command of No.2 WS, G/C Owen (foreground) was host to HRH The Duke of York, c.1944. *DAVID L. OWEN*

school, he served four years at RCAFHQ in various administrative positions and was promoted wing commander.

Specialist knowledge enabled Owen to obtain an RCAF commission in 1924, but became a handicap as he advanced in rank. His annual appraisals during the 1920s and 1930s extolled his abilities, keenness, work habits and appearance. His 1930 report, for example, commented, "As officer commanding the photographic section he has carried out his duties with zeal and efficiency. He is highly qualified in all branches of aerial photography except the field operations which he has been unable to carry out owing to medical reasons." The assessments were less enthusiastic in later years. As senior air staff officer of No.3 Training Command, he was praised for his achievements and promoted group captain, but his lack of staff experience "owing to a long period of employment in specialist work" was noted. After posting to No.2 Wireless School in Calgary, his work was again praised, but A/V/M G.R. Howson, No.4 Training Command, proposed that, as he was not a pilot, he be posted to a non-flying school. Howson, contrary to other reviewing officers, was very critical of Owen's lack of tact, administrative ability and unsound judgement. Nevertheless, Owen continued to command No.2 WS until the British Commonwealth Air Training Program started to wind down in 1944.[52]

In one of the most colourful ceremonies of the 1942 Calgary Stampede, G/C Owen was inducted with all the pomp and ceremony of ancient tribal rites as an honorary chief of the Blackfoot Tribe in recognition of the expanding role of the RCAF.

Owen was retired on 18 October 1944 because of "the contraction of the

52 Ibid.

Air Training Plan, the limited possibilities for promotion and the difficulty of finding advantageous employment for him at his present rank." In arriving at this decision, the Air Council Retirement Committee noted that he was "a keen, conscientious officer, although handicapped by his lack of Station, Staff and Flying Experience caused by his long employment on specialist duties up to the outbreak of the present war." Owen requested promotion to air commodore on the basis of his 31 years of exemplary military service, but was refused.[53]

After retirement, Owen lived in Rockcliffe Village where a large property allowed him full exercise of his gardening skills. He served on the village council for about ten years. Not one to be inactive and with a need to supplement his pension, he was manager of the Ottawa Humane Society for several years and then manager of the Mayfair Apartments on McLaren Street, Ottawa. He died at Rockcliffe on 29 September 1970.

53 Ibid.

Charles Roy Slemon

ROY SLEMON was born on 7 November 1904 in Winnipeg, where his father worked for the streetcar system. He was among the RCAF "Originals," served for 42 years, and was one of Canada's most distinguished airmen.

Recognizing the imminent need for younger, "new blood" with technical training to take over from the World War I veterans who had launched the embryo air force, the CAB initiated a pilot training program for university students in 1923. Slemon had the prerequisite Canadian Officers Training Corps "A" Certificate, obtained at the University of Manitoba, and signed on as a provisional pilot officer in 1923. He undertook pilot training at Camp Borden and received his wings in December 1924. The general practice was for provisional pilot officers to continue their university studies during the winters when little flying was undertaken. This Slemon did and obtained a degree in civil engineering. By 1927, two of the nine cadets recruited in 1923 had died in aircraft accidents and six had not made the grade or had resigned, leaving Slemon as the only serving member of the original group.

Slemon went to the flying boat school at Vancouver before being posted to Winnipeg. He was confirmed pilot officer on 1 April 1925, flying officer one year later and flight lieutenant in 1929. With the exception of four months on a navigation course with the RAF at Calshot and three months on a meteorology course with the RAF, the next eight summers were spent as a photographic pilot in Northern and Western Canada. The winters were fully occupied instructing and preparing for the next season's program. Over 100,000 miles of northern flying, often over unmapped territory, earned him the sobriquet of "GNP"—Great Northern Pilot.

Slemon's good friend and colleague for many years, A/V/M Ralph McBurney, a member of the 1924 intake for pilot training, remembers him in those days as "a rather solemn, serious young man, but with a quiet sense of humour." He was not exactly the life and soul of the party, but obviously the leader of his group and was the first to go solo. Evidencing characteristics that ultimately, it could be argued, saved the world from nuclear disaster, Slemon

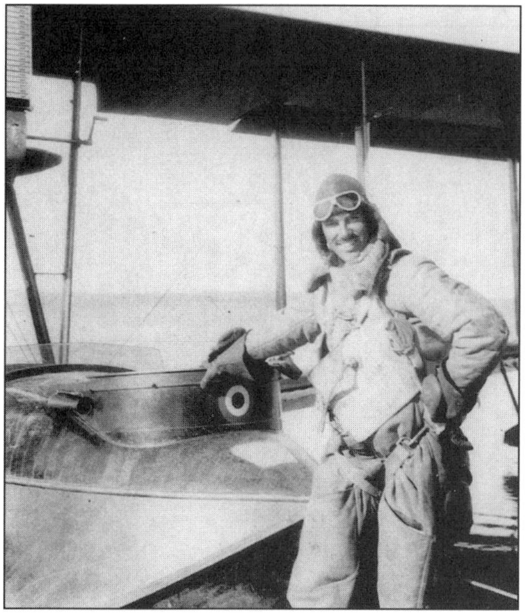

Slemon with a Viking. Ralph McBurney recalls: "He always looked like a movie star dolled up to rescue his best girl." *DAVID R. SLEMON*

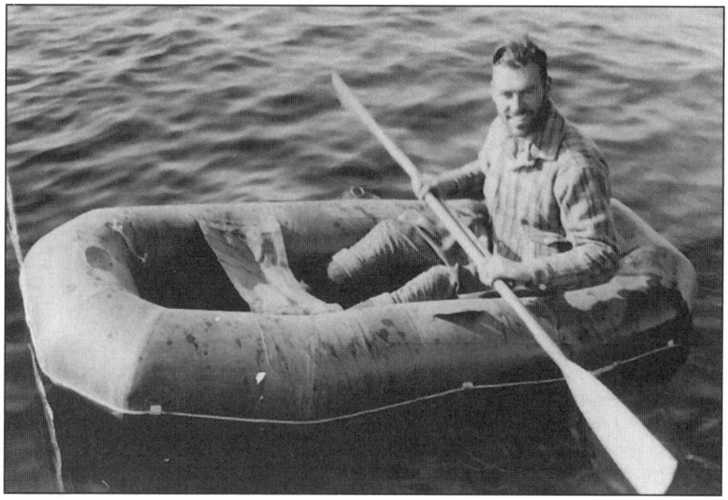

In a more relaxed moment at Lake Grand Rapids in 1928. *DAVID R. SLEMON*

Slemon set an impeccable example with his shined shoes and buttons and pressed uniform. Even in the Barren Lands, where civilian clothes more suitable for the environment were the normal dress, he was immaculate. This "spit and polish" image was maintained throughout his career, not always endearing him to subordinates whom he expected to do the same.

was a stickler for detail. He was always careful to make absolutely sure that he—and all other participants—thoroughly understood any proposal, situation or problem that arose. He would go over the task many times, sometimes to the frustration of other participants, but would then execute the decision firmly, quickly and with precision.[54]

Tempering this stern image was a tendency to eccentricity. McBurney remembers Slemon's fondness—"if not addiction"—to canned corn that he used to order by the case for his detachment, often to the disgust of his crew who did not share his passion. Although always on time for duty assignments, he was invariably late for social engagements, to the frustration of his hostess, to the extent that Gert McBurney once made him eat burned steak and onions. For reasons he never explained, he always identified himself to the McBurneys as "Captain Schlieyminck." He took great delight in surprising his friends by arriving at social events with a different, but always attractive, lady on his arm. "He had a way with women, enhanced by the glamour of being a pilot."

Early evidence of Slemon's diplomatic skills was demonstrated in 1928 while he was photographing near Lake Athabasca. The local RCMP constable brought him an invitation to visit the local Indian chief. After the social courtesies were exchanged, the very dignified elderly chief explained to Slemon that his were the first aircraft he had seen. He thought they had great potential and had a proposition to make. If Slemon would stay over the winter and use his "Big Birds" to scout for game, in exchange he would receive the

54 Ralph McBurney, personal communication.

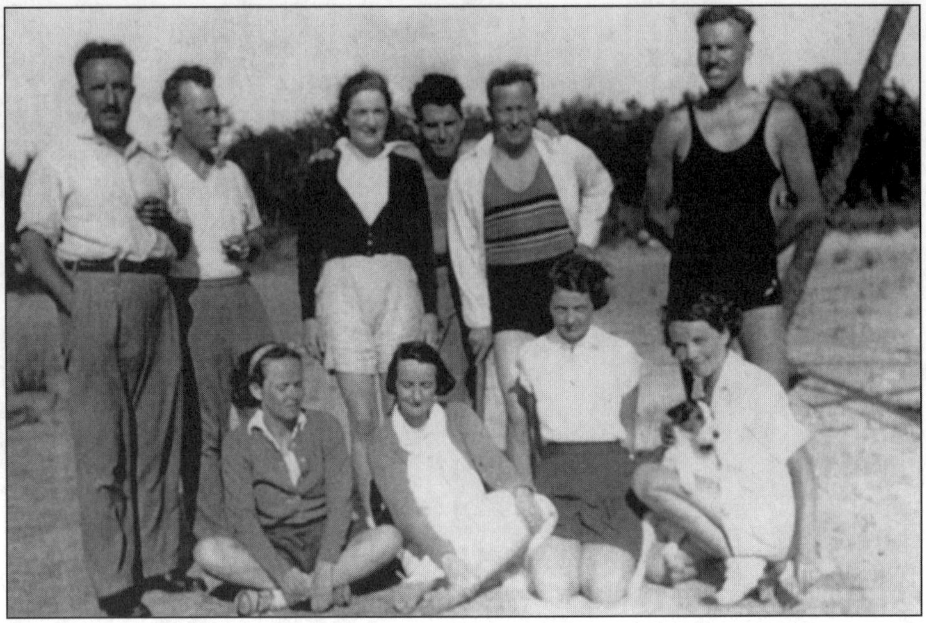

Slemon and the McBurneys shared a cottage on Cawaja Beach near Penetanguishene on Georgian Bay. They had an agreement that the married couple would act as chaperones—"a social nicety more important in those days"— when Slemon brought a companion. Pilots, wives and girlfriends were frequent visitors. In 1934, standing (L-R): Bill Butler (left the RCAF to be a furniture salesman), Art James (crashed in a Tiger Moth and was badly injured), Audrey James, Ralph McBurney (wrote the foreword and provided much of the information for this book), Brian Carr-Harris (killed in an air accident) and Roy Slemon. Sitting: "Butler's girl," Mary Carr-Harris, Muriel Meech and Gert McBurney.
R.E. McBURNEY

The McBurneys' chaperone service was not required for very long as Slemon's procession of beautiful ladies ceased with the arrival on the scene later in 1934 of the last, his second cousin Marion Slemon, daughter of Bowmanville physician Dr. Slemon. The Slemons were married in 1935. The wedding in Bowmanville was enhanced by an "air raid" with toilet paper bombs led by F/O Gordon Kerr. The Slemons had one son and two daughters.

hand of the chief's daughter in marriage. The Mountie was very concerned that the Indians would be offended by a refusal. Slemon carefully explained that the aircraft were not his, but belonged to the Great White King in England. Honour was satisfied until Slemon returned the next summer to be welcomed by the chief who thought the young pilot had come to his senses and decided to accept the deal. Slemon quickly told the chief how appreciative the king was of both the offer and for allowing his aircraft to operate in the chief's territory, but he had instructed Slemon to advise the chief that the aircraft were committed elsewhere. All ended satisfactorily, but Slemon did not return the next year.

Slemon was a pilot to be relied on. The Wolsley Viper engines in the Avro biplanes used for forestry patrol were notorious for throwing connecting rods, with sometimes disastrous consequences. In *Roundel*, Jan/Feb 1959, W/C T.F. Cooper recollected being taken into tiny Stormy Lake in central Manitoba when he was a sergeant fitter to investigate the results of one such incident in 1925. F/O Bill Weaver had an interesting few minutes with a dead-stick landing on the one-mile-long lake, only 100 yards wide and surrounded by trees. The engine was unrepairable and F/O Frank Wait flew Cooper in to check the situation, but stalled and crashed on his departure, making two aircraft stranded in the "pothole." Another mechanic, Cpl. Joe Maskell, was flown to nearby Beresford Lake and hiked through the bush to Stormy Lake. A winch was improvised and the crashed aircraft towed ashore; the engine, magnetos and carburettor were removed, cleaned of mud and slime, and installed in F/O Weaver's aircraft. Flying out the repaired aircraft was the next problem to solve. Everything moveable was jettisoned and fuel reduced to the minimum. With the tail tied to a tree, F/O Roy Slemon, who had walked in from Beresford Lake, ran the engine to full throttle before Cooper cut the rope with a sharp axe. Everyone held their breath and the aircraft just cleared the trees.

From 1935 to 1938, Slemon was a staff officer at RCAF Headquarters in Ottawa, then he attended the RAF Staff College. On return from the United Kingdom in January 1939, he spent two months in Ottawa before being posted as a squadron leader to Western Air Command in Vancouver and Victoria. In 1941, he was a wing commander and director of operations in Ottawa. He moved to Britain in 1942 to help organize the famed Canadian 6 Group of Bomber Command. For two-and-one-half years he was senior air staff officer of 6 Group, which grew to 11 stations with 13 squadrons of Halifax and Lancaster heavy bombers. His forethought and meticulous planning is credited with forestalling last-gasp efforts by the Luftwaffe to attack returning bombers over their home bases, saving many lives. He commanded 6 Group for a short period before he was appointed deputy commander-in-chief of RCAF Forces Overseas. He was named to command the Canadian Tiger Force destined for the Pacific, but this was disbanded after the Japanese surrender in 1945.

The honours awarded Slemon included Commander, Order of the British Empire in 1942; Companion of the Order of the Bath in 1946; three Mentioned in Dispatches; the US Legion of Merit in 1946; and the French Chevalier of the Legion of Honour and Croix de Guerre with Palm in 1947. Two lakes were named in honour of his photo pilot days, one near the "kink" in the Ontario–Manitoba boundary and a second north of Great Slave Lake in the Northwest Territories. His CB citation sums up the later stages of his career:

> As one of the senior RCAF staff officers, A/V/M Slemon performed outstanding service at Air Force Headquarters and Western Air Command. He was specially selected for overseas service when No.6 (RCAF) Bomber Group was in its formative stages. By his unceasing endeavour and his complete understanding, many of the obstacles which face newly created organizations were overcome with little or no difficulty. As Senior Air Staff Officer, he was to no small degree responsible for bringing No.6 Group into early effectiveness and he contributed very greatly to making No.6 Group one of the outstanding formations in Bomber Command. Although his main duties dealt with actual operations, he nevertheless found time to ensure that the welfare of all personnel in 6 Group was maintained at the highest possible level. During his tour of duty at Bomber Command his genial personality and admitted ability contributed to the splendid relations between Bomber Command and his old Group and earned for him the highest respect from his Royal Air Force confreres. As Deputy Air Officer Commanding-in-Chief, this officer was charged with the responsibility of organizing the RCAF participation in Tiger Force. He rendered

admirable guidance in the planning phases and due to his outstanding qualifications of leadership and his proven administrative ability, he was selected to command the RCAF effort in the Far East. A popular leader and an understanding officer, his career has been marked with zeal and energy, and an exceptionally keen sense of responsibility.

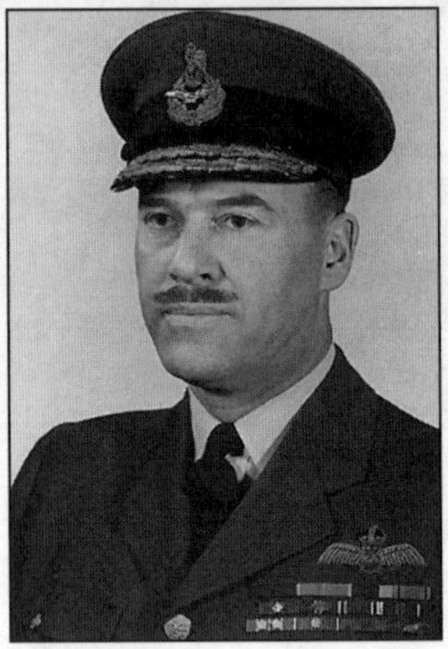

Air Marshal C. Roy Slemon, CB, CBE, CD.
DAVID R. SLEMON

After the war, A/V/M Slemon served as air member for supply and organization in Ottawa, then went to Trenton as air officer commanding training command in 1949. During 1952 he was temporarily assigned to Allied Air Forces Central Europe at the request of the commander-in-chief, General Lauris Norstad, to assist in planning a series of war training manoeuvres. He returned to Ottawa as air marshal and chief of the air staff in 1953.

In August 1957, Slemon was appointed to the key post of deputy commander-in-chief of NORAD with headquarters in Colorado Springs. He was in the "hot seat" on 5 October 1960 when the new ballistic missile early warning system (BMEWS) in Thule, Greenland, gave positive notice of an overwhelming number of nuclear missiles on their way to North America. With only minutes in hand, Slemon weighed the information from other sources and decided merely to put Strategic Air Command on alert. Anything more could have initiated a disastrous series of events leading to World War III. He was right; it was later realized that the new radar had not been programmed to discriminate between oncoming missiles and the moon rising over the USSR. The media made great play with the acronym BMEWS, but Slemon was certainly not bemused.

Following his retirement from the RCAF and NORAD in 1962, Slemon served as executive director of the US Air Force Academy Foundation, a fund-raising organization for flying scholarships. He also designed his own house on a mountainside at Colorado Springs.

Roy Slemon died at Colorado Springs on 12 February 1992. Attending his funeral was USAF Gen. Donald J. Kutyna, then in command of NORAD with the responsibility for warning of imminent attack. He remarked to Slemon's son David that the event "completed the circle" because, as a young lieutenant B-47 commander, he was one of those alerted by the false missile warning in 1960. He later wrote, "It was a wonderful personal opportunity to honor the man who some thirty years ago had, in his role in NORAD, precluded us from going to war. I felt a strong identification with him, and was pleased to be present as he launched on his next journey."[55]

In the June 1935 issue of *Canadian Aviation*, F/L Slemon wrote of the Photo Detachments: "At season's end the detachment heads for the main base, personnel happy at the prospects of returning home and satisfied with the knowledge of a job well done and one which will serve a useful purpose after they themselves are long forgotten."[56]

55 Don Kutyna, personal communication.
56 C. Roy Slemon, "RCAF Photographic Detachments," *Canadian Aviation* 13, no. 6 (June 1935): 20, 21, 27.

Harry Johnstone Winny

HARRY WINNY was born in Vancouver on 17 April 1904. His father was a marine engineer serving in Canadian Pacific ships to the Orient, but he left the sea in 1907 to help his brother Reg run his ranch, the Nicola Stock Farm, at Merritt, British Columbia. There, Harry and his sister Alice attended the local public and high schools. Winny's service record shows that he was also tutored for two years by the Rev. J. Thomson. From 1921 to 1924, he was employed at Nicola Stock Farm, while also serving as a corporal in the British Columbia Light Horse Militia Regiment.

In July 1924, Winny applied to join the RCAF, volunteering to do "anything for which I am considered fitting." His father added a prophetic note to the application: "My son is very keen to join the RCAF and is quite willing to start from the bottom and work his way up." Despite endorsement by Maj.-Gen. J.H. McBrien, chief of staff, the best that could be done for the inexperienced Winny was employment in September 1924 as night watchman at Jericho Beach.[58] However, during three enlistments he rose from AC2 to wing commander.

Anxious to expand his experience and improve his promotional opportunities, he remustered to AC1 "Engine Fitter Learner" in 1926, and volunteered to sign on for three years if he could be transferred to a station in Eastern Canada in order to broaden his experience. Headquarters had no use for him in the east, so Winny purchased his discharge in July. During his time out of the RCAF, he may have worked for a technical training school as he obtained his engineer's certificate in September 1926. He re-enlisted in the RCAF as an AC2 Labourer on 14 June 1927 and was promoted the same day to AC1 Engine Fitter.

All the non-commissioned RCAF pilots were required to maintain their basic trades. A/Sgt Winny just scraped through his engine fitter course in July 1929 with a mark of 65 percent, which caused him problems, as pilots were expected to achieve at least 80 percent. Nevertheless, he was confirmed as sergeant on 1 April 1930 and flew on historic expeditions to the Northwest

57 NAC Personnel File.
58 bid.

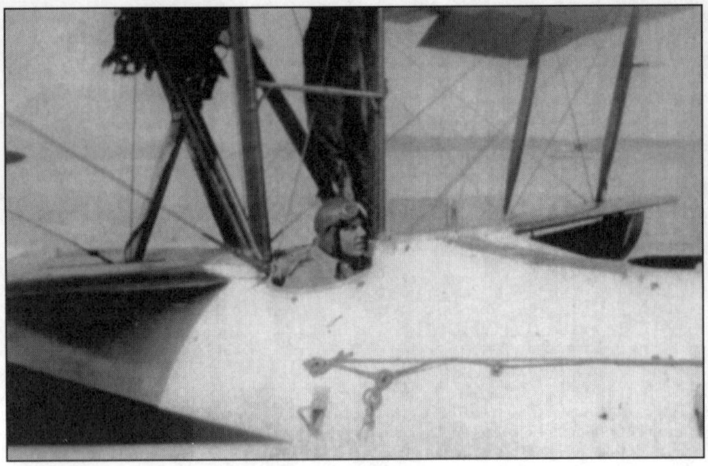

Above: In October 1927, LAC Winny was a grateful and enthusiastic member of the second airman pilot's course at Camp Borden. He was awarded his wings on 23 February 1928, made acting sergeant and sent back to Vancouver for seaplane training. *J. O. RENNIE*

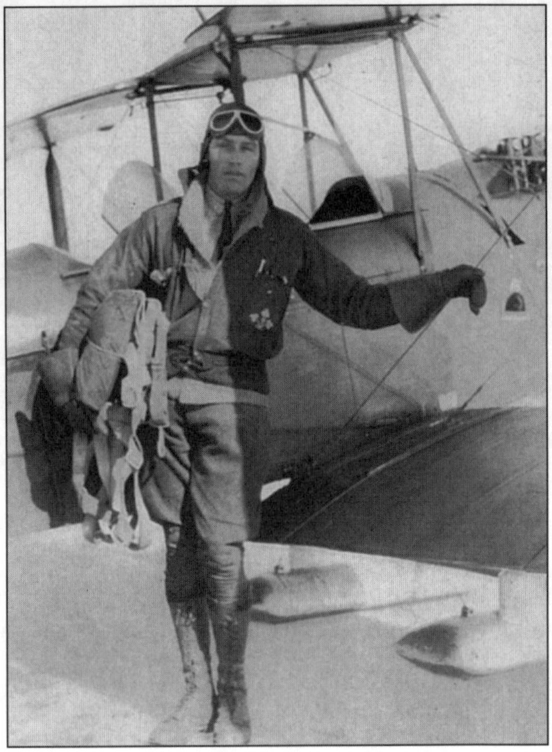

Right: Winny was posted to Winnipeg in April 1928. Here, he is about to set out on forestry patrol in Cirrus Moth 'YR from Ladder Lake, Saskatchewan, in the spring of 1930. For the next two years he was a busy pilot, flying from Norway House, Ladder Lake, Cormorant Lake and other northern bases on the forest patrols, photo surveys and other diverse tasks expected of pilots in those days. *R.E. McBURNEY*

Territories in 1930 and 1931. "Exemplary" ratings earned him promotion to acting flight sergeant on 1 July 1930. His high marks at the advanced photographic course in Ottawa during 1931 received an accolade from W/C A.E. Godfrey: "Flight Sergeant Winny proved to be one of the most capable pilots on the course. He showed a keen interest in the work on the ground and unusual ability in the air. This NCO was the outstanding member, earning the highest marks in the examination set by F/L Cairns and F/L Jellison, and by showing exceptional ability in the air."[59]

In 1935, Winny had the rare distinction of being awarded the Medal of the Military Division of the Most Excellent Order of the British Empire. This came about through an initiative of Prime Minister R.B. Bennett who enquired about outstanding achievements in Canadian aviation that might be recognized in the January 1935 Honours List. Maj.-Gen. A.G.L. McNaughton recommended one civilian—W.R. "Wop" May; one officer—S/L Roy S. Grandy; and to join this illustrious pair, one NCO—F/Sgt. H.J. Winny. The memorandum nominating him chronicles his performance as pilot of a Vedette dispatched in 1930, under the command of F/L F.J. Mawdesley flying a Fairchild 71, to explore and photograph the main waterways of the Northwest Territories and to investigate fuel caches laid down in the area. Also cited was his 1931 photographic reconnaissance with F/L R.E. McBurney of the unmapped area of the NWT situated between the

59 Ibid.

Mackenzie River system and Hudson Bay.

In a strange quirk of military protocol, Winny could not use the usual abbreviation for the award—MBE—because he was below the rank of warrant officer. He was entitled to use British Empire Medal, abbreviated to BEM. These letters, however, could not be used after his name, creating confusion throughout his career. He did wear the distinctive purple with narrow red stripe ribbon common to both the MBE and BEM, but further confusion resulted from its rarity after this ribbon was radically changed in 1937 because the civil version lacked the red stripe and could be mistaken for the ribbon of the Victoria Cross.

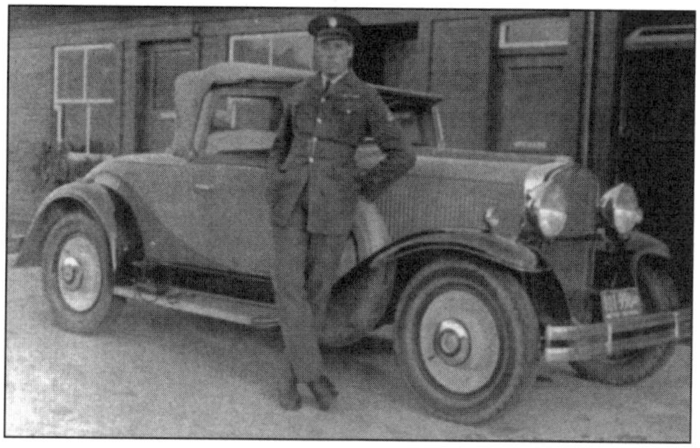

Flying was, by this time, Winny's life. He bit the bullet at Winnipeg in 1931, sat the engine fitter examination again, and passed with more than the 80 percent necessary to stay on the flying list. From 1932 to 1935, Winny operated from Jericho Beach, flying preventive, immigration and fishery patrols along the west coast, and making contacts that stood him in good stead a decade later. Photo 1933.
J.O. RENNIE

Apprehensive that he was about to lose his flying status, as happened to several other non-commissioned pilots in the austere days of the Depression, Winny requested permission on 3 July 1936 to purchase his discharge. His release was approved, with some reluctance as expressed by W/C A.B. Shearer: "It Is very much regretted that a commission could not be granted to such a thoroughly competent pilot as F/Sgt. Winny has proven to be."[59] Winny's payment for discharge was waived. He signed on as a pilot with Canadian Airways Limited in August 1936, engaged in general charter work and the Vancouver–Victoria–Seattle passenger run. Attracted by the opportunity to fly a new Stinson Reliant, and do some prospecting, Winny transferred to Territorial Explorations Co. Ltd in March 1937, the same month he married Jessie Amelia Rennie.

In June 1937, Winny's Reliant was destroyed by fire in the Northwest Territories. Left without an aircraft, he rejoined Canadian Airlines in Edmonton until April 1939 when he moved to the Hudson's Bay Company, attracted by its new twin-engine Beech-18, fitted for wheels, floats or skis. Pilots Winny and Paul Davoud with engineer Duncan McLaren were welcome visitors in CF-BMI throughout the fur trading districts of the Yukon, Saskatchewan, Manitoba and Ontario. In *Canada's Flying Heritage*, Frank Ellis describes one flight transporting no less a personage than Patrick Ashley Cooper, who was honoured with the historic title, Governor of the Company of Adventurers Trading Into Hudson's Bay, on an inspection tour of the company's activities and holdings. Boarding 'BMI at Edmonton on 18 June accompanied by P.A. Chester, Canadian general manager, and R.H. Chesshire, assistant manager of fur trade, Cooper covered 4,685 miles and visited 28 posts

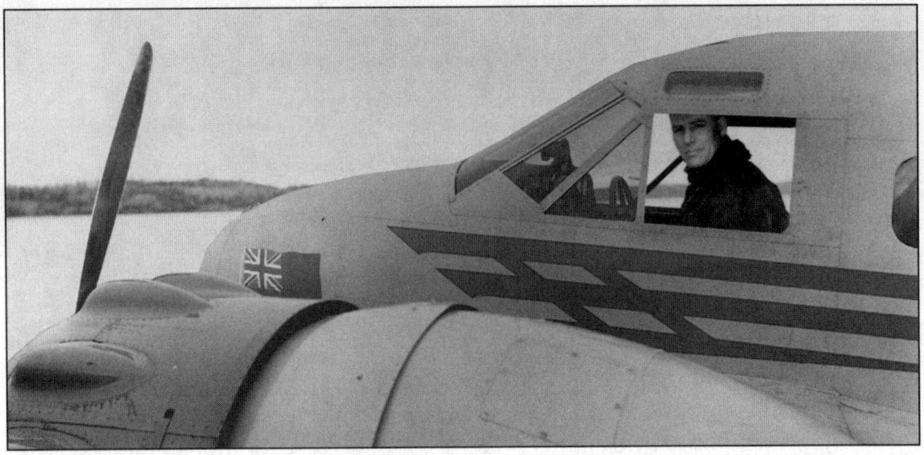

Harry Winny in the cockpit of Hudson's Bay Company Beech-18 CF-BMI, August 1939.
HBC ARCHIVES/D.D. MCLAREN COLLECTION 1984/39/40

in one week before disembarking at Winnipeg. In nine days the following month, 'BMI carried fur trade commissioner Ralph Parsons 3,200 miles from North Bay to visit posts along the shore of Hudson Bay.

Winny had flown many miles with engineer Duncan Dingwall McLaren before he left the HBC to join the RCAF in November 1939. By this time, McLaren had his pilot's licence and took over Winny's job. He later wrote his autobiography, *Bush to Boardroom*, much of which recounts the experiences of the two friends. Winny was commissioned as a pilot officer on 29 November 1939 and posted to No.120 (BR) Squadron at Vancouver. He moved to command No.6 BR Squadron of Supermarine Stranraers at Alliford Bay in the Queen Charlotte Islands as a squadron leader in March 1942.

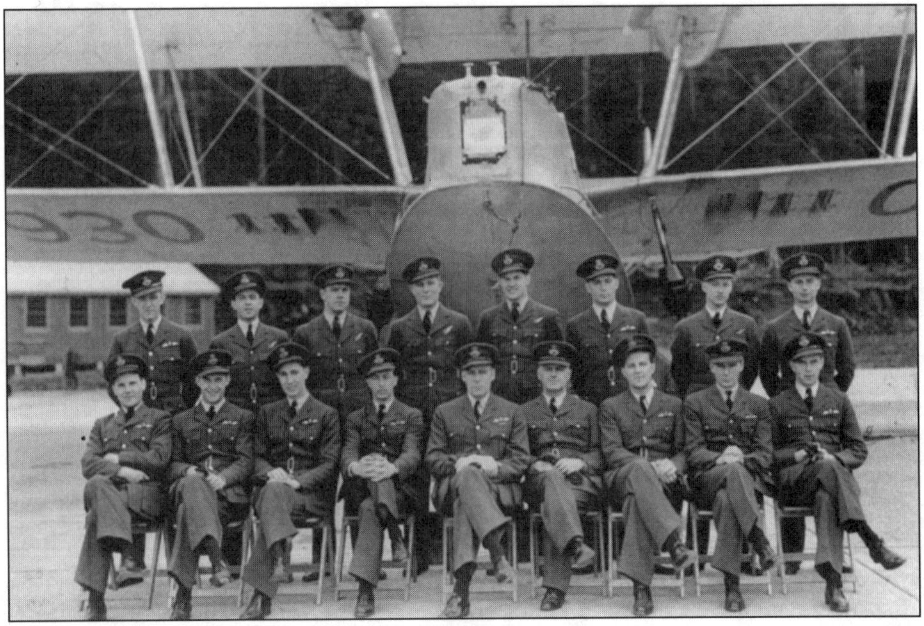

H.J. Winny (front centre) was promoted to wing commander and CO of Alliford Bay in August 1942. He went to command at Ucluelet on the west coast of Vancouver Island in May 1943.
HUGH M. MILLER

Winny was Mentioned in Dispatches in 1943:

> Wing Commander Winny commands an isolated station on the Pacific Coast. He is an outstanding pilot, a resourceful commander and an exceptional leader. His diligence, fine judgement and personal example have brought his station to a high state of efficiency. Wing Commander Winny has flown 1,066 operational hours since the outbreak of war.

A more personal tribute came from retired W/C Hugh Miller in a speech to the Air Force Officers' Association of Vancouver in 1994 on the subject of leadership skills. Miller was a pilot officer wireless air gunner at Alliford Bay when Winny took command: "This was not a happy camp: morale was poor; the Japanese ruled the Pacific Ocean; our Stranraers were ancient and we were a very thin line of defence . . . we were a sorry bunch." Then Winny took command and one of his first actions was to organize a station dance. He contacted all the friends he had made on the coast during the 1920s and dispatched everything that would fly or float to import that rare commodity—women—in quantity from Skidgate Village, Queen Charlotte City and fish canneries and logging camps along the coast. The dance lasted two days, did a lot to restore morale, and was recorded in a lengthy poem, which started:

Way out in the West,
Where the fishing is best,
Is a place called Alliford Bay.
And the boys out there,
Just tear their hair,
For a blonde in a mound of hay.

Now there came one day,
To Alliford Bay,
A man who is known as Harry.
He swore by God,
That every sod,
Would have his chance to marry.

The rest of the poem is best left to the imagination. Improved morale had an immediate effect on both operational effectiveness and the station's sense of community with other residents of the Charlottes. Miller recalls Winny's great rapport with all ranks, "a touch which made us all feel needed and important."

Winny's west coast experience was put to the test in 1942 when he advised on the construction of a radar station at a difficult site on tiny Marbel Island, Canada's first line of defence against feared Japanese attacks. Moving to Edmonton in 1944, he helped organize the Northwest Staging Route to Alaska. He returned to the HBC after release from the RCAF in April 1945.

The Winnys' only daughter, Diane, was born soon afterwards and the family moved to Winnipeg in 1946. Winny and Duncan McLaren had the task of selecting new aircraft and had an exciting time checking out war surplus machines in Canada, and evaluating the new Mark V Norseman at Noorduyn Aircraft in Montreal and the Widgeon at Grumman on Long Island. McLaren recalled that Winny had little interest in the "office work" or even in putting in regular hours, greatly preferring the bush pilot's independent lifestyle. He was seldom in the office after collecting a new Norseman, CF-BHT, from Montreal. One of his first jobs was a protracted flight from Winnipeg to evacuate David Drysdale, HBC post manager, and his expectant wife from Repulse Bay, off the northwest tip of Southampton Island. In bitter December weather, surmounting mechanical problems and repairing the aircraft after it caught fire while warming the engine at Chesterfield Inlet, Winny and his engineer successfully ferried the Drysdales to Winnipeg.

Veteran pilot Fred "Brick" Bradford flew co-pilot with Winny in the Norseman and recalled that his idea of supreme happiness was to sit behind "anything that was attached to a prop" and map read from about 3,000 feet. "He could do this all day; he never got hungry; he just carried a chocolate bar with him."

As chief pilot for the HBC, Winny flew most of the company aircraft, including Canso CF-FOQ. With co-pilot Fred Bradford, engineer Arnold Wallace and crewman Len Fowler, he left Winnipeg in July 1947 to rescue crew and passengers from the Arctic supply ship *Nascopie* wrecked off Cape Dorset in Hudson Strait. The round trip of 2,400 miles was completed in three

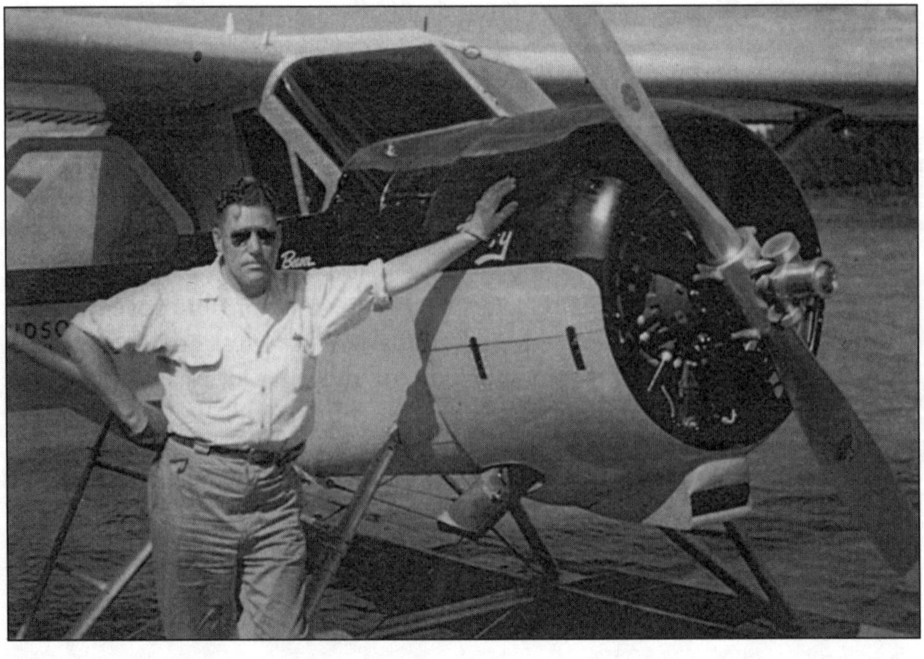

Leaving the HBC in June 1950, Winny moved back to Edmonton to fly de Havilland Canada Beaver CF-GQD for the Hudson Bay Oil and Gas Company. His brother-in-law Jack Rennie remembers: "He was in his element, flying his *Betsy* all over the northern part of Alberta, BC, and the Northwest Territories with geologists searching for oil and gas, then servicing the camps and well sites once they found something promising. He operated on wheels, skis and floats and had the time of his life."
DIANE MacDONALD

days. A remarkable achievement considering that it was over largely unmapped territory with few navigational aids, no reliable weather forecasts and Churchill the only refuelling point.

The only cloud on Winny's horizon was a back injury in 1953 that necessitated an operation, leaving him with a weakness in one leg. He moved to Calgary in 1955 when Hudson Bay Oil and Gas relocated the base for *Betsy*. Unfortunately, by 1959 he had to stop flying due to his back injury and subsequent blood pressure problems. Restricted to a ground job scheduling supplies to mine and drill sites, Winny succumbed to a fatal heart attack on 18 December 1960.

Harry Winny contributed many hours teaching navigation and other subjects to air cadets of No. 604 Moose Squadron, where he is remembered by a memorial trophy awarded to the outstanding cadet each year. The trophy is proudly presented at the annual inspections by his daughter Diane Macdonald, a nurse at the Alberta Children's Hospital in Calgary.

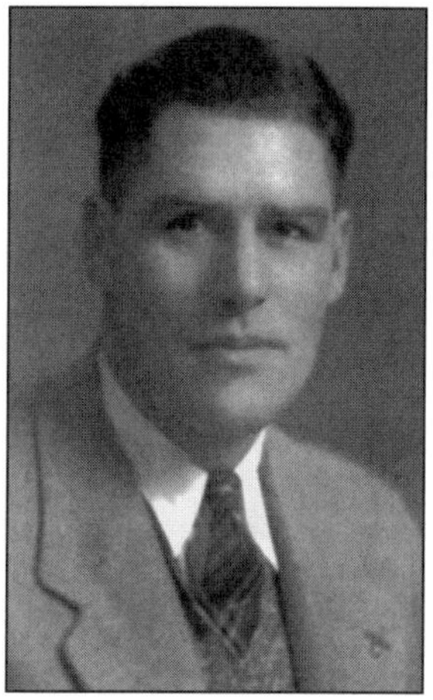

In addition to RCAF wings Harry Winny held a public transport pilot's licence and qualified as both a radio operator and air engineer. He died a highly respected man at 57 years of age, having achieved more than most who live much longer. Winny Lake, north of Great Slave Lake, Northwest Territories, was named in recognition of his outstanding service to aviation.
J.O. RENNIE

Part III

THE MACHINES

Photographic aircraft of the Canadian Air Board,
Canadian Air Force and Royal Canadian Air Force, 1919–1939.
Listed by date of introduction.

de Havilland D.H.4

Engine:	One 345 hp Rolls-Royce Eagle VIII
Dimensions:	Span 42'4½" Length 30'8" Height 11'0"
Weight:	Empty 2,387 lbs. Max. 3,472 lbs
Speed:	Cruise 90 mph Max. 143 mph

The D.H.4 was the first British aircraft intended specifically for the bomber role. Designed by Capt. (later, Sir) Geoffrey de Havilland and powered by a 250 hp Rolls-Royce Eagle, its performance was comparable to many fighters in 1916 when it was operating in squadron strength in France. The Imperial Gift of 115 aircraft delivered to Canada in 1919 included 12 D.H.4s, each powered by a 345 hp Rolls-Royce Eagle VIII engine.

By the end of 1923, the D.H.4 variants were beginning to show their age, at least three having been written off in accidents. The change in environment from high moisture content in Britain to the low humidity of prairie winters resulted in serious shrinkage of wooden parts. The adjacent layers of laminated spars shrank in opposite directions to the extent that glued joints could fail. Wing fabric soaked up moisture, increased weight and warped the structure. High River was kept open during the 1923–24 winter and a wide variety of cold weather testing was undertaken, including tests of a D.H.4B on skis, the heaviest aircraft to be so equipped at that time.

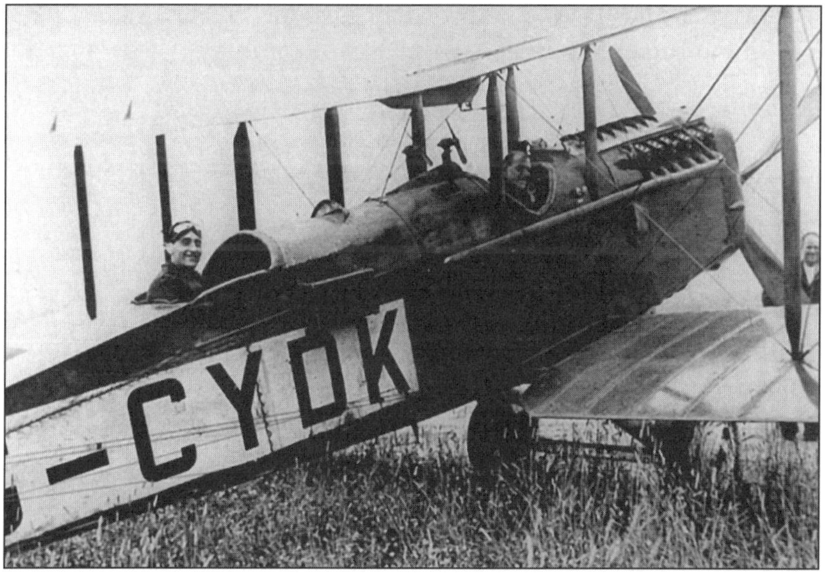

Excellent handling characteristics of the D.H.4 were somewhat limited by difficulty of communication between pilot and observer, separated by the centrally located fuel tank, as shown in this photograph taken at Rockcliffe in 1921. Two propellers driving fuel pumps can be seen behind the pilot. *J.A. GRIFFIN*

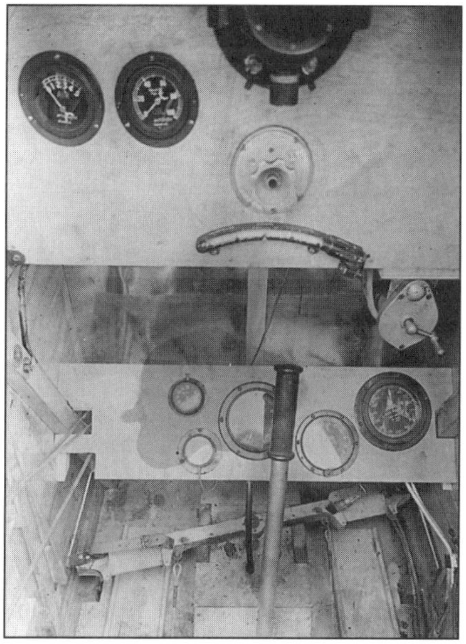

Pilot's cockpit of D.H.4 G-CYDM, High River, 4 August, 1922. *NAC PA52128*

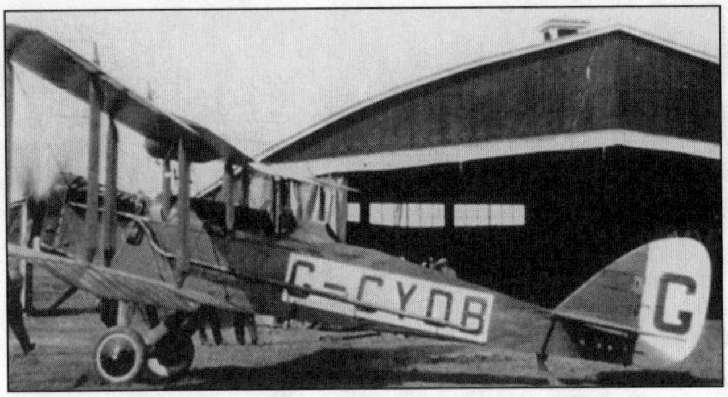

Several of the D.H.4s, designated D.H.4Bs, were modified by moving the pilot's seat rearwards and mounting the fuel tank between him and the engine, permitting communication between the crewmen. This D.H.4B was one of two at High River in 1923 used on photographic missions. The exhaust pipe was also extended past the observer's cockpit. The rear cockpits of some D.H.4Bs were enclosed to provide a degree of protection for the cameraman and his equipment. *J.A. GRIFFIN*

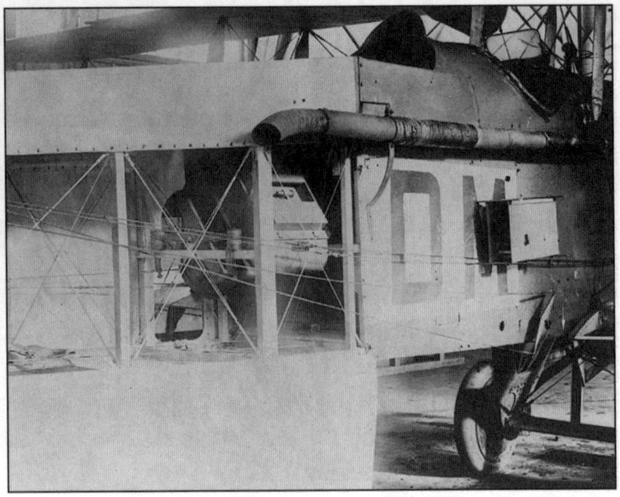

D.H.4B G-CYDM with a camera mounted for oblique photography, High River, 31 October 1922. *NAC PA52125*

Avro Viper 552As were introduced for the patrol task in 1925, gradually replacing the D.H.4s. Additional D.H.4s had been converted to D.H.4Bs and these venerable aircraft undertook the photographic duties. G-CYDM and G-CYEC were the mainstay of this work and by 1926 were the only serviceable D.H.4Bs. All versions of the D.H.4 were written off by 30 November 1929 after performing well in arduous flying conditions long past their expected life—a tribute to their maintenance crews. They had the lowest failure rate in the RCAF and probably gave the greatest value of any aircraft type in the Imperial Gift.

Curtiss HS-2L

Engine:	One 350 hp Liberty XII
Dimensions:	Span 74'1/2" Length 39'0" Height 14'7"
Weight:	Empty 4,300 lbs. Gross 6,432 lbs
Speed:	Cruise 76 mph Max. 82 mph

The Curtiss H-12 and H-16 flying boats were developed in the USA during World War I and employed with some success against German U-boats. Early in 1917, Curtiss designed and built the HS-1, a smaller and more economical flying boat, its designation derived from its H-Series ancestry: the "H" for hydroplane, "S" denoting single-engine, and "1" the mark. The HS-1 with a British-designed "Porte" hull and powered by a 200 hp Curtiss engine first flew at Buffalo, New York, in June 1917. Installation of the 350 hp Liberty V-12 engine developed by the Hall-Scott and Packard companies, and some other refinements, including inserting six-foot-long panels into each wing, resulted in the HS-2L (L for Liberty engine). This version, able to carry two 230-pound bombs, was in US naval service by July 1917.

A squadron of US Navy HS-2Ls was stationed at North Sydney and Dartmouth, Nova Scotia, to provide air cover for the convoys departing from and arriving at Halifax. At the termination of hostilities in November 1918, Canada purchased the US installations, and the US Navy donated the 12 HS-2Ls based there. The value of the gift was not immediately appreciated by the federal government, although two HS-2Ls were released in 1919 to the St. Maurice Forestry Protective Association, Grand Mère, Quebec (predecessor of Laurentide Aviation) for forest survey.

In 1920, the remaining ten HS-2Ls were taken out of storage and reassembled: six at Dartmouth, two at Jericho Beach, Vancouver, one at Ottawa, and one at Victoria Beach on Lake Winnipeg. Two assembled at Dartmouth were flown to Roberval on Lac Saint-Jean in central Quebec, primarily for forest patrol on behalf of the Quebec government. In deplorable weather conditions, one of the Dartmouth aircraft flew the first leg, Halifax to Ottawa, of the first Trans-Canada flight in October 1920.

Newly appointed as director of technical services in October 1920, Lt.-Col. Ernest W. Stedman ordered stress tests of the wings that indicated the six-foot extensions added by the US Navy to increase payload and range of the flying boats were inadequate for the anticipated demands of Canadian operations. New sections were designed, manufactured in a rented workshop

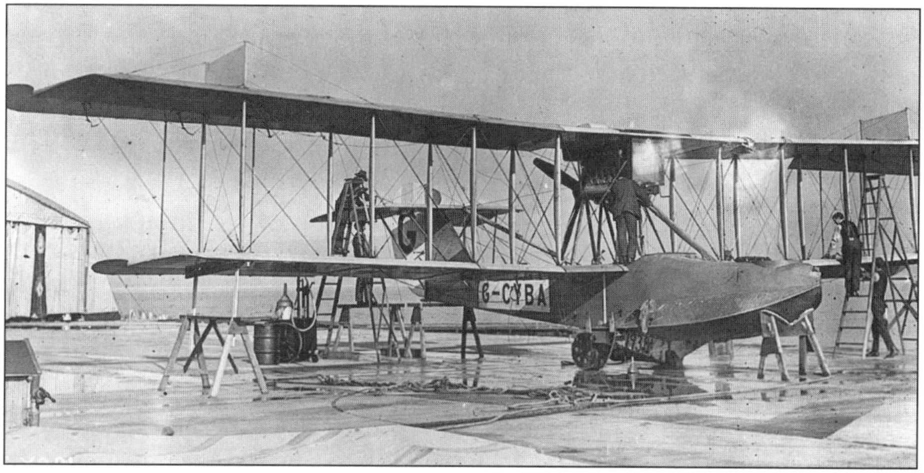

One of the first pair of Curtiss HS-2Ls to be shipped to Vancouver being readied for flight at Jericho Beach, 26 October 1920. The US Navy stripes are still visible on the tail. *NAC C46959*

A "Jericho Beach" nose fairing was designed at Vancouver, c.1922, for the HS-2L to protect the exposed photographer on vertical photography operations. Covered in rubberized fabric, it did limit the intake of water in rough seas but was inconvenient, hindered mooring and prohibited oblique photography. The sole installation, on G-CYBB, was soon removed.

The vertical camera installation, typical of all early flying boats, can be seen. A hole was cut into the floor of the front cockpit through which a camera could be aimed when airborne. When not on camera missions, the port was sealed by a plug inserted from outside the hull. While on the water on photo missions, the port was sealed by a "stovepipe" cylinder projecting above the water line and easily removed for photography. *J. A. GRIFFIN*

The remains of HS-2L G-CAAC are on display at the National Aviation Museum alongside a superb reconstruction in Laurentide Aviation colours. Pilot Don Foss and engineer Jack Caldwell made an emergency landing with 'AC in a heavy downpour on a small, unnamed lake en route to Remi Lake, Ontario on 2 September 1922. Despite jettisoning unnecessary items, the left wing tip float struck the water on takeoff and the aircraft crashed. The two crewmen were not seriously hurt and walked out. A salvage crew recovered the engine, but the fuselage was left to rot and lay forgotten for nearly half a century. The remains of G-CAAC were

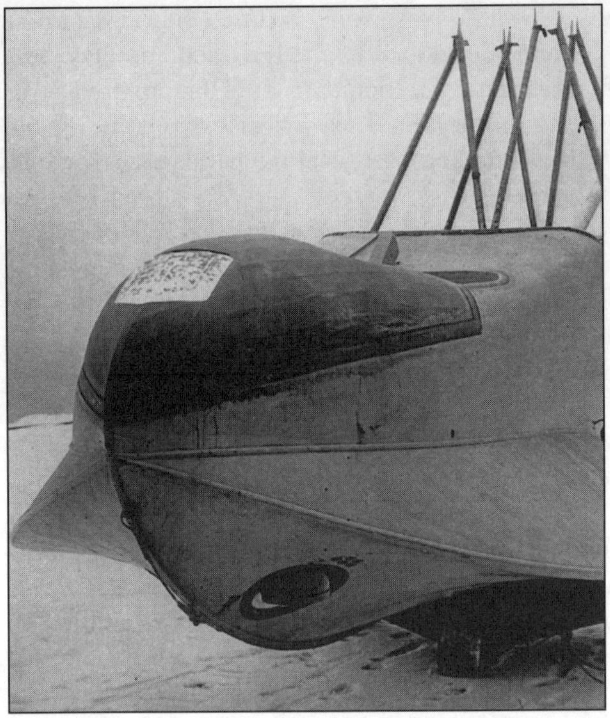

on Sparks Street, Ottawa, and installed, probably much to the relief of the aircrews. Other changes included enlargement of the front cockpit to make room for the photographer and his equipment, and modification of the front cockpit "machine gun ring" to mount an oblique camera.

Designed for salt-water conditions, the HS-2L performed well on the west coast and five were stationed at Jericho Beach in 1921. They were limited by an optimistic design ceiling of 6,500' for work over the Rocky Mountains, where most mountain passes were over 7,000'.

During 1921, 12 more ex-US Navy HS-2Ls were purchased at $1,000 each.

By 1924, Canadian Vickers Vikings were gradually replacing the HS-2Ls. The venerable flying boats had been flown in primitive conditions, frequently from isolated lakes infested with rocks, shallows, deadheads and floating debris. Their very survival was a tribute to their air and ground crews. The "boats" went indoors for winter overhaul and storage, emerging dry and performing up to specification, but as the year advanced, the wooden structures absorbed water to the extent that takeoff and climb performances were barely within safe limits. In a last attempt to extend the life of the HS-2Ls, G-CYEL was fitted with a high compression Liberty engine in 1924, giving greater altitude capability.

Ten "new" HS-2Ls were purchased and delivered in crates from the US Navy base at Norfolk, Virginia, in 1925. They turned out to have been built in 1917 and placed in storage after the war, and were just as weary as those the RCAF had recently retired. One, G-CYGU, in fact, was shipped to Camp Borden and was never taken out of its crates until sold to Canadian Airways in 1928.

Some clever juggling of parts kept one of the five HS-2Ls at Vancouver in the air and one in reserve into 1927, but mishaps and old age took their toll.

In 1928, G-CYGS was sold to Western Canada Airlines and a tidy up operation saw all the remaining HS-2Ls taken off the books; two—C-GYGQ and 'GR—at Vancouver were burned "to eliminate an eyesore at this station."

The HS-2Ls performed an invaluable service for Canada, pioneering civil applications of aviation with great success, aiding greatly in the evolution of the industry. Limited funds for new equipment stretched their use beyond reasonable limits and, with the benefit of hindsight, the purchase of the last batch of ten may have been a mistake. The experience gained, however, had a direct bearing on the design and development of their successors and could be seen as the beginning of a Canadian aerospace manufacturing tradition.

located in 1968 by Kapuskasing businessman Don Campbell. What turned out to be Canada's first bush plane, *La Vigilance*, flown by Stuart Graham in 1919, was recovered during the summers of 1968 and 1969 to be carefully transported to Rockcliffe. The lake in which it was found is now known as Foss Lake.

Canadian Vickers Viking

Engine:	One 360 hp Rolls-Royce Eagle IX.
Dimensions:	Span 50'0" Length 35'0" Height 14'2"
Weight:	Empty 3,800 lbs Gross 5,800 lbs
Speed:	Cruise 90 mph Max. 100 mph

Vickers Limited, Weybridge, England, introduced its first flying boat—the Viking—in 1919. R.K. Pierson's original design was refined over the next two years, culminating in the Mk IV in 1921, the first version to enter quantity production. The Mk IV, while not a true amphibian, incorporated a set of wheels that could be manually lowered after a water landing, allowing the aircraft to be taxied or towed ashore. A significant market was envisaged for the small, versatile and efficient flying boat at a time when few airfields existed.

Canada with its long coastline and huge landmass dotted with innumerable lakes and rivers was ideal territory for the Viking. In June 1922, the newly formed Laurentide Air Service Limited at Lac-à-la-Tortue, north of Three Rivers, Quebec, imported the first Viking, G-CAEB with a 450 hp Napier Lion engine.

Conscious that the days of its aging HS-2L flying boats were numbered, the CAB prepared a specification for a replacement able to carry a pilot and four crewmen at a minimum speed of 80 mph at a working ceiling of 12,000' for at least six hours. Cultivation of a domestic industry was an important consideration and manufacture had to be in Canada. Vickers's shipbuilding subsidiary in Montreal, Canadian Vickers Limited, won the competition with the Viking (the only other bid came from Laurentide Air Service). Eight Viking IVs were ordered for a total of $150,650, with the CAB to provide the

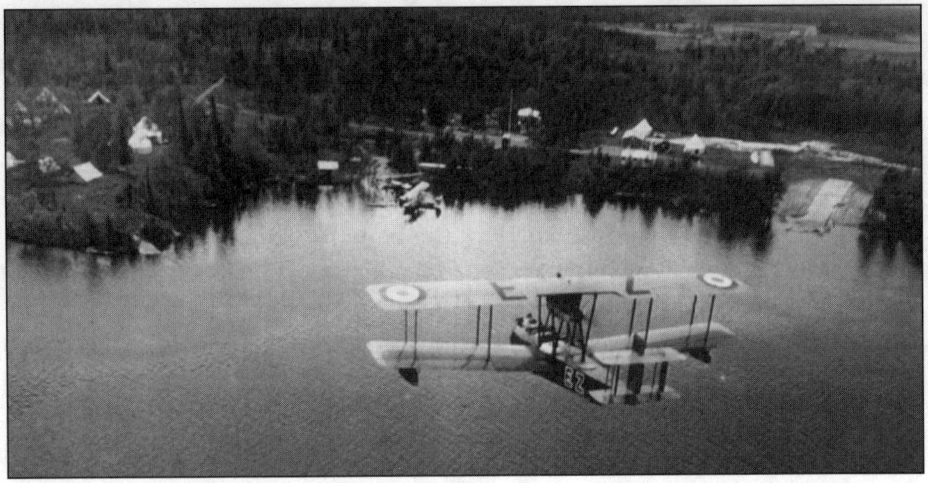

Canadian Vickers Viking, identified by its distinctive double tail, over Lac du Bonnet Air Station, Manitoba, 15 June 1927.
NAPL FA535-94

engines. A compromise was reached in order to accelerate delivery and the first two were delivered from UK production and assembled by the Canadian company.

The urgency for delivery was prompted by the need to have the Vikings available for the 1923 photographic season. Only half the original fleet of waterborne aircraft was operating and all were suffering the penalties of hasty wartime production, compounded by long service in a demanding environment. Time constraints prohibited the design, testing and manufacture of a completely new aircraft that would probably have come much closer to satisfying Canadian needs. The Viking, however, was an effective compromise.

The CAB had some unique requirements for its Vikings. Aware that the aircraft would spend much of their time in the water on detached photo operations, often with no provision for beaching, it specified a pump operable from inside the aircraft. In anticipation of rough usage, extra bulkheads were added, along with dual controls, rubber-covered fuel lines, and a radiator capable of supplying the large amount of cooling necessary in the hot Canadian summers. A wire screen mounted on the forward engine struts was specified to prevent loose items being sucked into the propellers. Skis were designed for winter operations and one installation was completed at Canadian Vickers, but not flown, as the additional weight made it impossible to carry a worthwhile payload.

Delivery of the six Canadian-built Vikings, G-CYEU to 'EZ, was completed by November 1923. Late, but still a remarkable achievement considering that building a complex, lightweight aircraft was new venture for Canadian Vickers, more used to shipbuilding and heavy construction. Problems with the engines also caused delays. G-CYEU, the first Canadian-built Viking, was christened in July by Mrs. G.J. Desbarets, wife of the deputy minister of national defence. Flight testing of 'EU was carried out by

J.L. Gordon and E.W. Stedman, both of whom rose to the rank of air vice-marshal.

The Viking was well mannered in the air and established a good reputation during its first full year of operation in the RCAF, despite being underpowered and having an occasional habit of "porpoising" on landing. The delicate trade-off between minimum weight and maximum strength of flying boat hulls, however, may have been cut too fine and several aircraft were damaged on the water.

Production Vikings built in the UK had a 450 hp Napier Lion engine as the standard installation. The Canadian Vikings initially had a 345 hp Rolls-Royce Eagle VIII engine, inadequate for the task, but a "bargain," as they were part of the Imperial Gift. They were also unreliable and an ongoing maintenance headache. During the winter of 1923–24, a more powerful 360 hp Eagle IX was installed in the Canadian Vikings and they performed excellently from there on.

By 1925, teething problems were over and the Vikings settled down as versatile workhorses. Pilots knew they had to be careful at high speed on the water when most mishaps occurred. By this time, the more economical Avro 552As were employed on fire patrol duties, but the Vikings flew in personnel and equipment to fight the fires, developing techniques still in use today. Photography was a major role and the Vikings were also used for general transportation duties. A series of mishaps experienced with the new type in very challenging operations occurred in the subsequent years, reducing the fleet to four by 1927, all of which were strengthened with an extra mahogany plank on the outside of the hull.

The four remaining Vikings, G-CYEU, 'EV, 'EX and 'EZ, in 1929 were reduced to supporting photographic detachments out of Lac du Bonnet, Manitoba. The sole survivor in 1931, G-CYEU, was eight years old with about 500 hours' flying time—a remarkable achievement for a wooden structure in very demanding operations—and the venerable aircraft was retired. Although not realized at the time, the last Viking, G-CYEU, was of some historical significance, being the first aircraft to be built in Canada to air force specifications since World War I.

The Canadian Vickers Viking was among the more successful of the many future fleet selections made by the RCAF. The RCAF was never entirely happy about its hull structure and the high degree of rot resulting from water absorption. Reinforcement, in the form of extra bulkheads, was added early in its life, and several hulls, wings and tail assemblies were replaced to the extent that little of the original airframes could have remained by the time they reached the end of their lives. Nevertheless, the Viking generated a high degree of respect and affection from experienced pilots able to manage its low power and difficult handling characteristics, especially on landing.

Canadian Vickers Vedette

Engine:	One 200 hp Wright J-4 Whirlwind or 215 hp
Armstrong Siddeley	Lynx IV
Dimensions:	Span 42'2 1/2" Length 32'11" Height 11'0 1/2"
Weight:	Empty 2,722 lbs Gross 4,000 lbs
Speed:	Cruise 76 mph Max. 92 mph

By 1924, Canadian experience had established a clear picture of the type of aircraft needed for forestry patrol and aerial photography. The Society of Forest Engineers in co-operation with the RCAF prepared an ideal aircraft specification for forest patrol and photography that defined a three-place flying boat with an air-cooled engine, cruising speed 70 mph and maximum speed 80 mph at 10,000', quick takeoff and good rate of climb with a crew and 1,000-pound payload. RCAF Specification C-6-24 went out for tender.

Canadian Vickers had the inside track by virtue of building the Viking and having the associated experience in dealing with the Canadian government procurement system. The Vedette was successfully proposed. This single-engine, pusher biplane was of conventional construction—laminated elm frames with the hull covered in cedar planking, double thickness on the bottom, wings fabric-covered wooden frames, and tail steel over wood. The large British Vickers Company evidently envisaged a profitable future in aircraft manufacture for its Canadian subsidiary, and established a design capability under the direction of Wilfred T. Reid, formerly chief designer for the British and Colonial Aircraft Company. The prototype Vedette, powered by a single Rolls-Royce Falcon III engine provided by the RCAF, took off from the Longueuil plant on 4 November 1924.The prototype was next flown on 22 November after a quick engine change for a lighter Wolseley Viper and the installation of balanced ailerons.

The engine was changed again, to a 200 hp Wright J-4 Whirlwind during the winter of 1924–25. The first flight by an RCAF pilot was by S/L B.D. Hobbs on 9 May 1925—evidently satisfactorily, as he took it on to the Ottawa seaplane base at Shirleys Bay. The prototype Vedette was accepted into the air force on 17 July, designated a Mk 1 and registered as G-CYFS. A season of experimental flights and testing of operational capabilities resulted in an enthusiastic comment in the RCAF annual report: "A complete success and a credit to her designer and builder, she is the equal, and for Canadian conditions the superior, of any flying boat in the world today."[1]

Refinements in the shape of enlarged cockpits, a vertical camera installation, improved flying controls and tailplane adjusting gear were incorporated into four Vedette IIs, G-CYGW to 'GZ, delivered

1 *Report on Civil Aviation for the Year 1925* (Ottawa: Dominion of Canada, 1925): 14.

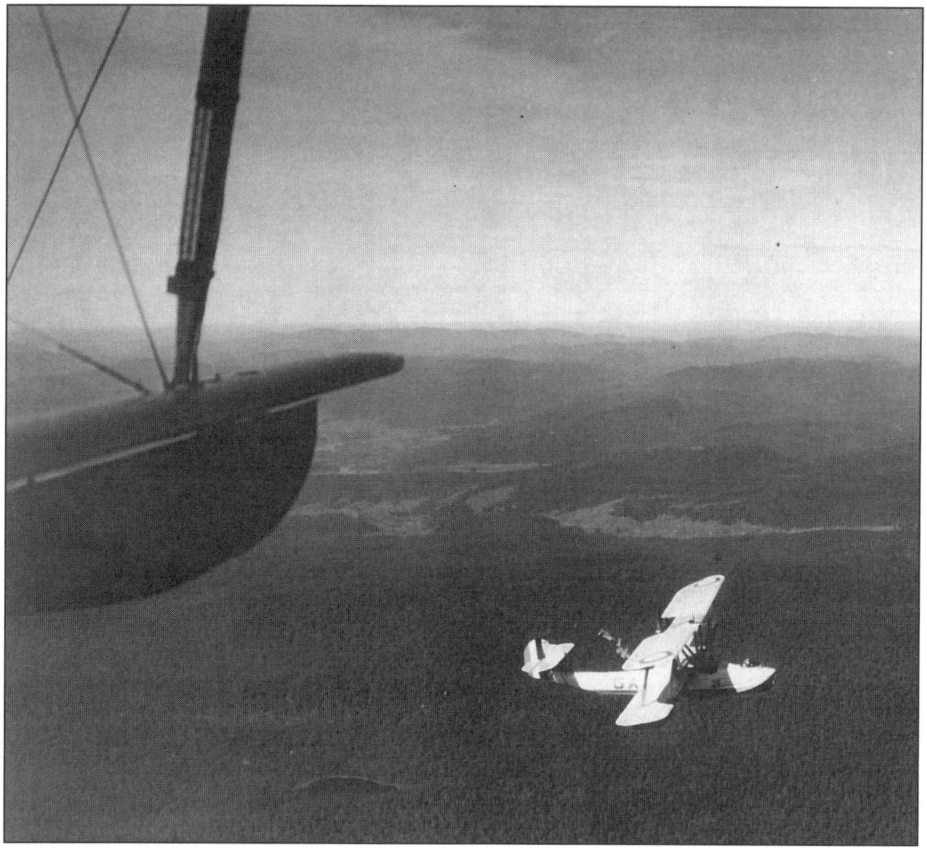

Vedette G-CYGA, pilot F/L J. Sadler (on exchange from the RAF) and cameraman Cpl. H. Diller, over St. Donat, Quebec. Photograph taken from Vedette 'ZL, pilot F/O A.L. Johnson, by Cpl. A.V. Green, 26 July 1928. *NAPL A-342-6*

August–September 1926. Two of the new Vedette IIs were hastily fitted with cameras and dispatched to photographic assignments in northern Manitoba and Ontario with commendable results, which established faith in the new type. Five additional Vedette IIs, G-CYZK to 'ZO, were ordered for 1927 delivery and an additional Vedette was acquired from the estate of J.D. McKee, who had been killed in the crash of his other Vedette.

Seven Vedette IIs were delivered in 1928, but experience and the crashes of two civilian Vedettes and a Varuna had raised doubts about the adequacy of the Canadian Vickers's hull design to withstand the rough treatment it was subjected to on detached operations. Accordingly, these aircraft, G-CYXZ to 'YF, had additional stringers and cross members plus a duraluminum cross-member between the wing spars, and the existing fleet was similarly retrofitted. The last three Vedette IIs were taken on strength by the RCAF in January 1929.

Canadian Vickers introduced a new model of the Vedette in 1929—the Mk V— with a stronger structure and improved control system, available in an amphibious version with a hand-cranked, wheeled undercarriage. The RCAF ordered ten flying boats for 1929 delivery.

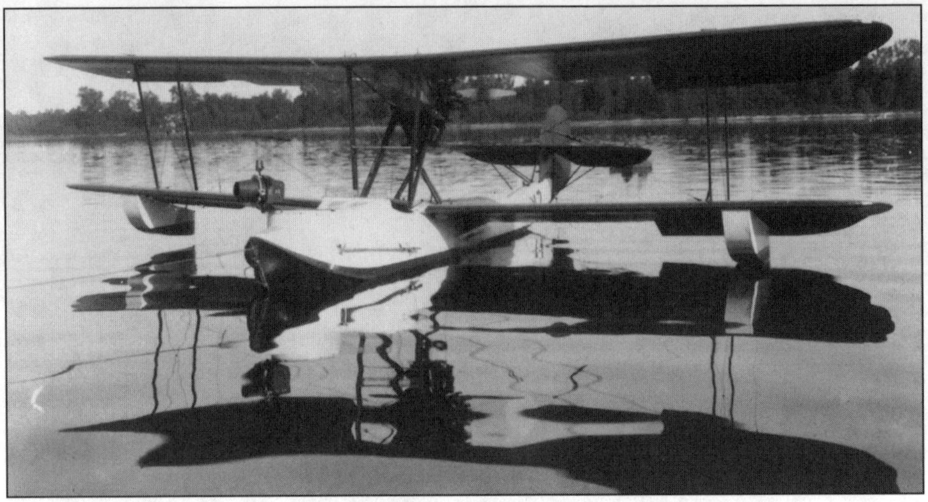

Vedette Mk V G-CYYZ equipped with camera for oblique photography, Rockcliffe, 24 August 1929.
NAC PA62564

Canadian Vickers incorporated Handley Page slots in the Vedette design, designating this version Mk VA. Three amphibian versions and eight flying boats were delivered to various customers during 1930. In the same year, the RCAF purchased one Vedette VA, G-CYVP, and one Vedette VI, G-CYWI, with a new-design metal hull 3'4" longer than the wooden version and an enclosed rear cockpit, both powered by the Wright J-6 engine.

G-CYVP had a short life, brought to an end in an effort to extend the use of the versatile Vedette past freeze-up. Skis were designed with the idea that the aircraft could, at least, continue flying from remote sites until freeze-up and then return to base on skis, rather than having to cease operations early to ensure return to open water. Failure of the ski tip restraining system, however, resulted in damage to the hull and lower wings, serious enough to write off the aircraft. The forward hull was used as a photographic crew training aid with the fortunate result that it is, today, the only surviving original Vedette airframe and is stored at the National Aviation Museum.

A Vedette is being meticulously built, using the remains of several different aircraft as a guide, by the Western Canada Aviation Museum in Winnipeg.

1930 saw a total of 36 Vedettes in RCAF service, reduced to 32 by year-end. The deepening Depression curtailed flying operations in 1931. No new Vedettes were delivered to the RCAF, but development work did continue at Canadian Vickers. Wing slots were fitted to G-CYYZ, and G-CYZD was fitted with a new metal hull, designated Vedette VAM.

The RCAF transferred five Vedettes to the province of Saskatchewan in 1933 and by 1935 the remaining 12 were primarily used in a training role, or held in reserve. Seven were ordered to Trenton, Ontario, in 1936 to equip a new training school. A tornado on October 29 greatly damaged the facility and sank Vedette 806. One venerable Vedette continued to serve at Vancouver and another at Ottawa until 1939, but stocktaking in 1940 reduced the inventory to four—two metal-hulled, 816 and 817, and two wooden-hulled, 809 and 812. One of each went to the Jericho Beach school to capitalize on the longer training season, but all four were written off in 1941.

While overall satisfaction was expressed with the design and operation of the Vedette, there was a continuing chronicle of complaints about the quality of workmanship. In common with other wooden-hulled flying boats, absorption of water remained a persistent problem, which detracted from performance to the extent they had to be taken out of the water and dried out several times each season.

The Vedette was undoubtedly the best aircraft produced by Canadian Vickers in the between-wars years. It enabled the RCAF to operate to the extreme limits of the Dominion, measure the extent of natural resources previously only guessed at, and to identify much of Canada hitherto virtually unmapped. Aircrews gained experience, particularly in northern operations, invaluable in the coming conflict. The Vedette also triggered the rebirth of a Canadian aircraft design and manufacturing capability now measured at over ten billion dollars annually.

The weather vane surmounting the Confederation Building to the west of the Parliament Buildings in Ottawa is a copper model of the Vedette, weighing 140 pounds. It was presented by the RCAF and installed in 1931. Looking small at its altitude of 265' above street level, wing span of the model is 5'6".

Canadian Vickers Varuna

Engine:	Two 180 hp Armstrong Siddeley Lynx
Dimensions:	Span 55'2 1/2" Length 38'3" Height 13'9 1/2"
Weight:	Empty 3,860 lbs Gross 5,850 lbs
Speed:	Cruise 81 mph Max. 94 mph

The Varuna was a qualified success, but only eight were built. RCAF Specification C-3-25 defined a small, twin-engine flying boat for photographic and transportation operations in more remote areas where twin-engine reliability could obviate the need for single-engine aircraft to travel in pairs.

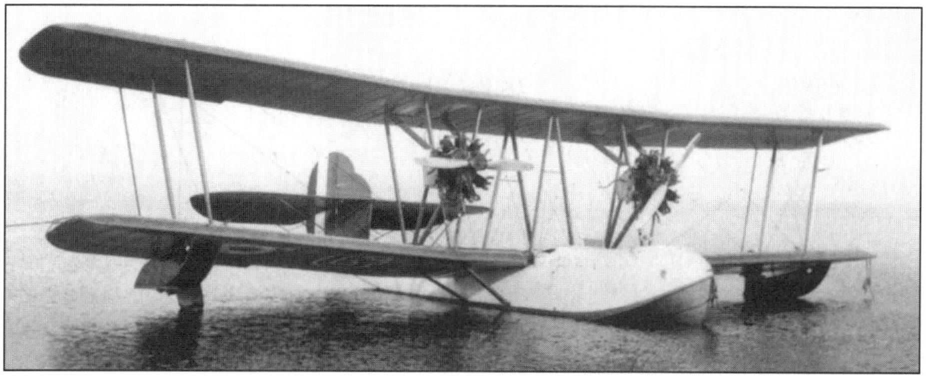

The prototype Varuna G-CYGV at Shirleys Bay, 18 November 1925. Equipped with Wright 220 hp J-5 engines it was accepted by the RCAF as a Mk 1 variant in June 1926.
NAC PA62316

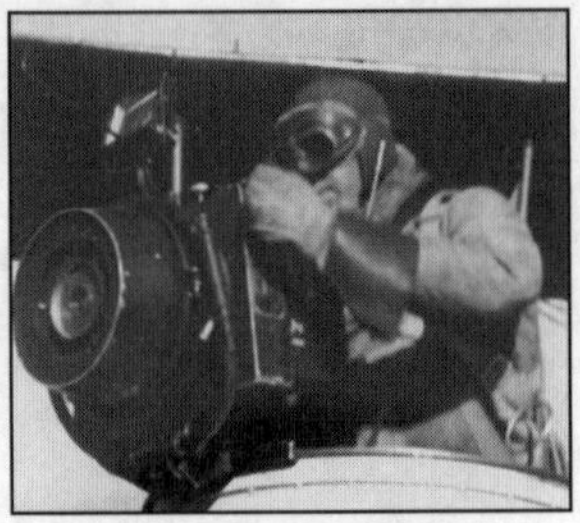

Photographer in the front cockpit of a Varuna, c.1928.
PAC PA62960

Seven Mk II Varunas powered by 180 hp Armstrong Siddeley Lynx engines were ordered for 1927 delivery. One went to Dartmouth for photographic operations, five went to Winnipeg for photographic and general duties, and one to Ottawa for evaluation. Three suffered hull damage early in their lives and were written off, resulting in the hulls of the remaining Mk II Varunas being reinforced with an additional frame nine inches forward of the existing bulkhead between the pilot's and photographer's cockpits. Spray shields for the rear cockpit were fitted, and the auxiliary fuel tank systems were removed to reduce weight in an attempt to improve the rate of climb. The bilge pumping system was ineffective, so that was also removed to reduce weight that remained some 700 lbs over the design objective.

Cost and additional weight considerations cancelled thoughts of installing more powerful engines in the four remaining Mk II Varunas. They continued flying, primarily in a transport role, until 1930 when two were holed by rocks, and the other pair showed evidence of wood rot. All four were written off, but the prototype Mk I remained in storage at Ottawa until November 1932 when the books were closed on an original Canadian design that did not measure up to expectations, but did fill the gap when no other aircraft were available.

The Varuna never did fulfill expectations as a photographic aircraft and the RCAF returned to a series of single-engine aircraft for photographic duties until the war years. The experience, however, helped Canadian Vickers to design the Vancouver, a more efficient twin-engine aircraft, that first flew in April 1929. The RCAF ordered five: three with 240 hp Armstrong Siddeley Lynx engines and two with 300 hp Wright J-6 engines. These aircraft gave excellent service and were armed with machine guns and bombs to carry out some of the first wartime coastal patrols.

Canadian Vickers Velos

Engine:	Two Wright Wasp-1340A de-rated to 304 hp
Dimensions:	Span 68'0" Length 41'0" Height 16'9"
Weight:	Empty 5,875 lbs Gross 7.918 lbs
Speed:	Cruise 80 mph Max. 85 mph

By the winter of 1925–26, it was clear that the RCAF had an important and continuing role in aerial photography. It was becoming obvious that the Varuna would not be the ideal aircraft for the job and experience had proved that adopting aircraft designed for other roles, while initially economical, was not the best solution in the long term. A specification was, therefore, drawn up to identify the desirable characteristics of a photographic aircraft. The current single-engine "puller" aircraft had obvious limitations, particularly with forward visibility. So, with safety also in mind, a twin-engine aircraft was specified, capable of operating with wheels, floats or skis. Accommodation for crew of three—pilot, navigator and photographer—would be required. Purchase Order 1348 was raised on 2 September 1926 for Canadian Vickers to design and build one such aircraft. Thus, with great optimism was the ill-fated Velos born.

A unique design, the Velos accommodated photographer, navigator and pilot seated in tandem in an enclosed cockpit with excellent forward visibility. In order to get the necessary field of view for oblique photography, however, the lower wing of the biplane had a small span and contributed little lift. The stub wing did provide a platform for access to the twin engines and, supposedly, for mooring the aircraft when in a float configuration, although a significant design shortcoming was the inadvisability of standing on the float while the propellers were rotating. In a unique departure from traditional materials, steel tubing was used as wing bracing instead of wire, and the fuselage structure was welded steel tubing with only the upper wing made of wood and covered with fabric. Original plans were to use a pair of 180 hp Armstrong Siddeley Lynx engines, but this was soon changed to 304 hp Wright J-6 engines. A three-blade, fixed-pitch metal propeller with a diameter of 8'3" was selected.

Construction must have commenced in anticipation of design completion as became apparent when S/L R.S. Grandy and S/L Allen Ferrier were called to test fly the float-equipped Velos at Montreal on 19 November 1927. The pilots discovered that the aircraft weighed 5,875

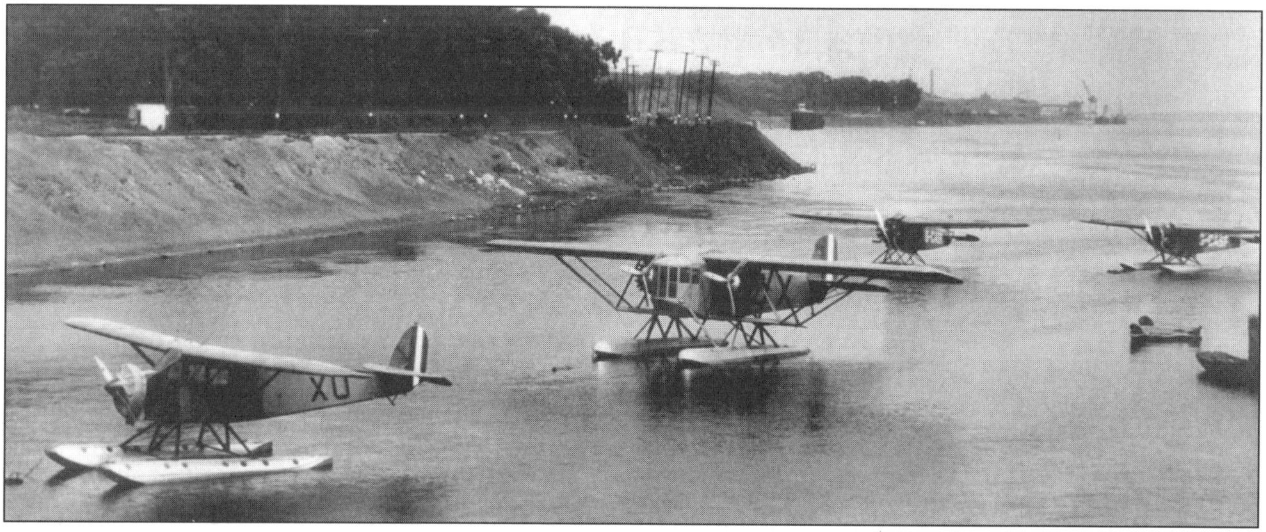

Twin-engined Velos at Canadian Vickers "Air Harbour," Montreal, 17 August 1928. Fairchild FC-2 in foreground, two Fokker Universals behind Velos. *DNDPU RE11710-33*

pounds empty compared with a design objective of 4,184 pounds. They refused to fly the aircraft and departed, leaving Canadian Vickers to resolve the problem.

S/L Grandy returned to fly the rebuilt Velos on 28 August 1928. He reported an initial rate of climb to 3,000' of only 250 feet per minute at an indicated air speed of 85 mph. He concluded a litany of shortcomings with the recommendation that it was most unsuitable for any operation carried out in the Royal Canadian Air Force. S/L Roy Grandy was one of the most experienced and respected officers in the RCAF. Nevertheless, W/C E.W. Stedman, chief of Aeronautical Engineering, directed him to re-fly the tests. This he did, under protest, with similar results.

Canadian Vickers, however, continued to work on the Velos without funding in the vain hope that its fortunes could be restored. The company test pilot flew it on 30 November 1928, reporting reaching 4,800' in under 14 minutes before running into a snowstorm, which forced him to discontinue the test. That night the Velos sank into the St. Lawrence River under a heavy load of snow. Salvaged the following day, it was dismantled and the engines dried, oiled and put in storage.

So ended the sad story of the first Canadian aircraft design intended specifically for aerial photography. Vickers supplied two Vedettes to the RCAF in compensation for progress payments made during development of the Velos. Better planning might have given the RCAF a useful aircraft, but the magnitude of the design shortcomings in conjunction with the looming Depression effectively brought an end to original aircraft design in Canada for a decade.

The Fairchild Family

Sherman Mills Fairchild (1896–1971) was a versatile inventor and machinist who held 30 US patents, many covering his aerial photography innovations. Following the Armistice in 1918, he turned his attention to civil applications of the cameras he had developed during the war. With encouragement and financial backing from his father, one of the founders and the first chairman of IBM, he established the Fairchild Camera & Instrument Corporation in 1924 to design and build aerial cameras. His cameras became the most commonly used by the RCAF between the wars.

Frustrated at his inability to persuade any manufacturer to meet his specifications for a photographic aircraft, except on an open-ended, cost-plus basis, Fairchild decided in 1925 that he would build his own. He defined an economical, reliable, single-engine, five-place, high-wing monoplane with an enclosed cabin to protect cameras and crew. The rear fuselage was constructed around three longerons, giving a characteristic triangular cross-section, known as "razorback." The FC-1 (FC for Fairchild Cabin), which had a wingspan of 44' and folding wings for ease of shipment and storage, evolved into the production version—designated FC-2—first flown on 20 June 1927.

Fairchild FC-2

Engine:	200 hp Wright Whirlwind J-5
Dimensions:	Span 44'0" Length 30'11" Height 9'0"
Weight:	Empty 2,166 lbs Gross 3,650 lbs
Speed:	Cruise 103 mph Max. 122 mph

The RCAF, with a major role photographing Canada, had been watching events at the Farmingdale, New York Fairchild plant with great interest. It was an early customer, ordering a total of 14 with deliveries starting in the fall of 1927.

G-CYYV, the RCAF's first Fairchild FC-2, was taken into the RCAF inventory on 20 October 1927. It was essential that photographic aircraft could operate from water, so 'YV was immediately scheduled for flight tests with Fairchild floats at the Shirleys Bay, Ottawa, base—with unfortunate results. On 12 November, a sudden gust of wind flipped the lightly loaded aircraft on its back. Pilot F/L A.T.N. Cowley and Cpl. R. Roberge escaped unharmed, but the brand-new FC-2 sank, and repairs, including a new fuselage, took until mid-May the following year to complete.

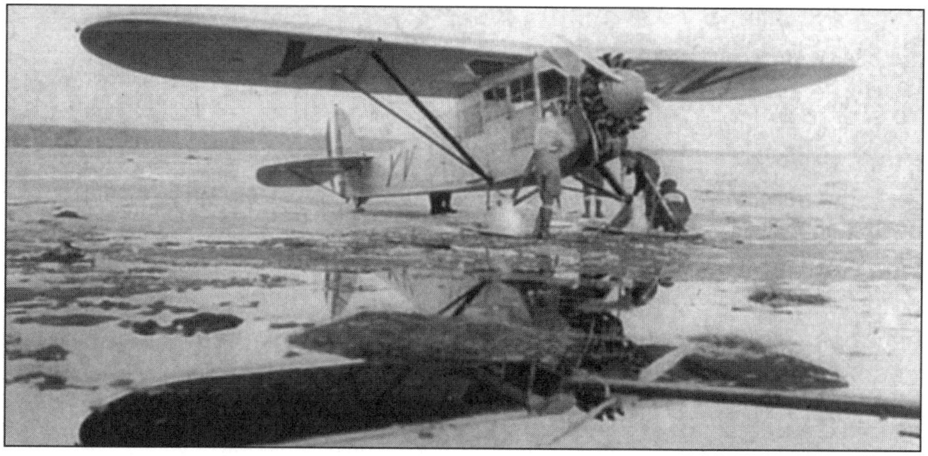

G-CYYV arriving at Ladder Lake, Saskatchewan, from Winnipeg on skis just before ice breakup in 1930. The skis were temporarily replaced with wheels for ease of handling on land until ice breakup when floats could be installed. F/L G.F. Mason-Apps at left, Sgt. Pilot "Rip" Collins kneeling with F/Sgt. "Red" Ramsden standing.
R.E. McBURNEY

Fairchild FC-2L

Resulting from a power play within the RCAF over the source of supply for aircraft and parts, it was decided in 1930 that the Canadian Vickers Vedette, Varuna and Vancouver, Avro 504N and the Fairchild FC-2 should all standardize on the British 238 hp Armstrong Siddeley Lynx engine. With Fairchild engineering approval, the Canada Car Company re-engined an initial four FC-2s, designated FC-2Ls.

Fairchild FC-2W

The RCAF was, in general, satisfied with the performance of the Fairchild FC-2, but required an increased rate of climb and improved high altitude performance for photographic operations. Fairchild responded by integrating the best features of the FC-2 with the wingspan increased from 44' to 50' and incorporating Frise-type ailerons to give better lateral control; 160-gallon fuel tank for increased range; four-longeron square-section rear fuselage for greater capacity; and a more powerful 410 horsepower Pratt & Whitney nine-cylinder Wasp "A" engine. This approach required a minimum of new development and test work, and was designated FC-2W1—"FC-2" for its origins, "W" for Wasp, and "1" for the first variant of the type.

S/L R.S. Grandy, commanding the Ottawa station, test flew the prototype FC-2W1 at Farmingdale, NY, in March 1928. As a result of his

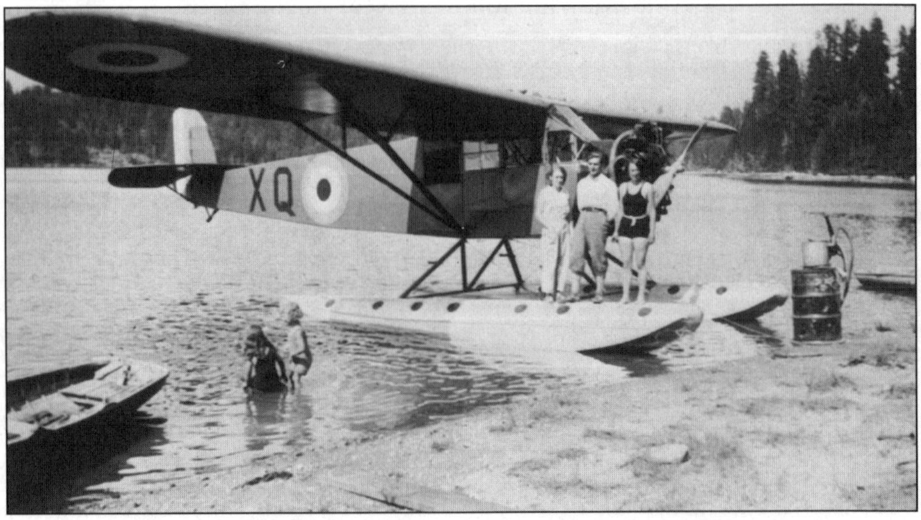

F/O C.R. Dunlap with Fairchild FC-2W 'XQ and friends at Sproat Lake, BC, 1931.
NAC PA202942/C.R. DUNLAP COLLECTION

suggestions, Fairchild lengthened the fuselage by two feet and improved lateral control by extending the length of each aileron by almost three feet, designating this variant the FC-2W2. Six FC-2W1s were delivered to the RCAF in 1928, three subsequently modified to FC-2W2s. Two were lost in crashes (one was repaired and sold to a US company), and the hangar roof literally fell in on three more at Vancouver in 1935. One of the damaged aircraft was repaired, but only two were operational by 1936. One was destroyed in a crosswind landing and the last was declared beyond economical repair in June 1936.

As often happens, the last survivor of the original six RCAF FC-2Ws had been the first to arrive. G-CYYU went on strength in May 1928, was renumbered 619 in 1933 when converted to a FC-2W2, and accumulated 1,521 hours and 40 minutes before the aircraft logbook was ruled off on 11 June 1936. The FC-2Ws were reliable, economical and effective aircraft, particularly for aerial photography, with an exceptionally low accident rate for the high utilization they enjoyed. Much of the servicing was on detached photographic operations, attesting to the efficiency of the maintenance crews and flight engineers.

Fairchild 51 and 51A

Influenced by the volume of orders from Canada, Fairchild made a most interesting offer in 1929 that meshed well with the RCAF's desire for more modern equipment and domestic sources of supply. A factory was proposed at Longueuil, on the shore of the St. Lawrence River at Montreal, to maintain, repair and build Fairchild aircraft. Modification to the RCAF's FC-2s was included as part of the deal—the fuselage changed from "razorback" triangular cross-section to a more commodious four-longeron, square cross-section, wings fitted with Frise-type ailerons, and a more powerful engine.

The new design was designated Model 51 when powered by a Wright Whirlwind J-6 engine, or a Model 51A when powered by a Pratt & Whitney Wasp Junior, both engines having nine cylinders and rated at 300 hp. Plans for the standardization on the Armstrong Siddeley Lynx had been abandoned and agreement was reached for Canadian manufacture. G-CYXK and No.29 went to the new factory on 2 July 1930, for conversion to, respectively, a Model 51 and a Model 51A.

Six more FC-2s were converted during 1931, two to 51s and four to 51As. Another conversion to a 51 in 1932 brought the total to four 51s and five 51As. The versatile and reliable aircraft were used as general transports, training aircraft and on photographic missions until 1938 when the surviving seven were all employed in training duties at Trenton. Wartime years saw them displaced by the vast intake of new aircraft and they were gradually written off or sold, the last disappearing from the books in 1946.

Fairchild 71 and 71C

Design of the Model 71 followed Fairchild's established practice of modest improvement on previous models, with no radical changes necessitating expensive and time-consuming flight test and certification programs. The square cross-section Model 51 fuselage was modified with a gently curved dorsal surface faired into the wing root, cabin windows were enlarged, and a semi-elliptical window aft of the cabin doors added a modern touch then the vogue in automobile design. The power plant was a 420 hp nine-cylinder radial Pratt & Whitney Wasp.

While Fairchild and the RCAF had previously shared the same design philosophy, they diverted during 1928 on the Model 71. Fairchild designers looked for greater market penetration by offering increased comfort and improved appearance, while the RCAF had a much more utilitarian concept in mind. The result was a Model 71B, eliminating much of the cabin lining to reduce weight and to improve access for inspection and maintenance. The rear bench seat doubled as a tool and storage box and the other four passenger seats were easily removed for cargo operations. The weight saved was invested in a cabin roof hatch with transparent panel for emergency use and to improve the pilot's rearward visibility, a larger diameter propeller, larger battery, electric engine starter, electrical generator, and additional instruments. The net result was a Model 71 with an empty weight of 3,130 pounds and a 71B at 3,076 pounds, both equipped with floats.

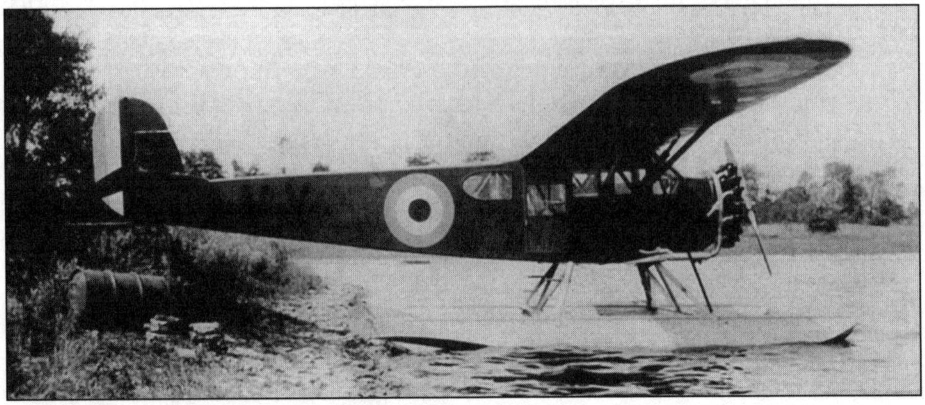

Fairchild 71 showing the distinctive rear window.
JOHN F. McNULTY

The RCAF directed additional changes in 1932 that resulted in the Model 71C. The major modifications were a strengthened fuselage, increased payload by 500 pounds, improved cabin heating and seating arrangements, and a cleaned up engine cowling that eliminated the traditional Fairchild "notch" in the upper cowling into which the side lights had been fitted. G-CTVE was sent to Fairchild Canada for conversion to a 71C in 1932.

The RCAF accepted its first four US-manufactured 71s in 1929 and ten more were delivered from Farmingdale in 1930. In the same year, the first four 71Bs were delivered from the Fairchild Aircraft Limited plant at Longueuil, Quebec. The 71Bs had welded steel fuselages imported from the parent plant and engines from Pratt & Whitney, but were otherwise of Canadian manufacture.

The photographic detachments were eagerly awaiting the new aircraft; after flight testing at Ottawa, they were assigned in pairs to PDs 2, 6, 7, 8, 10 and 11. Four more 71s were delivered from the Canadian plant in 1931, two assigned to the training squadron at Camp Borden and two for aerial photography from Fitzgerald, Alberta.

The Depression limited use of all Model 71 variants during the early 1930s, but increased funding in 1934 brought them back into service, all registered in the numerical list after overhaul, primarily for patrols on behalf of the RCMP.

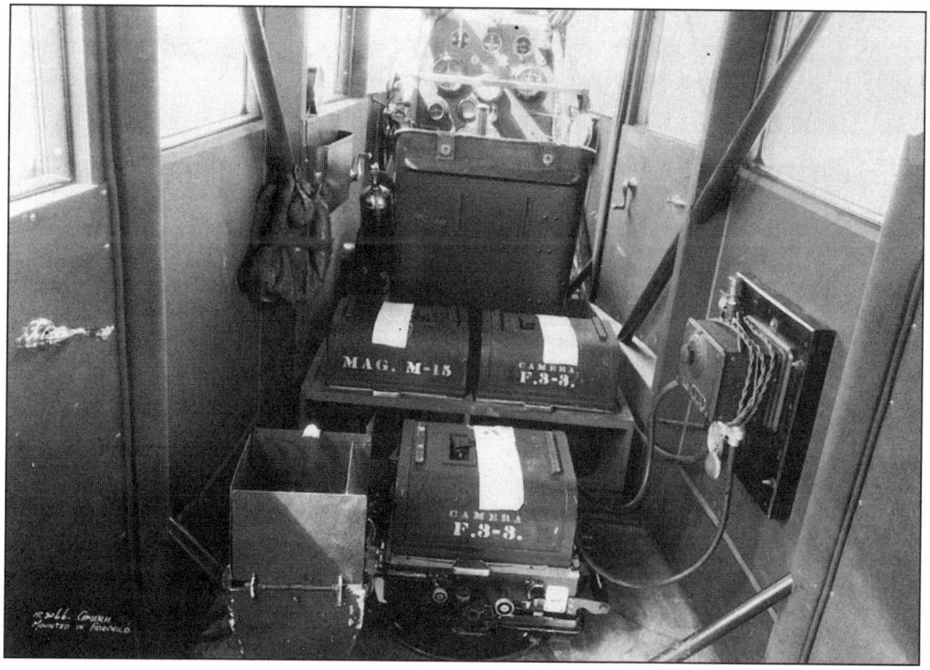

Vertical photography installation in a Fairchild 71. The viewfinder and camera are in the foreground with spare magazines at the ready behind the pilot's seat. Rockcliffe, 24 August 1929.
NAC PA62562

Fairchild Super 71

Although still carrying the "71" designation, the Super 71 was a significantly different design from its predecessor, with a larger, all-metal fuselage and a 520 hp Pratt & Whitney Wasp engine. Designed at the Canadian plant as an advanced bush plane offering long-range and high-altitude capability for photo operations, only three were built, two for the RCAF and one with a partitioned passenger cabin for commercial use. The RCAF Super 71s were delivered in the fall of 1936. Serial 666 crashed a year later in Manitoba and Serial 665 continued on photo operations until struck off the register in April 1940.

The Bellanca Pacemaker had gradually replaced the 71s for photographic duties, but the reliable workhorses continued general duty and training operations well into World War II. The fabric-covered aircraft, many with over 1,000 hours of hard use, had by this time exceeded their

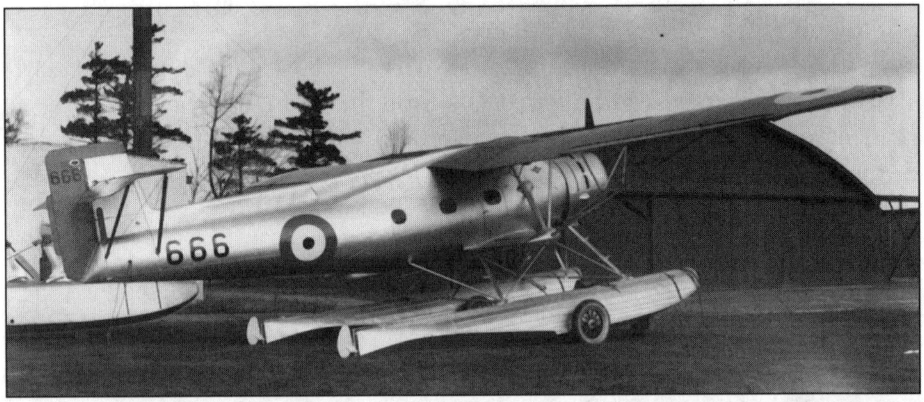

Fairchild Super 71 #666.
NAC PA63184

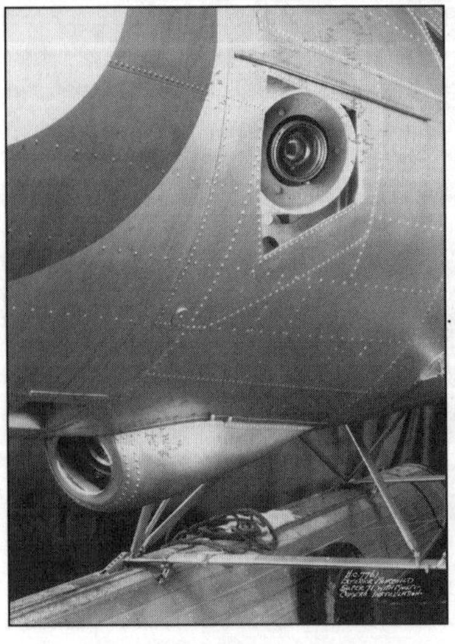

Three-camera oblique photography installation in Fairchild Super 71, Serial 665, Rockcliffe, 3 December 1936.
NAC PA63180

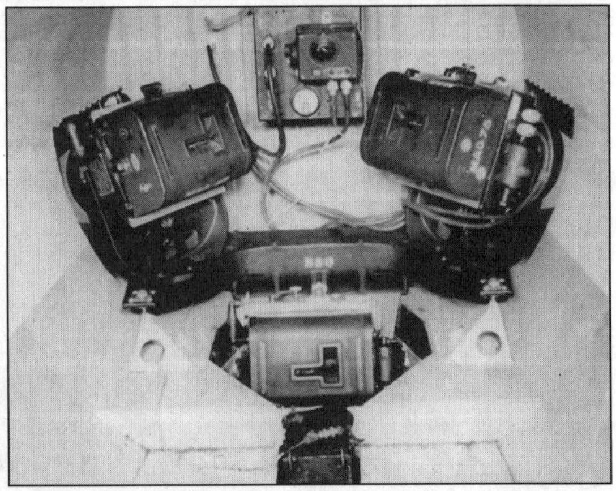

Interior view of a three-camera oblique photography installation in Fairchild Super 71, Serial 665, Rockcliffe, 3 December 1936. *NAC PA63181*

expected lifetime and several were used as instructional airframes for the enormous intake of raw personnel. Four 71s were attached to No.112 (Army Co-operation) Squadron in 1940 for pilot conversion to Lysander aircraft. During 1941, 13 of the original 22 Fairchild 71 variants were still on the books, spread from coast to coast, but most were donated to flying schools. The last survivor, Serial 642 (ex-G-CYWG), was recalled from Cartierville School to be sold to Grey Rocks Air Service on 15 April 1942 with 1,520 hours on the airframe. This brought to an end the long service of an outstanding aircraft that had contributed greatly to the effectiveness of the RCAF between the wars.

Bellanca CH-3000 Pacemaker

Engine:	One 300 hp Wright J-6
Dimensions:	Span 46'4" Length 27'9" Height 8'4"
Weight:	Empty 2,647 lbs Gross 4,300 lbs
Speed:	Cruise 110 mph Max. 143 mph

Giuseppe M. Bellanca had attracted considerable attention with his aircraft designs. One of them, a model WB-2, was selected by Charles Lindberg for his transatlantic attempt in 1927, but disagreements resulted in his choosing the Ryan Aircraft Corporation's *Spirit of St. Louis*. Clarence D. Chamberlain with Charles A. Levine as passenger flew a WB-2 from Long Island to Berlin only two weeks after Lindberg's epic flight to Paris. This achievement did not receive the same publicity as Lindberg's flight, but did underscore the reliability of Bellanca's designs and attracted sorely needed financial support from the du Pont family. Among the models produced by the busy Bellanca Aircraft Corporation in New Castle, Delaware, was the CH-300 Pacemaker, a high wing, cabin monoplane powered by a single Wright J-6, 300 hp, nine-cylinder radial engine.

Experience with the Fairchild aircraft had proven to the RCAF that a high-wing, cabin monoplane, adaptable to skis, wheels or floats, was the best configuration for aerial photography and light transport duties. The Pacemaker, like all single-engine aircraft, had poor forward visibility, but its modern design promised better performance than the Fairchilds then in service. Positive reports on three Pacemakers operated by Commercial Airways Ltd. of Edmonton and the establishment of Bellanca Aircraft of Canada Ltd. in January 1929 encouraged the RCAF to order its first Pacemaker in the spring of 1929.

The first Pacemaker, G-CYWT, performed well during a winter evaluation program, but crashed in the Gatineau River Valley while

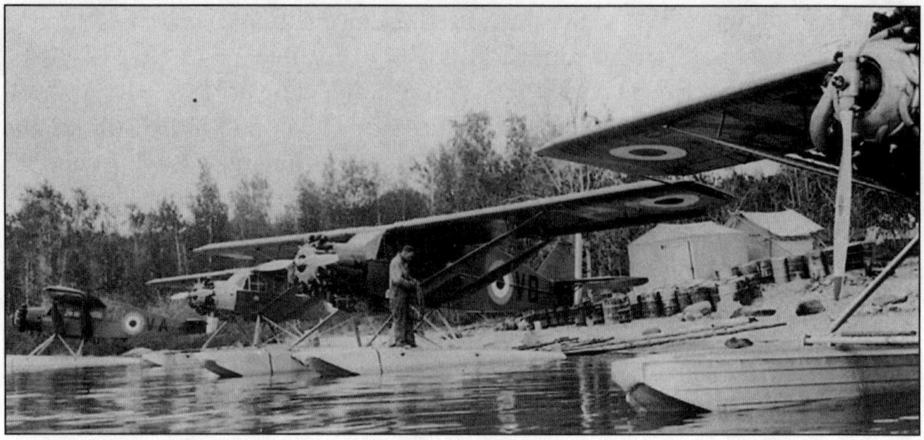

Bellanca Pacemakers of Nos 3 and 4 Photo Detachments at Slemon Lake, NWT, 1934.
DNDPU PMR74-468

Slemon Lake was named for photo pilot C. Roy Slemon who joined the CAF in 1923 and became an air marshal in the RCAF.

attempting to land in thick fog. Pilot F/L A. Carter, mechanic AC1 Ordidge and F.P. Steers of the Geodetic Survey Branch were shaken, but unhurt, as the aircraft disintegrated in some trees. The accident had a positive aspect in endorsing Giuseppe Bellanca's contention that his fuselage construction was strong enough to survive and protect the crew if they were well strapped in. The RCAF decided that the reliable and stable Fairchilds already equipped for photography should concentrate on that role, supported by Pacemakers that could carry an additional 500 pounds and cruise at least 7 mph faster than the Fairchild 51.

Six more Pacemakers were ordered from Bellanca Canada for delivery in 1930. The RCAF agreed to accept them from the US parent on condition that a second batch of six, for 1931 delivery, would be built in Canada. Canadian Vickers assembled these aircraft, which probably contained a significant portion of US components. Two, G-CYVJ and 'VK, were fitted for photography in 1930 and assigned to No.7 PD for the 1931 season. Most of the other Pacemakers were subsequently equipped for photography, but their utilization was spasmodic during the next few Depression years, and several remained in storage.

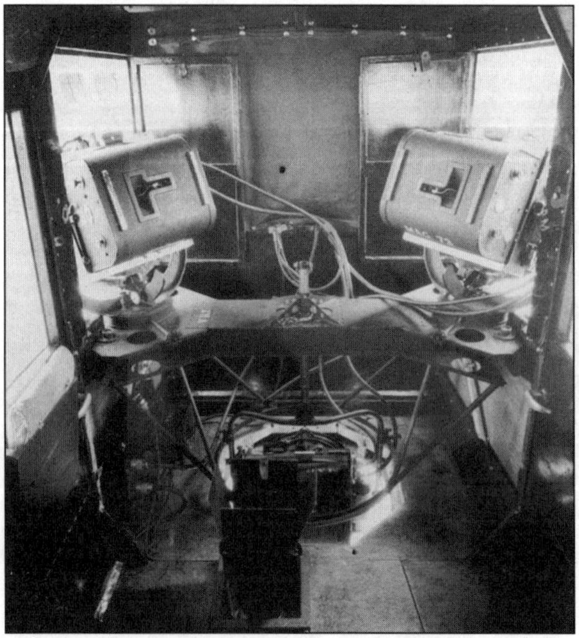

A semi-permanent, three-camera, electrically driven installation housed in the enclosed cabin of Pacemaker No.612 was a vast improvement for oblique photography over the single camera operated from the open cockpits of the early flying boats. The operator's viewfinder is in the foreground. Rockcliffe, 13 May 1935.
NAC PA63077

The economic situation improved a little in 1933. The modern and efficient Pacemakers were in demand, only two of the twelve being retained in reserve or undergoing maintenance. By the end of 1935 all 12 were available for duty, seven in Ontario and five in Manitoba.

Four Pacemakers were sold in March 1937 to Canadian Airways. The remainder continued photographic and general duties with No.7 and No.8 Squadrons. The threat of war, and the unsuitability of the Pacemakers for a military role, hastened the demise of the by now venerable aircraft in favour of more modern equipment, notably the all-metal Northrop Delta introduced specifically for photographic duties in 1937. The last year of active flying for RCAF Pacemakers was 1939, although three airframes were used for ground instructional training of aircraft riggers and fitters until advances in airframe technology made them completely redundant. The last was written off the RCAF inventory in February 1944.

The Bellanca Pacemaker was rugged, and reliable, even on detached photographic operations, where maintenance was rudimentary. The type suffered less than the average accident rate and no fatalities resulted. It is probable that they would have continued to make a valuable contribution, especially in the photographic role, if events had not combined to make the essentially civil aircraft obsolete in an air force that was gearing up for war.

Northrop Delta

Engine:	775 hp Wright SR-1820F-52 Cyclone
Dimensions:	Span 48'0" Length 33'2" Height 10'6"
Weight:	Empty 4,566 lbs Gross 7,350 lbs
Speed:	Cruise 170 mph Max. 205 mph

John Knudson Northrop developed a series of innovative, low-wing monoplanes commencing with the Vega in 1926, followed by the Alpha, Beta and Gamma in the early 1930s. He pioneered stressed metal skin fabrication techniques and applied experience gained with the previous aircraft to the dual-control, eight-passenger Delta, built in two halves and joined on the horizontal centre line at final assembly. Control surfaces were metal structures covered with fabric. The US Coast Guard purchased the first Delta and it became the personal transport of the secretary of the treasury in 1935.

The RCAF was attracted to the Delta as a high-performance photographic aircraft, and also welcomed an opportunity to introduce stressed-metal skin fabrication techniques into the Canadian manufacturing industry. The RCAF cut a deal with the California-based Northrop Aircraft Inc. to provide engineering support for Canadian Vickers Limited,

Montreal, to build a modified version of the Delta to Specification C/10/35. Known as the Model 75, the monocoque fuselage was strengthened and an oversize door hinged at the top improved cargo handling. A cast aluminum beam spanned the rear fuselage to accommodate three Fairchild A3 Automatic Aerial Cameras, one pointing out of each side and one downwards, all tilted slightly aft—a system known as tri-camera projection. The centre camera could be adjusted for vertical photography. The rear cabin was lined with aluminum alloy sheet and sprayed with a flock finish that somewhat reduced vibration and noise. Vickers Type 75 floats and Type F streamlined skis were interchangeable with the standard fixed-wheel undercarriage. The Wright Cyclone was retained, matched with a two-blade Hamilton Standard variable-pitch propeller.

Three Model 75s, RCAF designation Mk 1, were ordered from Canadian Vickers in 1936. Canadian Vickers's aircraft division manager R.J. Moffett had problems in satisfying the Canadian government procurement policy of the time that insisted, "buy British." On the one hand, Canadian industry, recovering slowly from the Depression, had very limited capability to provide complex parts such as forgings and castings. On the other hand, Britain was concentrating all its resources on rearmament and had little to spare for Canada. Canadian government hurdles had to be surmounted to purchase materials and parts from the USA. However, the experience gained and contacts made with US industry were important factors in Canada's own subsequent massive and unprecedented expansion of aircraft manufacture.

The three Delta Mk 1s were delivered to Ottawa in the fall of 1936. The first, Serial 667, was used for experimental photography and pilot training. Pilots, used to the excellent side visibility from high-wing monoplanes with highly visible wing struts, took some time to adapt to the low wing Delta with no external bracing. The Delta's wide cockpit restricted the pilot's side view of the ground, making it more difficult to follow photo "lines" than with the narrow cockpits of earlier aircraft. The solution was to add a triangular window at both sides of the cockpit and a rectangular window in the floor. Four more Deltas, Serials 670 to 673, designated Mk II, were ordered in 1937 to this configuration and four more, Serials 674 to 677, in 1938. For good measure, and a sign of the times, hard points were added under the wings of these aircraft for bomb racks and guns. On floats, the Delta could carry one 250-pound bomb under each wing and the land and ski versions an extra four 20-pound bombs under the centre section. The original three Deltas were modified to the new configuration during the 1937-38 winter and designated Mk IA.

The Deltas were assigned to No.8 (GP) Squadron at Ottawa. The general procedure was for a self-contained detachment of two aircraft and six crewmen to work their way north to the Northwest Territories as the ice melted on lakes and reverse the procedure in the fall. Deltas assigned to the west coast flew with wheeled undercarriages through the USA and had floats installed at Vancouver.

F/O H.H.C. Rutledge, OIC No.3 (GP) Detachment, was the first to experience a characteristic of the Delta that caused some nervousness for the rest of its career. At 15,000' over the prairies in 1938 he executed a tight turn and the Delta went into a spin. Rutledge recovered, but a similar incident a few weeks later raised suspicions of a tendency towards involuntary spinning. The National Research Council conducted one-eighth scale-model spinning tests without any definite result, but recommended the installation of an anti-spin parachute mounted in a modified tail cone. It appears, however, that the parachute was never used in an emergency.

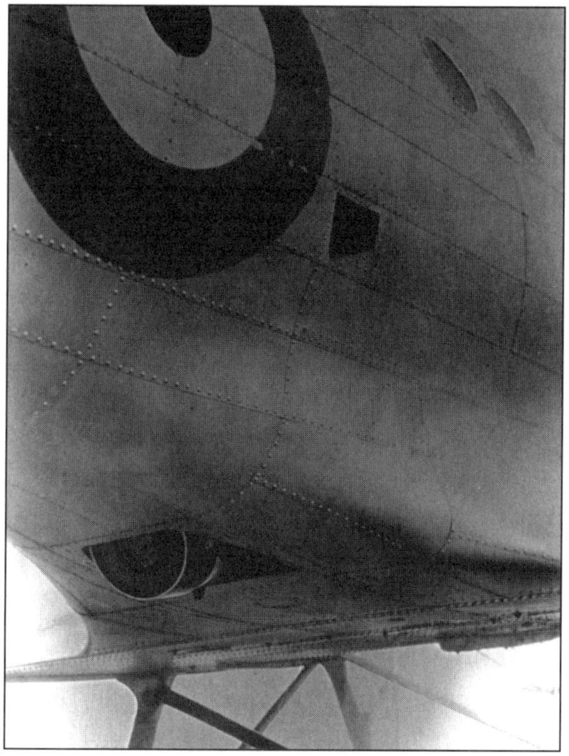

Ventral oblique camera installed in a Delta, 1938. The opening for the starboard camera is adjacent to the roundel. All three cameras pointed rearwards. The first photographic flight was made over Algonquin Park on 1 October 1937 when it was discovered that carbon monoxide from the exhausts was drawn in through the camera openings. The exhaust system was quickly modified to vent over the wing, rather than under the fuselage.

One Delta, Serial 682, was acquired in 1939. Although eight more were on order for 1940 delivery, the threat of war raised doubts as to the advisability of continuing domestic photographic operations. The aircraft were unsuitable for wartime operations and the crews lacked the military skills they would need should war be declared.

Delta Serial 673 disappeared during the move eastwards in September and, despite intensive search for two months, no trace was found until the wreckage was located in July 1958, 40 miles north of Fredericton, New Brunswick. No trace of the crew members, F/S Doan and LAC Rennie, was ever found and they were listed among the first of ten thousand RCAF aircrew casualties during World War II.]

Efforts were made to utilize the Deltas in wartime roles. Serial 675 was assigned for a brief period in 1939 to No.1 (Fighter) Squadron at Calgary to aid in the conversion of pilots to Hawker Hurricanes. Serial 672 and 670 were fitted with a "mid-upper" machine-gun position and a machine gun in each wing by Canadian Vickers in 1939. The Deltas, then with No.8 (Bomber Reconnaissance) Squadron, at Sydney were taken off photography and sent on anti-submarine patrol, but with some hesitation because of the

The urgent demand to update maps of the east coast in anticipation of hostilities resulted in the transfer of operational Deltas from No.8 (GP) Squadron to the maritimes during 1939. Serial 676 is being loaded prior to departure from Rockcliffe for Sydney, Nova Scotia, 26 August 1939.
NAC PA63532

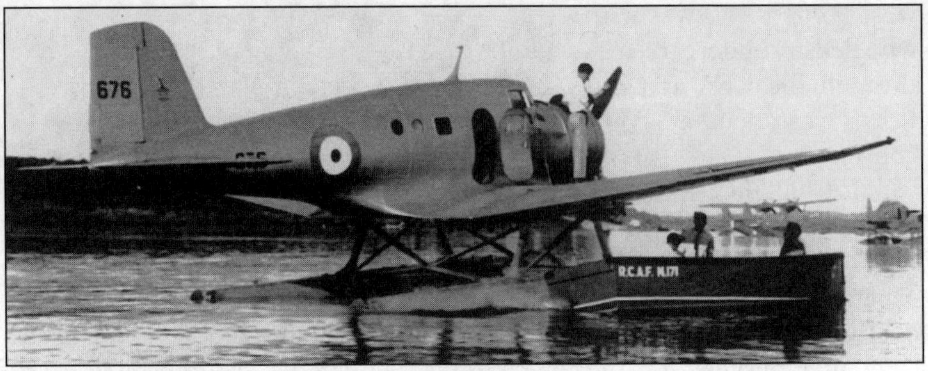

dangers inherent in flying a single-engine aircraft for long periods over water. The introduction of twin-engine, land-based Bristol Bolingbrokes in 1940, and the destructive corrosive effects of salt water on the float-equipped Deltas terminated this role.

Continuing concern regarding the spinning tendency of the Delta led to the eight delivered in 1940, Serials 683–690, having a redesigned fin and rudder and a designation of Mk III. All were dispatched to equip No.120 (Bomber Reconnaissance) Squadron on the west coast. Serials 675 and 676 were also modified to the Mk III configuration.

By 1941, accidents and corrosion had taken their toll. Of the original 20 Deltas, nine remained on the west coast and three on the east coast with another three on the books at Ottawa and one with No.4 Training Command. The decision was taken in the fall of 1941 that the Deltas had exceeded their useful life and should be withdrawn from active flying duties. Most were transferred to Training Command for use as instructional airframes, although Serial 689 did continue, briefly, flying into 1942; it was placed in storage without repair after a crash at No.8 Bombing and Gunnery School, Lethbridge, Alberta. On 14 February 1945, the remaining seven airframes were formally written off.

The Northrop Deltas manufactured by Canadian Vickers represented a significant technological advance, at the time, for both the RCAF and the Canadian manufacturing industry. Events conspired to limit their intended use as advanced photographic aircraft and their term of useful service was significantly less than any other aircraft type that had been adopted by the RCAF. The aircraft had no real military application—too fast for army co-operation, too slow for combat, unsuitable for over-water operations, too small for realistic transport duties, too complex for training purposes, and subject to saltwater corrosion. With its retirement and the RCAF's concentration on wartime priorities, it would be many years before the systematic aerial photography of Canada could be resumed.

Bibliography

Association of Ontario Land Surveyors Annual Report #40. Proceedings of the 33rd Annual Meeting (Toronto, 1925).

The Aviators of Hudson Strait, National Film Board video, 1973, 28m.

Babington-Smith, Constance. *Air Spy*. New York: Ballantine, 1957.

Bagley, James W. *Aerophotography and Aerosurveying*. New York: McGraw-Hill, 1941.

Barker, Ralph. *Aviator Extraordinary: The Sidney Cotton Story.* London: Chatto & Windus, 1969.

Brandt, Anthony S. "The First Canadian Fairchild Company, 1922–1929." *CAHS Journal* 9, no. 2 (1971): 38–51.

"Building Report 89-75: Photographic Establishment." Ottawa: Federal Heritage Review Office, c. 1991.

Canada From the Air. Photo album, National Aviation Museum. Unattributed, c.1924.

Canadian Air Board. *Report of the Air Board for the Year Ending March 31, 1920, 1921 and 1922*. Ottawa: Dominion of Canada.

Chesneau, Roger. *Aircraft Carriers of the World*. London: Arms and Armour Press, 1984.

"The Civil Aviation Branch." *Canadian Aviation* 8, no. 12 (1935): 6–10, 18–19.

Cooper, T.F. "Trials and Tribulations of the Ground Crew." Ottawa: *Roundel* 11, no. 1 (1959): 8–10.

Craig, J.D. *Canada's Arctic Islands: Log of Canadian Expedition 1922*. Ottawa: Department of the Interior, 1923.

Davidson, R.D. "Development of Mapping from Air Photography to 1930." *The Canadian Surveyor* 13, no. 8 (1957): 499–506.

Dunlap, C.R. Lecture to the Canadian Aviation Historical Society 3 June 1972. Willowdale, Ontario: CAHS, 1972.

Ellis, Frank H. "Arctic Airfield Survey." *The Beaver* 1, outfit 276 (1945): 22–25.

———.*Canada's Flying Heritage*. Toronto: University of Toronto Press, 1954.

———. *In Canadian Skies*. Toronto: Ryerson Press, 1959.

Fuller, G.A., J.A. Griffin and K.M. Molson. *125 Years of Canadian Aeronautics: A Chronology 1840–1965*. Willowdale, Ontario: CAHS, 1983.

Fullerton, Elmer J. "Pioneering Flying in the Sub-Arctic." Canadian Aviation 7, no. 5 (May 1934): 4–8.

———. "Pioneering Flying in the Sub-Arctic." Canadian Aviation 7, no. 6 (June 1934): 12–15, 23.

———. "Pioneering Flying in the Sub-Arctic." Canadian Aviation 7, no. 7 (July 1934): 10–11, 24.

Gardiner, E P. "The 'Delta' Story." *CAHS Journal* 3, no. 2 (1965): 33–39.

Graham, Ron and Roger E. Read. *Manual of Aerial Photography*. London: Butterworth, 1986.

Graham, Stuart. Audiotaped interview by Dr. Margaret Mattson, 1973. Robert Graham.

Greenhous, Brereton. *The Crucibles of War 1939–45*. Toronto: University of Toronto Press, 1994.

Griffin, J.A. *RCAF Aircraft Types and Markings 1920-1939*. Ottawa: Unpub., 1999.

Hitchins, F.H. *Air Board, Canadian Air Force, Royal Canadian Air Force*, Paper No.2, Mercury Series. Ottawa: Canadian War Museum, 1972.

———. Paper: "Photographic operations." Ottawa: Directorate of History, Department of National Defence.

Hobbs, B.D. "Report on Reindeer Lake and Churchill River Photographic Operations 1924." NAC Accession No. 912040. Winnipeg: n.p., 1924.

Hunter, J.D. "Flying in the Prewar RCAF." *CAHS Journal* 29, no. 1 (1991): 12–22.

———. "Maudie." *CAHS Journal* 33 no. 2 (1995): 50–51, 65

Lambart, H.F. Collection MG30. Ottawa: National Archives of Canada, n.d.

Lawrence, T A. *Report to the Director Civil Government Air Operations, Department of National Defence, on the Hudson Strait Expedition 1927–28*. Ottawa: National Archives of Canada, 1928.

Link, Theo. A. "Aerial Photography Applied to Geology." *Imperial Oil Review* 14, no. 4 (1930): 40–44.

Logan, R.A. Letter to Don W. Thomson. *The Canadian Surveyor* XVIII (1964): 427.

Macdonald, C.S. "Thirteen Thousand Miles by Air." *The Canadian Surveyor* V no. 1 (1934): 4–8.

MacLeod, E.L. "Early Flying in British Columbia." *CAHS Journal* 12 no. 4 (1974): 100–127.

Malayney, Norman. "Canada's Liberator Conversion Program." *CAHS Journal* 38 No. 4 (2000): 139-143.

McLaren, Duncan D. *Bush to Boardroom*. Winnipeg: Watson & Dwyer, 1992.

Milberry, Larry. *Air Transport in Canada*. Toronto: CANAV Books, 1997.

———. *Sixty Years: The RCAF and CF Air Command 1924–1984*. Toronto: CANAV Books, 1984.

——— and Hugh Halliday. *The Royal Canadian Air Force—At War 1939–45*. Toronto: CANAV Books, 1990.

Milne, Alex J. "Pioneer Survey Flights in Northern Manitoba and Saskatchewan." *CAHS Journal* 21 no. 4 (1983): 100–109

———. "Pioneer Survey Flights in Northern Manitoba and Saskatchewan. *CAHS Journal* 22 no. 1 (1984): 3–11.

Miller, Hugh. *Harry Winny*. Vancouver: Unpub., 1994.

Miller, John M. "A Timber Survey for the Chicago Tribune." *Aviation* 8, no. 11 (1922): 319.

Molson, K.M. "Canada's First Air Survey." *Aircraft* 24 no. 11 (1959): 50, 72.

———. "A History of Laurentide Air Service—Canada's First Scheduled Air Service." *CASI Journal* 16 no. 7 (1970): 269–284.

———. "Aircraft Manufacturing In Canada During The First Great War." *Canadian Aeronautical Journal* 5 no. 2 (1959): 47–50.

—— and H.A. Taylor. *Canadian Aircraft Since 1909*. Stittsville, Ontario: Canada's Wings Inc., 1982.

Mulligan, Paul. "En Route Charts to be Revamped." *Canadian Flight* (March 1999): 3.

Narraway, A.M. Report: Natural Resources Canada 1 no. 9 (October 1922): 1.

Newhall, Beaumont. *Airborne Camera: The World From the Air and Outer Space*. New York: Hastings House Pub Inc., 1969.

Pearce, Harold P. "Canada and the Airborne Camera." *Canadian Aeronautical Journal* 5 no. 2 (1959): 55–66.

———. "Development of Photography, Cameras, Aircraft and Camera Mounts." *The Canadian Surveyor* 13, no. 10 (1957): 635–643.

Pickler, Ron and Larry Milberry. *Canadair: The First Fifty Years*. Toronto: CANAV Books, 1995.

Pierce, J.W. "The Adaptation of Aerial Photographs to Surveys and Maps." *Proceedings of the Association of Ontario Land Surveyors* (February 1925): 168–179.

Ralph, Wayne. *Barker VC*. Toronto: Doubleday, 1997.

Rayburn, Alan. *Dictionary of Canadian Place Names*. Toronto: Oxford, 1999.

Reeves, A.I. "Wood That Could Fly." *CAHS Journal* 37 no. 2 (1999): 66–74.

Report on Civil Aviation for the Year 1923, 1924, 1925, 1926, 1927, 1928, 1929, 1930 and 1931. Ottawa: Dominion of Canada.

Report on Experimental Aerial Survey at Ottawa, 1920. Ottawa: Air Board Bulletin #2, March 1921.

Sandwell, Arnold H. "The Camera Takes to the Air." *Canadian Geographical Journal* 1 no. 1 (1930): 61–73.

Slemon, C. Roy. "RCAF Photographic Detachments." *Canadian Aviation* 13 no. 6 (1935): 20, 21, 27.

Stedman, E.W. *From Boxkite to Jet*. Toronto: University of Toronto Press, 1963.

Stewart, N.C. "Surveying from the Air." *Journal of the Royal Astronomical Society of Canada* 25, no. 4 (1931): 153–165.

Stretton, W.D. "From Compass to Satellite." *The Canadian Surveyor* 36, no. 4 (1982): 123–126.

Thomas, R.I. *Stories for a Grandson*. Unpublished, c. 1980.

Thomson, Don W. *Men and Meridians*. Vol. 3. Ottawa: Department of Energy, Mines and Resources, 1969.

———. *Skyview Canada*. Ottawa: Department of Energy, Mines and Resources, 1975.

Vincent, Carl. "Canadian Classic: The Canadian Vickers Vedette in the RCAF." *High Flight* 2, no. 4 (1982): 131–139.

———. "Viking—the Story of the Air Force's First New Aircraft." *High Flight* 1, no. 4 (1981): 134–142.

Weicht, Chris. *Jericho Beach and the West Coast Flying Stations*. Chemainus, BC: MCW Enterprises, 1997.

Wheeler, A.O. "Application of Photography to the Mapping of the Canadian Rocky Mountains." *Canadian Alpine Journal* 11 (1920): 76–96.

——— and H.F. Lambart. "Mountain Reconnaissance by Aeroplane." *Canadian Alpine Journal* 13 (1923): 146–157.

Wikene, Ingwald. "The Curtiss HS-2L in RCAF/CAF Service." *CAHS Journal* 10, no. 1 (1972): 13–14.

Wise, S.F. *Canadian Airmen and the First World War*. Toronto: University of Toronto Press, 1980.

Wood, E. P. "Northern Skytrails." *The Roundel* 1, no. 1 (Nov. 1948): 28–32.

———. "Northern Skytrails." *The Roundel* 1, no. 2 (Dec. 1948): 22–27.

———. "Northern Skytrails." *The Roundel* 1, no. 3 (Jan. 1949): 26–31.

———. "Northern Skytrails." *The Roundel* 1, no. 4 (Feb. 1949): 22–29.

———. "Northern Skytrails." *The Roundel* 1, no. 5 (Mar. 1949): 22–28.

———. "Northern Skytrails." *The Roundel* 1, no. 6 (April 1949): 24–27.

———. "Northern Skytrails." *The Roundel* 1, no. 7 (May 1949): 24–32.

———. "Northern Skytrails." *The Roundel* 1, no. 8 (June 1949): 18–24.

———. "Northern Skytrails." *The Roundel* 1, no. 9 (July 1949): 24–32.

Abbreviations

2Lt.	Second Lieutenant
A/	Acting
AC1	Aircraftsman 1st Class
AC2	Aircraftsman 2nd Class
A/C/M	Air Chief Marshal
AFC	Air Force Cross
AFHQ	Air Force Headquarters
AFR	Air Force Reserve
AHQ	Air Headquarters
A/M	Air Marshal
AMICE	Associate Member, Institute of Civil Engineers
ARCSc	Associate of the Royal College of Science
A/V/M	Air Vice-Marshal
BA	Bachelor of Arts
BEM	British Empire Medal
B&GTS	Bombing and Gunnery Training School
CAB	Civil Air Board
CAF	Canadian Air Force
CAHS	Canadian Aviation Historical Society
Capt.	Captain
CASI	Canadian Aeronautics and Space Institute
CB	Companion (of the Order of) the Bath
CBE	Commander of (the Order of) the British Empire
CCG	Canadian Coast Guard
CD	Canada Decoration
CEF	Canadian Expeditionary Force
CGAO	Civil Government Air Operations
CGS	Canadian Government Ship
CMG	Companion of the Order of St. Michael and St. George
CNR	Canadian National Railway
CO	Commanding Officer
Col.	Colonel
Cpl.	Corporal
CWM	Canadian War Museum
CVSM	Canadian Volunteer Service Medal
DCGAO	Director, Civil Government Air Operations
DCM	Distinguished Conduct Medal
DFC	Distinguished Flying Cross
DHist	Directorate of History
DLS	Dominion Land Surveyor
DND	Department of National Defence
DNDPU	DND Photo Unit
DOT	Department of Transport
DSC	Distinguished Service Cross
DSO	Distinguished Service Order
DTS	Dominion Topographical Surveyor
e.a.	enemy aircraft
F/L	Flight Lieutenant
F/O	Flying Officer
FRGS	Fellow of the Royal Geographical Society
FRPS	Fellow of the Royal Photographic Society
F/Sgt.	Flight Sergeant
F/S/L	Flight Sub-Lieutenant
G/C	Group Captain
GPD	General Purpose Detachment
GSGS	Geographical Section of the General Staff
KC	King's Counsel
KCMG	Knight Commander of (the order of) St. Michael and St. George
LLD	Doctor of Laws
Lt.	Lieutenant
Lt.-Col.	Lieutenant Colonel
Maj.	Major
MB	Bachelor of Medicine
MBE	Member of (the Order of) the British Empire
MC	Military Cross
MEIC	Member of the Electrical Institute of Canada
MG	Machine Gun
MID	Mentioned in Dispatches
MP	Member of Parliament
NCO	Non-Commissioned Officer
NORAD	North American Air Defence Command
NP	Non-Permanent
OBE	Officer of (the Order of) the British Empire
OIC	Officer In Charge
PD	Photographic Detachment
PLS	Provincial Land Surveyor
P/O	Pilot Officer
PPO	Provisional Pilot Officer
RAF	Royal Air Force
RCAF	Royal Canadian Air Force
RCAMC	Royal Canadian Army Medical Corps
RCCS	Royal Canadian Corps of Signals
RCN	Royal Canadian Navy
RCNAS	Royal Canadian Navy Air Service
RD	Repair Depot
RFC	Royal Flying Corps
RNAS	Royal Naval Air Service
Sgt.	Sergeant
S/L	Squadron Leader
TCA	Trans-Canada Airlines
USAAF	United States Army Air Force
W/C	Wing Commander
W/O	Warrant Officer
W/O1	Warrant Officer, 1st Class
VC	Victoria Cross
WS	Wireless School

Index

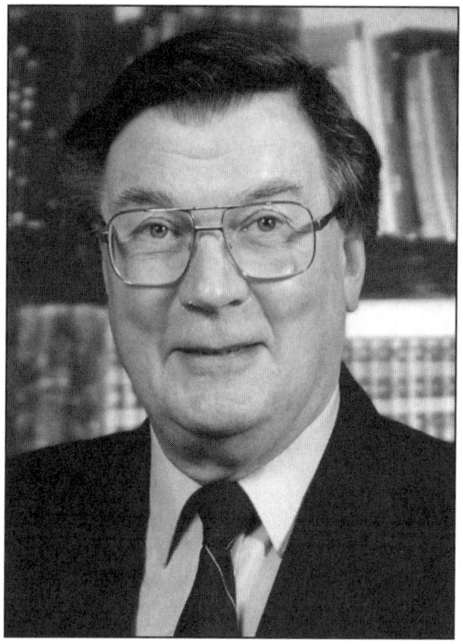

About the Author

S. Bernard Shaw is no stranger to readers of Algonquin Park and Ottawa Valley historical lore. In his first book, *The Opeongo: Dreams Despair and Deliverance* (1994), he traced the colonization road from the Ottawa River westward toward the Algonquin highlands and Lake Opeongo. He followed that with his comprehensive accounts of two famous lakes—*Canoe Lake, Algonquin Park: Tom Thomson and Other Mysteries* (1996), and *Lake Opeongo: Untold Stories of Algonquin Park's Largest Lake* (1998). He has also written more than 200 magazine articles on historical, aviation, and business topics. He found vintage aerial photographs of great help in his research, which prompted his interest in the men who took them and the conditions in which they worked to lay the essential foundations for the development of Canada.

A native of Derbyshire, England, the author emigrated to Canada with his family in 1957 and currently resides near Ottawa. During a varied career he has served in the Royal Air Force as a meteorological assistant; designed dairy equipment, and subsystems for automobiles, aircraft engines, navigation equipment and research rockets; and has been a business development officer with the federal government. For the past decade, his business consultancy has been increasingly neglected in favour of historical research and the publication of heritage stories that he hopes will help illuminate the achievements of pioneers and improve the chances of their patriotic dreams being realized in a united Canada.